Bioprobes

AF793367

Hiroyuki Osada
Editor

Bioprobes

Biochemical Tools for Investigating Cell Function

Second Edition

Springer

Editor
Hiroyuki Osada
Chemical Biology Research Group
RIKEN Center for Sustainable Resource Science
Wako, Saitama
Japan

ISBN 978-4-431-56813-1 ISBN 978-4-431-56529-1 (eBook)
DOI 10.1007/978-4-431-56529-1

© Springer Japan KK 2000, 2017
Softcover reprint of the hardcover 2nd edition 2017
This work is subject to copyright. All rights are reserved by the Publisher, whether the whole or part of the material is concerned, specifically the rights of translation, reprinting, reuse of illustrations, recitation, broadcasting, reproduction on microfilms or in any other physical way, and transmission or information storage and retrieval, electronic adaptation, computer software, or by similar or dissimilar methodology now known or hereafter developed.
The use of general descriptive names, registered names, trademarks, service marks, etc. in this publication does not imply, even in the absence of a specific statement, that such names are exempt from the relevant protective laws and regulations and therefore free for general use.
The publisher, the authors and the editors are safe to assume that the advice and information in this book are believed to be true and accurate at the date of publication. Neither the publisher nor the authors or the editors give a warranty, express or implied, with respect to the material contained herein or for any errors or omissions that may have been made. The publisher remains neutral with regard to jurisdictional claims in published maps and institutional affiliations.

Printed on acid-free paper

This Springer imprint is published by Springer Nature
The registered company is Springer Japan KK
The registered company address is: Chiyoda First Bldg. East, 3-8-1 Nishi-Kanda, Chiyoda-ku, Tokyo 101-0065, Japan

Preface

The first edition of this book *Bioprobes* was published in 2000. The term "bioprobe" was introduced then as the name for a chemical tool, a probe, to investigate biological functions. Eventually, the concept of the research inspired by the bioprobe was similar to that of chemical biology and it became familiar to scientists. The epoch-making research results, namely, that FK506 and cyclosporin inhibit the activity of phosphatase calcineurin by forming the ternary complexes FK506- FKBP12- calcineurin or cyclosporine-cyclophilin-calcineurin, opened a new door in chemical biology. It was typical research in basic chemical biology.

The new tide of chemical biology has flowed from the Institute of Chemistry and Cell Biology (ICCB) at Harvard Medical School and has initiated the building of chemical biology-related centers in other parts of the world. One of the main activities of the centers is the screening of small bioactive compounds from chemical libraries. This tide merged into chemical genomics, which deals with the systematic screening on a large scale from chemical libraries to discover lead compounds of drugs.

Based on this background, the present volume describes bioprobes that are mainly isolated from microorganisms, along with some synthetic compounds. Most of the bioprobes are not developed for clinical use yet, but they are useful as chemical probes as well as lead compounds of drugs.

The book consists of six chapters. After a brief introduction of the development of bioprobes, the biological fields in which bioprobes affect the molecular targets are surveyed. These reviews cover the broad, advancing areas of the cell cycle, epigenetics, apoptosis–autophagy, and immunological responses. A final chapter contains important bioprobes that are useful for investigating biological studies.

In this second edition of *Bioprobes*, we deal with small molecules isolated from marine organisms in addition to microbial metabolites. I hope this book will contribute to the work of both natural product chemists and cell biologists, eventually resulting in the progress of bioprobe research.

Wako, Japan Hiroyuki Osada
March 25, 2017

Contents

Chapter 1
Trends in Bioprobe Research

Hiroyuki Osada

Keywords Screening • PPI • Target Identification • Cell Morphology • Profiling • Proteome • Affinity Beads • DARTS • CETSA

1.1 Introduction

The term "bioprobe" was used for the first time in the previous version of this book [1]. Bioprobes were defined as small molecules which are useful not only for biochemical research but also for the source of drug candidates with diverse activities. Historically, microbial metabolites have been quite useful not only as medicines but also as bioprobes. As a typical example, penicillin was originally discovered as a therapeutic agent against bacterial infectious diseases. Afterwards, research on the mode of action of penicillin gave us the insight on the structure and the biosynthesizing mechanism of the bacterial cell wall [2, 3]. Another well-known example is that the reverse transcriptase was discovered by the aid of daunomycin and actinomycin D. In the twentieth century, the central dogma of molecular biology suggested that the genetic information flows from DNA to RNA, and then finally to protein. However, some tumor viruses transcribe DNA from RNA by reverse transcriptase which is RNA-dependent DNA polymerase [4, 5]. Specific inhibitors of DNA and RNA synthesis were used to prove the template of the polymerase.

Based on the above mentioned background, we especially focused on the microbial metabolites in the first version of *Bioprobes* and attempted to merge the interests of organic chemists who are interested in the isolation of inhibitors, and molecular biologists who are interested in the biological function of enzymes, into the same stream. The attempt was fulfilled as chemical biology. There are many success stories to prove the usefulness of microbial metabolites [6, 7]. The most recent glorious example is the development of the antiparasitic agents avermectin and ivermectin by Omura and Campbell [8]. Unfortunately, pharmaceutical

---

H. Osada (✉)
Chemical Biology Research Group, RIKEN Center for Sustainable Resource Science,
2-1 Hirosawa, Wako, Saitama 351-0198, Japan
e-mail: hisyo@riken.jp

© Springer Japan KK 2017

H. Osada (ed.), *Bioprobes*, DOI 10.1007/978-4-431-56529-1_1

companies gave up natural products screening and shifted to high-throughput screening using small molecules synthesized by combinatorial chemistry. However, the Nobel Prize 2015 recalled the usefulness of the natural products.

1.2 Screening of New Bioactive Compounds

1.2.1 Cell-Based Screening

To screen for bioactive small molecules, it is essential to establish a simple rapid bioassay system. Screening systems are roughly categorized into enzyme-, cell-, and animal-based assays, each with its own pros and cons. When constructing a new bioassay system, we make a point of verifying whether it meets the 4S (simple, speedy, sensitive, and specific) and 1D (distinctive) criteria [9]. In another words, a good screening system provides an easy-to-follow assay protocol, a quick turn-around time for assay results, high sensitivity, high specificity, and easy interpretation of assay results. Regarding these criteria, enzyme-based assay is easy to transform to a high-throughput screening system. Conversely, cell-based assays are generally time consuming and laborious, therefore, cell-based screening is difficult to transform to high-throughput screening. However, phenotypic screening by using cells is suitable for natural products screening [10].

Recently, high-content screening using mammalian cells is becoming popular because automatic microscope systems such as IN Cell Analyzer (GE Healthcare Life Science), Cell Voyager (Yokogawa Electric Corporation), Cell Insight (Thermo Fisher Science), Opera Phenix (Perkin Elmer), etc., are commercially available. With these visualization tools, it is now possible to analyze qualitative and quantitative cellular phenotypes such as morphological changes in a high-throughput manner.

More interestingly, we found that mammalian cancer cells also showed specific morphological changes in response to the mode of action of a drug. Thus, we constructed a phenotypic screening system based on the high-content cell morphological database "MorphoBase" [11]. After several experiments using various cell lines, we decided to use *src*^{ts}-NRK [12] and HeLa cells for the assay system. Both cell lines exhibited a variety of changes in shape in response to various small molecules. To eliminate bias of individuals during observation, we performed automated observations using IN Cell Analyzer. Specifically, we designed a high-content imaging algorithm that segments an individual cell into three fragments (cell, nucleus, and granule) and quantified 12 morphological parameters. This system enables us to classify the mode of action of typical anticancer drugs and profile various molecular targets of bioactive compounds and even crude extracts [13, 14].

1.2.2 Target-Based Screening

Receptor–ligand binding (RLB) assay and protein–protein interaction (PPI) assay are typical examples of target-based screening [15, 16]. In general, the target-based screening is constructed as a cell-free assay. However, RLB and PPI can be constructed as a cell-based assay, if thc aimed protein can be expressed on the cell surface.

Membrane proteins are difficult to obtain as a soluble form, therefore, it is a challenging subject to screen the inhibitors of membrane proteins. The screening of agonists or antagonists of G-protein coupled receptor (GPCR) is one of the applicable RLB assay systems. Yanai et al. [17] identified that the novel inverse agonists of super conserved receptor expressed in brain (SREB) consists of GPR27, GPR85, and GPR173 from the compound library of NPDepo. They constructed the assay system in Sf9 cells expressing the GPR173-Gsα fusion protein and tested the binding of [^{35}S] GTPγS to the receptor in the presence of the screening compounds.

Recognition of phosphorylated protein is one of the key events in the signal transduction pathway [18]. There are many cases where the recognition of the phosphorylated domain is involved in protein–protein interaction such as the SH2 domain in Src family proteins, polo box domain (PBD) [19] in the polo-like kinase (PLK) family, etc. Small molecules modulating these interactions are attractive lead compounds for the treatment of various diseases including tumors. Watanabe et al. established a screening system to explore inhibitors of phosphorylation-dependent PPI between the polo box domain of PLK1 and its recognition peptide in 96-well plates. Purpurogallin, a benzotropolone-containing natural compound derived from nutgalls deposited in the NPDepo library, was identified as a strong inhibitor of PBD-dependent binding [20].

A chemical array platform which can immobilize thousands of compounds on a glass slide is applicable for the high-throughput screening to detect the interaction of a protein of interest and its binding ligands. For the chemical array screening, we developed two kinds of detection methods; one is based on the fluorescent tagged protein fused with red fluorescence protein (RFP) [21, 22]. Instead of RFP, GST, His, or FLAG can be used to probe proteins. In this case, a fluorescent tagged antibody against GST, His, or FLAG is required. The other detection method is based on surface plasmin resonance (SPR) imaging [23, 24]. For SPR imaging, library compounds should be attached on the slide glass coated with gold. In this case, native protein without a tag can be used and the interaction of protein and a small molecule can be detected by SPR detector. These methods can be used to find binding compounds even when the structure or function of the protein is not known.

DNA consists of only four kinds of bases and a phosphate group; therefore, it was not so difficult to attach the DNA chain on the slide glass by a uniform method. However, the structural variation of small molecules is very wide; therefore, it was difficult to attach the variety of the small molecules on the slide glass by a uniform method. Before we established the chemical array platform, several research groups reported the immobilization methods of small molecules. In general, small molecules with reactive functional groups are attached to the linker on the surface of glass slides by a selective coupling reaction. MacBeath et al. immobilized small molecules containing free thiols to glass slides coated with maleimide groups by the Michael addition reaction [25]. Hergenrother et al. activated the slide glass with thionyl chloride and prepared chlorinated slides. Small molecules containing a primary alcohol can be attached to the slide glass [26]. By activating the glass surface with diazobenzylidene, Barnes-Seeman et al. developed a versatile method to attach small molecules containing phenols, carboxylic acids, or comparably acidic groups [27]. These immobilization methods require a specific group such as -SH or -OH on the molecule. On the other hand, we reported the photo-cross-linking method for immobilization of small molecules in a functional group independent manner [28]. Trifluoromethyl aryldiazirine is used as a photo reactive group and linked with a spacer such as polyethylene glycol. UV irradiation forms a highly reactive carbene species which covalently binds to any kinds of small molecules without requirement of the specific functional group. Our method is suitable for screening complex natural products [21, 24].

1.3 Target Identification of Bioactive Compounds

There are mainly two approaches to identify cellular targets of bioactive small molecules, namely, direct and indirect approaches. Affinity beads are used to pull down the target protein depending on the direct interaction between small molecules and protein. Moreover, several new methods such as drug affinity responsive target stability (DARTS) [29, 30] and cellular thermal shift assay (CETSA) [31, 32] based on the stability of the protein were also reported. When the protein binds with the small molecule, the protein becomes resistant to heat and protease digestion.

Phenotype-based approaches compare the biological profiles of small molecules of interest and known reference drugs, such as the cell line panel [33, 34], the proteomic profiling [35], and the cell morphological profiling [11]. In addition to the approaches above mentioned, genetic-based approaches using genetically engineered yeasts are useful [36], but the method cannot be described here due to space limitations.

1.3.1 Cell Morphology–Based Profiling

The National Cancer Institute [33] and Japan Foundation for Cancer Research [34] developed the cell line panel assays using various types of cancer cell lines for target speculation. These assays are based on the difference of the drug sensitivity of each cell against small molecules.

Moreover, we found that cells exhibited a particular morphological change depending on the mode of action of the small molecule before death. Based on this observation, we developed a database of the morphological changes of two cell lines, *src*^{ts}-NRK and HeLa cells, named MorphoBase [11]. To construct MorphoBase, we established a high-content imaging method that segments the cells and quantifies 12 morphological parameters of the cells. We also developed a data analysis program that incorporates multivariate statistical tools. Principal component analysis is applied to visualize the phenotypic responses, and target prediction of a small molecule is performed by using two statistical computations: "similarity ranking" determined by Euclidean distance metrics and "probability scores" which are based on the z-scores of a test compound relative to the 14 target classes.

1.3.2 *Proteomic Profiling*

As described above, bioactive compounds induced morphological change of the cell. Upon the morphological changes, the protein expression should be changed in the cells. Actually, some protein levels were augmented and some protein levels decreased responding to the small molecule treatment. Then, we analyzed the proteomic change induced by the small molecule by using two-dimensional difference gel electrophoresis (2D-DIGE). The expression data that were presented by ca. 300 spots were reproducibly detected in images of HeLa cells treated with well-known inhibitors. The proteomic pattern in 2D–DIGE was deposited in the database named ChemProteoBase [35]. The pattern was classified by cluster analyses according to their mechanism of action. The mechanism of action of a small molecule can be accurately established if it induces a proteomic profile that is similar to a profile induced by a reference small molecule in the database. To obtain more information about the effects of small molecules linking to the gene ontology or biological pathway, about 90% of the spots in ChemProteoBase were identified by mass spectrometry. Spot information and a subset of the expression data for 19 inhibitors are shown on our web site; http://www.npd.riken.jp/csrs/xtra/ProteomePage/.

1.3.3 *Affinity Beads*

The affinity beads approach is a direct approach to find target proteins. As a typical successful example of chemical biology, the affinity beads method was applied to identify the target protein of FK506 [6]. The small molecule is chemically attached to the beads thorough a linker. The binding protein for the affinity beads is precipitated from the crude cell extract and identified by mass spectrometry (MS) or *N*-terminal peptide sequencing.

The affinity purification using affinity beads is powerful; however, there are several drawbacks. The first drawback is the derivatization of small molecules. The small molecule must be immobilized in an appropriate manner that maintains its target recognition properties; however, this approach is often time consuming and requires chemical expertise to elaborate the structure–activity relationship (SAR). It is difficult to apply this method for small molecules with complicated structures or without suitable functional groups. The second drawback is nonspecific interactions. For example, the proteins may bind to the resin itself or linkers, because such proteins are detected frequently in independent experiments using different affinity beads. The third drawback is the preparation of cell extracts. In many cases, whole cell lysates are used as a source of target proteins. The salts and detergents used for preparing cell lysate often affect protein–small molecule interactions. Moreover, it takes time to find suitable conditions. The fourth issue is the scale or sensitivity of detection. Detection of small molecules targeting low-abundance proteins may be difficult. When the desired results are not obtained, it is often difficult to judge whether this is due to scale or to other factors. To overcome these limitations, a variety of approaches have been developed [37–39].

In general, the affinity beads consist of the small molecule of interest, the linker or linker with a functional tag, and the beads. Therefore, improvements in the probes can be achieved by making changes in a matrix, linker, or functional tag. Sugar-based matrices such as agarose or Sepharose beads, which are commercially available, have been frequently used. They have a highly hydrophilic property that reduces nonspecific protein binding, but are easily denatured under organic synthesis conditions.

Length and properties of the linker are important to reduce the nonspecific binding with proteins. Polyethylene glycol (PEG) is mainly used as a linker. Depending on the depth of the binding pocket of the protein, the length of PEG should be optimized. When the longer length of the PEG linker reduces the binding efficacy of small molecules with the target protein, the combination of PEG and polyproline is effective as the linker [40, 41].

1.3.4 DARTS and CESTA

Efficacy of therapeutics is dependent on a drug binding to its target molecule. Therefore, it is very important to measure the interaction between a small molecule and its target; in most cases, the targets are proteins. Recently, two methods, being dependent on the protein stability enhanced by the binding with a compound, have been reported to measure the small molecule binding to target proteins in cells. The methods are named DARTS (drug affinity responsive target stability) [29, 30] and CETSA (cellular thermal shift assay) [31, 32].

DARTS was reported to identify potential protein targets for small molecules. The principle of this method is based on protection against proteolysis when the protein binds with a small molecule. When the cell lysate with a small molecule of interest is treated with proteases, most proteins are digested but the binding target proteins are protected by the small molecule from the proteolysis. The merit of this method is that protein tagging or small molecule immobilization are not necessary.

Recently, Chin et al. applied this method to probe that alpha-ketoglutarate binds to ATP synthase subunit beta and the inhibition of ATP synthase extends the life-span of *Caenorhabditis elegans* [42]. DARTS is thought to be applicable not only to identify the target of small molecules but also to detect protein–small molecule interactions in general.

CETSA is based on the thermal stabilization of target proteins depending on the binding with a small molecule. When the cell lysate is treated at various temperatures, proteins are denatured and aggregated. However, the protein binding with its ligand small molecule becomes resistant to the thermal denaturation. We applied this method combined with a ChemProteoBase method to identify the hidden molecular target of TH287 which was recognized as an inhibitor of MTH1 [43].

1.4 Trends and Prospects

We are searching for candidate bioprobes mainly from microbial metabolites. To find the possible candidates, it is important to establish a screening/bioassay system. According to our experience, the cell-based/phenotypic screenings are suitable for the screening of natural products, even crude extracts, and the target-based screening is suitable for the screening of the synthesized small molecules focusing on the specific target. According to the recent trends, pharmaceutical industries are giving up natural product screening and are carrying out high-throughput screening using the synthesized small molecules. However, the recent advance of biosynthetic studies enables us to diversify the chemical space of the natural products and refocus the spotlight on natural product screening. Especially, phenotypic screening using iPS cells established from patients with incurable diseases is a promising approach for drug discovery from natural products [44].

A target specific bioprobe as well as siRNA can be used to analyze the protein function. Moreover, the bioprobe can be a lead compound in drug development [45–47].

References

1. Osada H (2000) Trends in bioprobe research. In: Osada H (ed) Bioprobes. Springer, Tokyo/New York, pp 1–14
2. Park J, Strominger J (1957) Mode of action of penicillin. Science 125:99–101
3. Trucco RE, Pardee AB (1957) Synthesis of Escherichia coli cell walls in the presence of penicillin. J Biol Chem 230:435–446
4. Gallo R (1972) RNA-dependent DNA polymerase in viruses and cells: views on the current state. Blood 39:117–137

5. Temin H, Baltimore D (1972) RNA-directed DNA synthesis and RNA tumor viruses. Adv Virus Res 17:129–186
6. Schreiber SL (1991) Chemistry and biology of immunophilins and their immunosuppressive ligands. Science 251:283–287
7. Osada H (1998) Bioprobes for investigating mammalian cell cycle control. J Antibiot 51:973–981
8. Crump A, Omura S (2011) Ivermectin, 'wonder drug' from Japan: the human use perspective. Proc Jpn Acad Ser B Phys Biol Sci 87:13–28
9. Osada H (2001) An overview on the diversity of actinomycete metabolites. Actinomycetol 15:11–14
10. Swinney DC, Anthony J (2011) How were new medicines discovered? Nat Rev Drug Discov 10:507–519
11. Futamura Y, Kawatani M, Kazami S, Tanaka K, Muroi M, Shimizu T, Tomita K, Watanabe N, Osada H (2012) MorphoBase, an encyclopedic cell morphology database, and its use for drug target identification. Chem Biol 19:1620–1630
12. Takahashi H, Osada H, Koshino H, Sasaki M, Onose R, Nakakoshi M, Yoshihama M, Isono K (1992) Reveromycins, new inhibitors of eukaryotic cell growth II. Biological activities. J Antibiot 45:1414–1419
13. Futamura Y, Kawatani M, Muroi M, Aono H, Nogawa T, Osada H (2013) Identification of a molecular target of a novel fungal metabolite, pyrrolizilactone, by phenotypic profiling systems. Chembiochem 14:2456–2463
14. Futamura Y, Muroi M, Osada H (2013) Target identification of small molecules based on chemical biology approaches. Mol BioSyst 9:897–914
15. Arkin M, Tang Y, Wells J (2014) Small-molecule inhibitors of protein–protein interactions: progressing toward the reality. Chem Biol 21:1102–1114
16. Watanabe N, Osada H (2016) Small molecules that target phosphorylation dependent protein–protein interaction. Bioorg Med Chem 24:3246–3254
17. Yanai T, Kurosawa A, Nikaido Y, Nakajima N, Saito T, Osada H, Konno A, Hirai H, Takeda S (2016) Identification and molecular docking studies for novel inverse agonists of SREB, super conserved receptor expressed in brain. Genes Cells 21:717–727
18. Watanabe N, Osada H (2012) Phosphorylation-dependent protein–protein interaction modules as potential molecular targets for cancer therapy. Curr Drug Targets 13:1654–1658
19. Elia AE, Cantley LC, Yaffe MB (2003) Proteomic screen finds pSer/pThr-binding domain localizing Plk1 to mitotic substrates. Science 299:1228–1231
20. Watanabe N, Sekine T, Takagi M, Iwasaki JI, Imamoto N, Kawasaki H, Osada H (2009) Deficiency in chromosome congression by the inhibition of PLK1 polo box domain–dependent recognition. J Biol Chem 284:2344–2353
21. Kondoh Y, Honda K, Osada H (2015) Construction and application of a photo-cross-linked chemical array. Methods Mol Biol 1263:29–41
22. Miyazaki I, Simizu S, Ichimiya H, Kawatani M, Osada H (2008) Robust and systematic drug screening method using chemical arrays and the protein library: identification of novel inhibitors of carbonic anhydrase II. Biosci Biotechnol Biochem 72:2739–2749
23. Kanoh N, Kyo M, Inamori K, Ando A, Asami A, Nakao A, Osada H (2006) SPR imaging measurement of photo-cross-linked small molecule microarrays on gold. Annal Chem 78:2226–2230
24. Kondoh Y, Osada H (2013) High-throughput screening identifies small molecule inhibitors of molecular chaperones. Curr Pharm Des 19:473–492
25. MacBeath G, Schreiber SL (2000) Printing proteins as microarrays for high-throughput function determination. Science 289:1760–1763
26. Hergenrother P, Depew K, Schreiber S (2000) Small-molecule microarrays: covalent attachment and screening of alcohol-containing small molecules on glass slides. J Am Chem Soc 122:7849–7850

27. Barnes-Seeman D, Park S, Koehler A, Schreiber S (2003) Expanding the functional group compatibility of small-molecule microarrays: discovery of novel calmodulin ligands. Angew Chem Int Ed Eng 42:2376–2379
28. Kanoh N, Kumashiro S, Simizu S, Kondoh Y, Hatakeyama S, Tashiro H, Osada H (2003) Immobilization of natural products on glass slides by using a photoaffinity reaction and the detection of protein-small-molecule interactions. Angew Chem Int Ed 42:5584–5587
29. Lomenick B, Hao R, Jonai N, Chin RM, Aghajan M, Warburton S, Wang J, Wu RP, Gomez F, Loo JA, Wohlschlegel JA, Vondriska TM, Pelletier J, Herschman HR, Clardy J, Clarke CF, Huang J (2009) Target identification using drug affinity responsive target stability (DARTS). Proc Natl Acad Sci U S A 106:21984–21989
30. Lomenick B, Olsen RW, Huang J (2011) Identification of direct protein targets of small molecules. ACS Chem Biol 6:34–46
31. Martinez Molina D, Jafari R, Ignatushchenko M, Seki T, Larsson EA, Dan C, Sreekumar L, Cao Y, Nordlund P (2013) Monitoring drug target engagement in cells and tissues using the cellular thermal shift assay. Science 341:84–87
32. Jafari R, Almqvist H, Axelsson H, Ignatushchenko M, Lundback T, Nordlund P, Martinez Molina D (2014) The cellular thermal shift assay for evaluating drug target interactions in cells. Nat Protoc 9:2100–2122
33. Shoemaker RH (2006) The NCI60 human tumour cell line anticancer drug screen. Nat Rev Cancer 6:813–823
34. Kong D, Yamori T (2012) JFCR39, a panel of 39 human cancer cell lines, and its application in the discovery and development of anticancer drugs. Bioorg Med Chem 20:1947–1951
35. Muroi M, Kazami S, Noda K, Kondo H, Takayama H, Kawatani M, Usui T, Osada H (2010) Application of proteomic profiling based on 2D-DIGE for classification of compounds according to the mechanism of action. Chem Biol 17:460–470
36. Arita Y, Nishimura S, Matsuyama A, Yashiroda Y, Usui T, Boone C, Yoshida M (2011) Microarray-based target identification using drug hypersensitive fission yeast expressing ORFeome. Mol BioSyst 7:1463–1472
37. Ohtsu Y, Ohba R, Imamura Y, Kobayashi M, Hatori H, Zenkoh T, Hatakeyama M, Manabe T, Hino M, Yamaguchi Y, Kataoka K, Kawaguchi H, Watanabe H, Handa H (2005) Selective ligand purification using high-performance affinity beads. Anal Biochem 338:245–252
38. Kanoh N, Nakamura T, Honda K, Yamakoshi H, Iwabuchi Y, Osada H (2008) Distribution of photo-cross-linked products from 3-aryl-3-trifluoromethyl-diazirines and alcohols. Tetrahedron 64:5692–5698
39. Kanoh N, Takayama H, Honda K, Moriya T, Teruya T, Simizu S, Osada H, Iwabuchi Y (2010) Cleavable linker for photo-cross-linked small-molecule affinity matrix. Bioconjug Chem 21:182–186
40. Sato S, Kwon Y, Kamisuki S, Srivastava N, Mao Q, Kawazoe Y, Uesugi M (2007) Polyproline-rod approach to isolating protein targets of bioactive small molecules: isolation of a new target of indomethacin. J Am Chem Soc 129:873–880
41. Saito A, Kawai K, Takayama H, Kanoh N, Sudo T, Osada H (2008) Improvement of photoaffinity SPR imaging platform and determination of binding site of p62/SQSTM1 to p38 MAP kinase. Chem Asian J 3:1607–1612
42. Chin RM, Fu X, Pai MY, Vergnes L, Hwang H, Deng G, Diep S, Lomenick B, Meli VS, Monsalve GC, Hu E, Whelan SA, Wang JX, Jung G, Solis GM, Fazlollahi F, Kaweeteerawat C, Quach A, Nili M, Krall AS, Godwin HA, Chang HR, Faull KF, Guo F, Jiang M et al (2014) The metabolite alpha-ketoglutarate extends lifespan by inhibiting ATP synthase and TOR. Nature 510:397–401
43. Kawamura T, Kawatani M, Muroi M, Kondoh Y, Futamura Y, Aono H, Tanaka M, Honda K, Osada H (2016) Proteomic profiling of small-molecule inhibitors reveals dispensability of MTH1 for cancer cell survival. Sci Rep 6:26521

44. Choi SM, Kim Y, Shim JS, Park JT, Wang RH, Leach SD, Liu JO, Deng C, Ye Z, Jang YY (2013) Efficient drug screening and gene correction for treating liver disease using patient-specific stem cells. Hepatology 57:2458–2468
45. Wada A, Hara S, Osada H (2014) Ribosome display and photo-cross-linking techniques for in vitro identification of target proteins of bioactive small molecules. Anal Chem 86:6768–6773
46. Aretz J, Kondoh Y, Honda K, Anumala UR, Nazare M, Watanabe N, Osada H, Rademacher C (2016) Chemical fragment arrays for rapid druggability assessment. Chem Commun (Camb) 52:9067–9070
47. Osada H (2016) Chemical and biological studies of reveromycin A. J Antibiot 69:723–730. doi:10.1038/ja.2016.57

Chapter 2
Cell Proliferation and Differentiation

Nobumoto Watanabe and Hiroyuki Osada

Abstract Cell proliferation and differentiation are highly coordinated by cellular regulatory proteins. These proteins receive and transduce signals from external and internal stimuli and determine cell fate accordingly. When one or more of these strict control systems are impaired, cells begin to grow disorderly and become malignant. Even in normal development, these systems regulate the capacity for differentiation—i.e., pluripotency—and modulate cell differentiation. The artificial induction of factors that are essential for pluripotency has recently been shown to render differentiated cells undifferentiated.

In this chapter, several essential systems for cell proliferation and differentiation are described, as are recently identified small molecules that regulate them. Small molecules that inhibit the activity of factors that mediate malignant tumor cell growth can be exploited for cancer therapy. Small molecules also render cells undifferentiated. In addition, there are many small molecules that regulate the activity of cellular regulatory proteins and can be valuable tools to study the signaling systems in growth and differentiation—i.e., bioprobes.

Keywords Cell growth • Cell cycle • Protein kinase • Small molecule • Signal transduction • Protein phosphorylation

2.1 Cell Growth Signaling Pathway

2.1.1 Growth Factors and Receptors

Cells receive growth signals from the environment, such as growth factors. Such signals are received by growth factor receptors, causing cells to enter the cell cycle.

N. Watanabe (✉)
Bio-Active Compounds Discovery Research Unit, RIKEN Center for Sustainable Resource Science, 2-1 Hirosawa, Wako 351-0198, Saitama, Japan
e-mail: nwatanab@riken.jp

H. Osada
Chemical Biology Research Group, RIKEN Center for Sustainable Resource Science, 2-1 Hirosawa, Wako, Saitama 351-0198, Japan

© Springer Japan KK 2017

H. Osada (ed.), *Bioprobes*, DOI 10.1007/978-4-431-56529-1_2

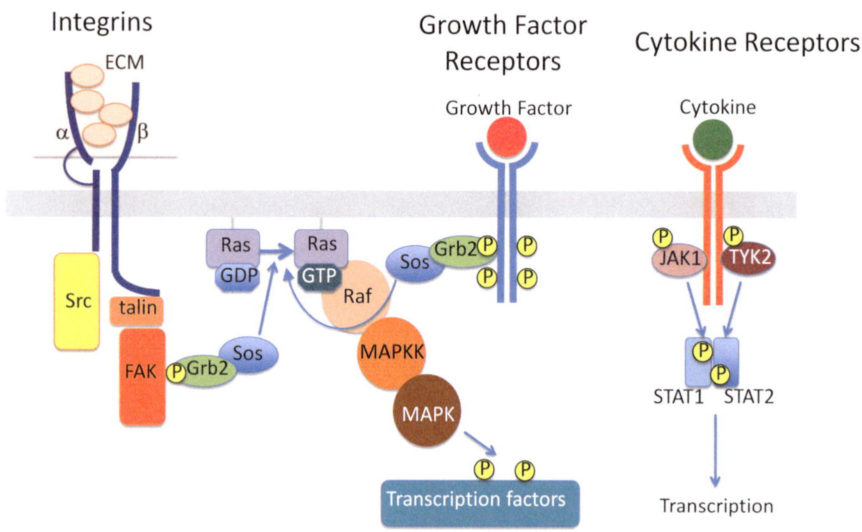

Fig. 2.1 Signaling pathways of membrane-bound receptors. Representative signaling pathways from cytokine receptor, growth factor receptor, and integrin family receptors are shown. All signals are transduced, culminating in the activation of nuclear transcription through various signaling molecules

Most growth factor receptors have protein tyrosine kinase activity [1]. On receiving growth factor signals, these receptors often autophosphorylate their tyrosine residues, which regulates the activity of the tyrosine kinase itself and affects its ability to associate with species that interact with phosphotyrosine-containing proteins, such as SH2 (Src homology 2) domain–containing proteins. Interactions with this domain effect the transduction of growth signals from the cell surface to the nucleus (Fig. 2.1.).

2.1.1.1 EGF Receptors

The study of signaling cascades by growth factors began with epidermal growth factor (EGF), the first growth factor to be discovered. EGF binds to its receptor on the cell surface and stimulates cell growth. Once EGF binds to EGF receptor, the receptor dimerizes and autophosphorylates its tyrosine residues [2]. Then, SH2 domain–containing proteins, such as Grb2/Sos complex, bind to the phosphotyrosines. For example, Sos can bind to an oncogenic protein, such as Ras, and facilitate its conversion from the GDP form to the active GTP form.

2.1.1.2 Ras-MAP Kinase Pathway

When Ras binds GTP, a domain in Ras, called the effector loop, is able to interact with many downstream signaling molecules—i.e., effectors. [3]. The first effector of Ras to be discovered was Raf kinase. Raf kinase is a cytoplasmic serine and threonine kinase that relocates to the plasma membrane on binding activated (GTP-bound) Ras. The association of Raf with Ras induces structural changes in Raf and activates Raf kinase activity.

Then, Raf phosphorylates and activates a second kinase, MEK (MAPKK), that can phosphorylate two MAP kinase members, extracellular signal–regulated kinases 1 and 2 (Erk 1, 2). The phosphorylation of Erk1 and 2 activates them, allowing them to phosphorylate several substrates and regulate cellular processes, such as transcription. This cascade is the typical signaling pathway that is stimulated on activation of Ras.

2.1.2 Cytokine Receptors Without Kinase Domains

Certain membrane-bound receptors lack a tyrosine kinase domain, including receptors for cytokines, such as granulocyte macrophage colony-stimulating factor (GM-CSF), G-CSF, interleukins (ILs), erythropoietin (EPO), and thrombopoietin (TPO). Instead, enzymes called Jaks (Janus kinases) associate with the cytoplasmic domains of these receptors through noncovalent links to transduce external signals to the inside of cells.

On binding their ligands, these receptors dimerize, the associated Jaks phosphorylate and activate each other, and the activated Jaks phosphorylate tyrosine residues in the C-terminal tails of the receptors. SH2-containing transcription factors called STATs (signal transducers and activators of transcription) bind to the resulting phosphotyrosines and are phosphorylated by Jaks. Phosphorylated STATs contain phosphotyrosine and SH2 domains that are specific for the phosphotyrosines in the surrounding sequence in STAT itself; as a result, STAT-STAT dimers form, in which each STAT uses the SH2 domain to bind to phosphotyrosines in the other STAT. STAT dimers migrate to the nucleus and function as transcription factors for genes that mediate cell proliferations, such as myc, cyclin D, and Bcl-XL.

2.1.3 Cell Adhesion Receptors

Integrins constitute a family of heterodimeric transmembrane surface receptors that comprise α and β subunits and transduce signals to Ras [4]. There are several types of subunits, and each extracellular domain in these subunits binds the extracellular matrix (ECM). α5/β1 integrin is the principal receptor for fibronectin. Another ECM molecule, laminin, binds over 10 distinct integrin heterodimers.

When cells attach to the ECM, integrins cluster and form multiple links to the ECM in localized areas, called focal adhesions. This clustering activates focal adhesion kinase (FAK), which associates with the C-terminal domain of integrins. FAK is a non-receptor-type tyrosine kinase and is phosphorylated on clustering by transphosphorylation [5]. FAK is phosphorylated at tyrosine residues, one of which becomes a docking site for c-Src. Subsequently, c-Src phosphorylates FAK at additional residues, rendering some of them ancillary binding sites for other SH2-containing molecules, such as Shc, Grb2, PI3K, and PLC-γ.

Grb2 recruits Sos into these complexes, and Sos activates Ras in a similar manner to that seen in other receptor-dependent pathways. This set of events is the typical signaling pathway of the response to extracellular adhesion with regard to cell growth through Ras activation. In certain cancer cells, in which Ras is activated by amino acid substitution, cells can grow without cell adhesion, because extracellular signaling is no longer required.

2.1.4 Dysregulation of Signaling Pathways in Cancer Cells

In other cancer cells, mutations in receptor tyrosine kinases allow the cell to grow independently of ligand binding. Because such growth relies completely on the mutated kinase activity, several therapeutic approaches have been developed. The inhibition of proteins that are essential for cancer cell growth is a very effective treatment. Among them, we will describe the development of small molecules that inhibit protein tyrosine kinase activity of cancer-specific tyrosine kinases.

2.1.4.1 Bcr-Abl Oncoprotein

In 1960, an abnormal, unusually small chromosome 22 was found in the majority of cells of chronic myeloid leukemia (CML) cells. This aberration was called the Philadelphia chromosome and turned out to be generated by reciprocal translocation between chromosomes 9 and 22 [6]. This short chromosome causes leukemia, due to generation of a fusion gene product between Abl and Bcr [7].

Abl, originally located on chromosome 9, is the human homolog of the mouse c-abl proto-oncogene. c-abl was discovered as an acquired oncogene in a rapidly tumorigenic retrovirus, Abelson murine leukemia virus. It encodes a cytoplasmic protein tyrosine kinase, and its regulatory region is lost by this breakage. The remainder of the catalytic domain of Abl is fused to Bcr (breakpoint cluster region). The resulting fusion between Abl and Bcr dysregulates normally well-controlled Abl kinase activity to produce a growth signal that activates the Ras-MAPK, phosphatidylinositol 3 (PI3K)-kinase-AKT, and JAK-STAT pathways [8].

2.1.4.2 Other Abl Fusion Genes in Cancer Cells

After the finding of the critical function of Bcr-Abl in leukemia, other types of c-Abl fusions on chromosome 9 were identified in leukemia: ETV6, RCSD1, SFPQ, ZMIZ1, NUP214, and EML1 on chromosomes 12, 1, 1, 10, and 14, respectively [9].

2.1.4.3 Inhibitor of Bcr-Abl Fusion Gene Products

Because the protein kinase activity of Bcr-Abl is essential for its transformation-inducing activity, a research program began to develop low molecular weight inhibitors of Bcr-Abl. As a result, a drug named Gleevec (CGP-57148, imatinib mesylate, STI-571, and Glivec) emerged [10]. Gleevec binds to the catalytic cleft between the two major structural lobes of the kinase domain of Bcr-Abl. Gleevec also inhibits other Abl fusion gene products, such as ETV6-ABL. Gleevec is a specific inhibitor for Abl and other tyrosine kinases, such as PDGF receptors and Kit receptors, and Arg protein.

2.1.4.4 EML4-ALK Oncoprotein

Another fusion protein with a tyrosine kinase catalytic domain was identified in a screen using a retrovirus-mediated complementary DNA expression system. To isolate novel transformation-related genes in non-small cell lung cancer (NSCLC), Mano's group in Japan generated a retroviral cDNA expression library from a lung adenocarcinoma specimen that was surgically resected from a 62-year-old male with a history of smoking. On infection of mouse 3 T3 fibroblasts with this library, they obtained many transformed foci, from which cDNA inserts were recovered by polymerase chain reaction (PCR), one of which encoded a fusion protein between EML4 (echinoderm microtubule-associated protein-like 4) and ALK (anaplastic lymphoma kinase), which was generated by a small inversion within chromosome 2p [11]. The resulting gene product was demonstrated to cause the cancer.

EML4 comprises a basic *N*-terminal region, a hydrophobic echinoderm microtubule-associated protein-like protein (HELP) domain, and WD repeats. In the predicted EML4-ALK fusion protein, the *N*-terminal section of EML4 that encompasses the basic region, the HELP domain, and a portion of the WD-repeat region are fused to the intracellular juxtamembrane region of ALK. ALK is a protein tyrosine kinase with a transmembrane domain that has been identified as a fusion partner in several tumors. The fusion point of ALK is conserved between most fusion products, including NPM1-ALK, TPM3-ALK, and EML4-ALK. The EML4-ALK fusion transcript was detected in 6.7% of NSCLC patients. Thus, this fusion kinase is a promising therapeutic target.

2.1.4.5 Inhibitor of EML4-ALK Fusion Gene Products

On identifying ALK fusion proteins in NSCLC, an inhibitor of ALK by Pfizer, crizotinib, was examined as therapy in patients who were positive for ALK gene reconstitution, as determined by fluorescent in situ hybridization [12]. The results were successful, and based on these results, other ALK inhibitors, such as AP26113 (ARIAD), CH5424802 (Chugai), ASP3026 (Astellas), and LDK378 (Novartis), were also in clinical trials and CH5424802 (Alectinib), LDK378 (Ceritinib, Zykadia) has been approved.

2.1.4.6 Drug-Resistant Mutation in ALK

As shown in the chemotherapy of Bcr-Abl-dependent and EGFR-dependent tumors, drug-resistant mutations of ALK have appeared in clinical trials of ALK inhibitors. The mutated amino acids in EML4-ALK are C1156Y, F1174L, and L1196 [13, 14]. L1196 lies in the regulatory region of the protein tyrosine kinase, which is a hot spot of mutations in the chemotherapy of other kinase-dependent tumors, such as Bcr-Abl and EGFR. Based on these findings, the development of second- and third-generation ALK inhibitors has been developed {Wu, 2016 #229}.

2.1.4.7 Other Fusion Protein Tyrosine Kinases

After the discovery of ALK fusion gene products, other such abnormalities were searched for in other types of lung cancer. KIF5B was found to fuse with ALK in at (2:10) translocation. Other fusions, such as KIF5B-RET, CCDC6-RET, SLC34A2-ROS1, CD74-ROS1, SDC4-ROS1, and TPM3-ROS1, were also identified and considered to be targets for cancer therapy [15–17].

2.2 Cell Cycle Regulation

2.2.1 Cell Cycle Regulators

The series of events that result in cell division is called the cell cycle, which begins with the duplication of cell components. During S phase (period of DNA synthesis), the genetic information in each chromosome is duplicated. Duplicated components are divided equally between two daughter cells in M phase (mitosis). This process must be conducted precisely. Preceding these steps, the cell cycle machinery, such as DNA replication enzymes, and chromosome segregation systems, must function accurately with regard to timing. To this end, many regulators effect precise advancement of the cell cycle. Abnormalities in the system, such as uncontrolled

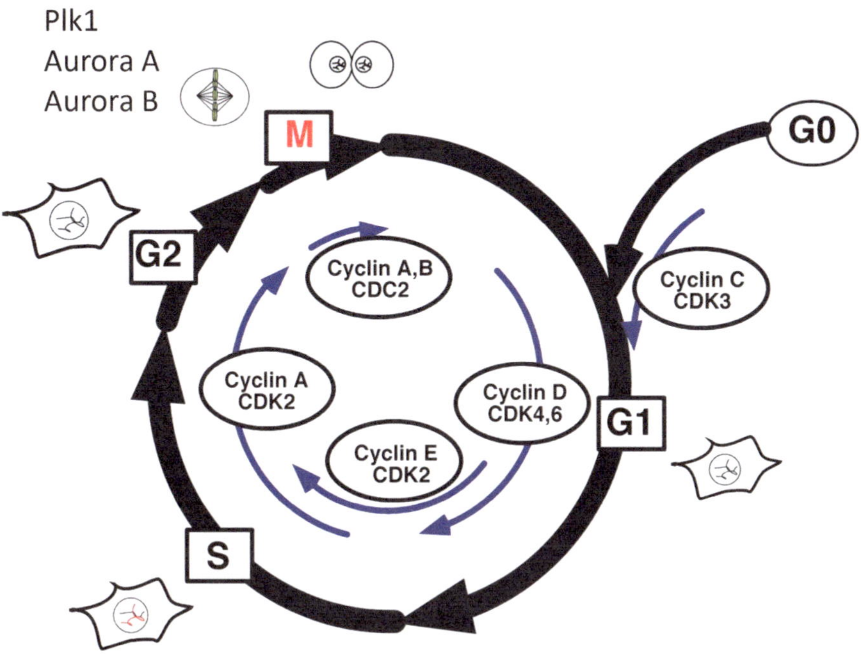

Fig. 2.2 Cell cycle-regulating protein kinases (cyclins) and cyclin-dependent kinases (CDKs) are major protein kinases that regulate the cell cycle. Cyclin A– and cyclin B–associated CDK1 has a major function in M-phase progression, as do other M-phase kinases, such as Plk1 (polo-like kinase 1) and Aurora A and B. Because these kinases are essential in cell cycle progression and cell growth, inhibitors of them are good candidate for cancer therapy

activation of regulators, lead to unrestrained cell growth, such as cancer. The small-molecule inhibitors that inhibit abnormally activated cell cycle regulators can be used as cancer therapies (Fig. 2.2).

2.2.2 Cyclins and CDKs

2.2.2.1 Cyclin and CDKs

The central components of the cell cycle system are the cyclin and cyclin-dependent kinase (CDK) complexes. Thus, CDK complexes are called the engine of the cell cycle. Initially, cyclin proteins were detected in M phase, complexing with Cdc2 (CDK1), and later, other cyclins and CDKs were identified in disparate phases of the cell cycle.

2.2.2.2 CDKs

CDKs are small proteins (approximately 35 kDa) with serine/threonine kinase activity. Activation of CDKs requires their binding to cyclin subunits and the phosphorylation of threonine in the catalytic site. In animal cells, at least nine CDKs are known, four of which function primarily during the cell cycle. CDK1 regulates M phase, whereas CDK4 and CDK6 controls G1 phase. CDK2 governs G1/S progression and M phase.

2.2.2.3 Cyclins

Cyclins were initially identified as proteins with concentrations that changed periodically during the cell cycle. Through this modulation, cyclins (cyclin B) activate CDKs (CDK1) in M phase, and inactivation of the kinase is induced by proteolysis of the cyclin at the exit of M phase. In animal cells, at least four cyclins are known to have important functions in cell cycle progression— cyclins D, E, A, and B—which bind and activate CDK family members in G1 phase (CDK4/6), G1/S phase (CDK2), S phase (CDK2,1), and M phase (CDK1).

2.2.2.4 CDK Inhibitors

Because CDKs have central functions in cell cycle progression, small-molecule inhibitors of them are expected to be effective cancer therapeutics. Thus, many small molecules with CDK inhibitory activity have been developed and examined for their clinical efficacy. Flavopiridol is a semisynthetic flavone derivative of the alkaloid rohitukine, an alkaloid that is isolated from the leaves and stems of *Amoora rohituka* and *Dysoxylum binectariferum*, plants that are indigenous to India [18]. It is a pan-CDK inhibitor, with activity against at least CDK1, 2, 4/6, 7, and 9, as well as other protein kinases.

Intravenous injection of flavopiridol induces selective apoptosis in the thymus, spleen, and lymph nodes, resulting in their atrophy, but not in the kidneys, liver, lungs, bone/bone marrow, muscle, or heart (except for intestinal crypts). Flavopiridol induces apoptosis in xenografts of human hematopoietic tumors, such as HL-60, SUDHL-4, and Nalm/6 cells. In addition to its antiproliferative effects, leading to cell cycle arrest in tumor cells, it effects apoptosis in hematopoietic cell lines [19].

Subsequent studies have revealed flavopiridol to be a global inhibitor of transcription, particularly of cell cycle and apoptosis regulators, which control mRNA turnover [20]. Flavopiridol was the first CDK inhibitor to enter clinical trials, but early results of flavopiridol monotherapy with 72-h continuous or 1-h bolus infusion schedules have generally been disappointing [21]. Recently, hematologic improvement after flavopiridol treatment was reported in a patient with hairy cell leukemia that was refractory to pentostatin and rituximab [22].

R-Roscovitine (seliciclib, CYC202) is the second CDK inhibitor to enter clinical trials. R-Roscovitine is a potent inhibitor of CDK2/cyclin E kinase activity, with an average IC_{50} value of 15.2 μM in a panel of 19 human tumor cell lines [23]. It selectively inhibits the growth of rapidly proliferating cells. In a study of the effects of CYC202 on the cell cycle in Lovo colorectal carcinoma cells, the major effect was the induction of cell death in all phases of the cell cycle—not merely in specific phases. Consistent with these results, R-Roscovitine inhibits the activity of CDK2 and CDK1, 5, 7, 8, and 9. Seliciclib exposure to human diffuse large B-cell lymphoma (DLBCL) cells results in G1- and G2/M-phase arrest and induction of apoptosis independently of any underlying chromosomal translocations [24].

Dinaciclib (Scheme 727965) inhibits CDKs 1, 2, 5, and 9 more potently than flavopiridol [25]. Dinaciclib completely suppresses retinoblastoma phosphorylation in cell-based assays, correlating with the onset of apoptosis and total inhibition of bromodeoxyuridine incorporation in >100 tumor cell lines of various origins and backgrounds. Further, dinaciclib induces regression in established solid tumors in several mouse models, inhibits the growth of AML (acute myeloid leukemia) and ALL (acute lymphoblastic leukemia) cell lines in vitro, and induces apoptosis. Based on these results, dinaciclib is in several phase II clinical studies [26].

Palbociclib (PD 0332991) is a highly specific inhibitor of CDK4 and CDK6, without any activity against a panel of 36 other protein kinases [27]. It is a potent antiproliferative compound against retinoblastoma (Rb)-positive tumor cells in vitro, inducing G1 arrest exclusively, with a significant reduction in phosphorylation at Ser780/Ser795 of Rb. Palbociclib is being examined in a phase III trial in breast cancer patients [28].

2.2.3 *M-Phase Kinases and Their Inhibitors*

As discussed above, CDK1 is the major protein kinase that is active during mitosis. But, several protein kinases are specifically expressed in mitosis that have important functions in the progression of mitosis. Two major families of these M-phase kinases are polo-like kinases (Plks) and Aurora kinases [29]. Mammals express four plks (Plk1, Plk2, Plk3, and Plk4). Plk1 is the only one that regulates mitotic progression. There are three Aurora kinases (Aurora A, B, and C) in mammalian cells, all of which mediated the progression of mitosis.

2.2.3.1 Plk1

Polo-like kinase-1 (PLK-1) is a conserved serine/threonine (Ser/Thr) kinase that was first reported to induce abnormal mitotic spindles in mutant *Drosophila* [30, 31]. During cell cycle progression, Plk1 levels increase in late G2 and decline rapidly by proteolysis on exit from mitosis. The localization of Plk1 also changes dramatically when cells transit through mitotic stages. Plk1 localizes to the centrosome during late G2 and mitotic prophase, during which it regulates centrosome maturation, separation, and microtubule nucleation [32–34]. Further, Plk1 regulates the activation of Cdc25 and degradation of Wee1 during mitotic entry [35–38].

Some Plk1 molecules relocalize to kinetochores in mitotic metaphase. Plk1 has significant functions in the spindle assembly checkpoint pathway and proper chromosome segregation. By sensitizing the tension between sister chromatids, Plk1 stabilizes kinetochore–microtubule interactions by phosphorylating kinetochore proteins, such as BubR1, and proteins that harbor 3F3/2 epitopes [33, 39–43]. The localization of Plk1 to chromosomal arms is essential for the loss of arm cohesion during M phase [44, 45]. Its localization is also important for the proper accumulation of kinetochore proteins, such as BubR1, Mad1, Mad2, Cdc20, and CENP-E [33, 40, 41, 43, 46, 47].

During the onset of anaphase, Plk1 relocalizes to the spindle midzone and regulates microtubule sliding. During telophase, when chromosomes segregate, Plk1 mediates the formation of the cleavage furrow in the central spindle and midbody [48–50]. This distribution of Plk1 is governed by its noncatalytic C-terminal half, the polo box domain (PBD), which comprises two polo box motifs [51, 52]. The PBD contains a phosphopeptide binding domain that binds with maximum affinity to phosphopeptides that harbor the S-pS/pT-P/*X* sequence [53, 54].

2.2.3.2 Plk1 Inhibitors

In the *N*-terminal region of Plk1, there is a highly conserved protein kinase domain. The kinase domain is one of the most promising drug targets [55], for which many inhibitors have been developed [56]. ON 01910.Na, a benzyl styryl sulfone analog, is an ATP-noncompetitive, multitargeted inhibitor of several tyrosine kinases and cyclin-dependent kinases. ON 01910 Na has strong potency (IC_{50} 9–10 nM) against Plk1 [57], although some results suggest that Plk1 is not its primary target [58, 59]. Microtubule dynamics are affected in ON01910.Na-treated cells, which experience mitotic defects with multipolar spindles and centrosomal abnormalities, inducing mitotic arrest and apoptosis [60].

ON01910.Na inhibits the growth of many cancer cell lines, including several drug-resistant cell lines, with IC_{50} values ranging from 50 to 250 nM in vivo and in vitro [57]. The phase I study of ON01910.Na has been completed, reporting several common adverse effects, such as fatigue, pain nausea, vomiting, and abdominal pain [61]. Additional phase I studies for patients with advanced cancers and B-cell chronic lymphocytic leukemia are also being conducted; one partial response was observed in a patient with ovarian cancer [61].

BI2536 is a dihydropteridinone derivative with inhibitory potency toward Plk1 that is more than 10,000-fold greater (IC_{50} 0.83 nM) than 63 other protein kinases [59]. It inhibits the proliferation of 32 human cancer cell lines, with EC_{50} values of

2–25 nM [59]. Because similar inhibitory effects have been observed in nontransformed, immortalized cells, BI 2536 might affect the growth of all proliferating cells. Cells that are treated with BI 2536 arrest in M phase with monopolar mitotic spindles, a phenotype that is similar to that induced by an siRNA against Plk1, suggesting that Plk1 is the main cellular target [59].

A phase I study of BI 2536 was conducted in 40 patients with refractory or metastatic solid tumors who were treated with a median of three chemotherapy regimens [62]. Patients who were administered the maximum tolerated dose (MTD) experienced grade 3/4 adverse events, including neutropenia (55.6%) and leukopenia (44.4%). Similar results were obtained in another phase I study of 70 patients with advanced solid cancers [63]. In the monotherapy phase II trial of BI 2536, 95 patients with advanced or metastatic NSCLC who had progressed after, or had failed, first- or second-line therapy received a single i.v. dose of 200 mg on day 1 or a single i.v. dose of 50–60 mg on days 1, 2, and 3 of each treatment course [64]. Neutropenia was the chief grade 4 adverse event, and most of the reported adverse events transiently affected the hematopoietic system.

GSK461364A, an imidazotriazine, inhibits Plk1 in an ATP-competitive manner with more than 1000-fold greater potency than most of the 48 other protein kinases [65]. The most apparent phenotype for GSK461364A-treated cells is prometaphase arrest, with a characteristic collapsed polar polo spindle. GSK461364A delays mitotic entry in G2, which is followed by gradual progression to terminal mitosis at high concentrations; in some cell lines, this phenomenon correlates with decreased apoptosis. Inhibition of cell culture growth by GSK461364A can be cytostatic or cytotoxic but leads to tumor regression in xenograft tumor models under proper dose scheduling. A phase I first-in-humans study of GSK461364 has been conducted in 40 patients with advanced solid malignancies [66].

Volasertib (BI 6727) is a dihydropteridinone derivative that targets Plk1 with selectivity and efficiency (IC_{50} 0.87 nM) similar to those of BI 2536 [67]. BI 6727 has a pharmacokinetic profile that favors sustained exposure of tumor tissues with a high volume and a long terminal half-life in mice and rats. BI 6727 has marked tumor inhibitory activity in several models of cancer, including a model of taxane-resistant colorectal cancer. With oral and i.v. administration, the total weekly dose of BI 6727 is most relevant for efficacy, supporting the use of a variety of well-tolerated dosing schedules. A phase I first-in-humans study of volasertib has been conducted successfully in patients with advanced solid tumors, including patients with NSCLC.

2.2.3.3 PBD-Dependent Binding Inhibitors

As discussed, Plk1 has many essential functions during mitosis [30, 31]. The localization of Plk1 alters dramatically when cells transit through mitotic stages; these changes are mediated by its noncatalytic C-terminal region, the PBD, which contains two polo box motifs [51, 52]. Yaffe and colleagues discovered that the PBD harbors a phosphopeptide binding domain that binds with maximum affinity to phosphopeptides that bear the S(pS/pT)(P/X) sequence [53, 54]. Crystallographic studies have confirmed that phosphopeptides are bound to the interface between the two polo boxes [54]. This PBD-dependent binding is important for subcellular localization and necessary for targeting Plk1 to specific substrates [52, 54].

Recently, we established a screening system to isolate small-molecule inhibitors of phosphorylation-dependent protein–protein interactions [68]. In this system, to quantitate the recognition of phosphorylation-dependent binding modules to their target peptides, the open reading frame of a fluorescent protein, such as GFP, is fused to the modules, and the fusion proteins are expressed in bacteria. Phosphopeptides of the binding sequence that are derived from the target proteins are synthesized chemically and bound covalently to maleimide-activated 96-well plates.

Using this screen, we examined compounds in RIKEN NPDepo (Natural Products Depository). Of approximately 2500 compounds, a benzotropolone-containing natural compound from nutgalls, purpurogallin (PPG), was identified as a robust inhibitor of PBD-dependent binding [68]. Several inhibitors of PBD-dependent binding, including ours, have been identified [69, 70]—some are small molecules and others are peptide-derived inhibitors. The detailed binding modes of PPG and another inhibitor have been solved by homology modeling [71]. It is expected that more potent inhibitors will be discovered.

The development of kinase domain–directed inhibitors for cancer therapy has long been pursued, but this class of inhibitors experiences high cross-reactivity. In this regard, targeting an essential noncatalytic domain in protein kinases is an effective strategy of isolating specific inhibitors. The PBD is an attractive example of such an approach. The phosphorylation-dependent protein–protein interaction is a significant binding mode of essential noncatalytic domains, and our screen to isolate inhibitors of this module will be an effective method of identifying such small-molecule inhibitors [72].

2.2.3.4 Aurora Kinases

The Aurora kinase family was discovered as a family of mitotic kinases during a study of mutant alleles that were associated with defective spindle pole organization in *Drosophila melanogaster* [73]. This family has been highly conserved during evolution. There are three paralogs of Aurora kinases in humans, designated A, B, and C [74]. Aurora A localizes to the centrosome during mitosis. Aurora B, a chromosomal passenger protein, localizes to the centromeres from the prophase to metaphase–anaphase transition. Then, it travels to the midzone of spindle microtubules during telophase and subsequently to the midbody during cytokinesis. Aurora C is another chromosomal passenger protein with localization similar to Aurora B [75].

Aurora kinases have multiple functions in mitosis. Aurora A localizes to centrosomes and regulates mitotic entry, separation of centrioles, bipolar spindle assembly, metaphase chromosomal alignment, and completion of cytokinesis.

Recently, Aurora A was also reported to govern the promotion of nuclear envelope breakdown. Aurora B has functions in chromosomal biorientation, regulation of the association between kinetochores and microtubules, and cytokinesis [76]. During chromosomal biorientation, Aurora B mediates the release of incorrect kinetochore–microtubule attachments. Through phosphorylation of a serine residue in histone H3, Aurora B regulates chromatin condensation and separation [77].

Compared with Aurora A and B, the functions of Aurora C are not as extensively described, but it is believed to act similarly to Aurora B. Aurora C associates with survivin and might be essential for cytokinesis [78]. Because these kinases maintain the genetic stability of cells, their aberrant expression leads to genomic instability or aneuploidy. Further Aurora kinases are overexpressed in many tumors.

2.2.3.5 Aurora Kinase Inhibitors

Because Aurora kinases are frequently overexpressed in cancer cells and have important functions in their growth, many small-molecule inhibitors of these kinases have been developed. Some are effective against all three members (pan-Aurora kinase inhibitors), whereas others are specific for Aurora A or B. Several compounds are being examined in clinical studies as cancer therapies.

2.2.3.6 First-Generation Inhibitors of Aurora Kinases

Hesperadin and ZM447439 are the first-generation Aurora kinase inhibitors that were developed by Boehringer Ingelheim and AstraZeneca, respectively [79, 80]. Both preferentially inhibit Aurora kinase B. Based on the use of these inhibitors, Aurora B was found to be important for chromosomal alignment through kinetochore–microtubule attachments in metaphase, activation of the spindle assembly checkpoint, and cytokinesis.

2.2.3.7 Pan-Aurora Inhibitors

Several small molecules inhibit Aurora kinases to a greater extent than first-generation compounds, some of which inhibit all Aurora kinases (pan-Aurora inhibitors).

One of the most frequently used pan-Aurora kinase inhibitors is VX-680 [81]. VX-680 inhibits Aurora A, B, and C with Ki values of 0.7, 18, and 4.6 nM, respectively [82] and suppresses the proliferation of several tumor cell lines, with IC_{50} values ranging from 15 to 130 nM. In many VX-680 treated cancer cell lines, mitosis is disrupted due to alterations in chromosomal segregation and cytokinesis, eventually inducing the accumulation of cells with 4n DNA content, activating checkpoints and subsequently effecting apoptosis [82].

Although clinical trials of this compound were suspended due to its toxicity profile, it has been tested in patients with ALL, myelodysplastic syndrome, and advanced CML, because it inhibits the T315I mutant of BCR-ABL, which is resistant to imatinib. VX-680 binds to wild-type BCR-ABL with a dissociation constant (K_d) of ~20 nM and to T315I with a K_d of 5 nM. In in vitro assays, VX-680 inhibits the activity of wild-type BCR-ABL and T315I BCR-ABL with IC_{50} values of 10 and 30 nM, respectively [83].

Other pan-Aurora kinase inhibitors, such as PHA739358, which was developed from PHA-680632 as a more potent compound, SNS-314, R763, GSK-1070916, AT9283, CYC116, and AMG-900, have been developed by pharmaceutical companies [46, 84–89]. Their activities have been examined with regard to the inhibition of enzymatic activities, tumor shrinkage in xenograft models, and inhibition of histone H3 phosphorylation during mitosis. The common phenotype that is induced by these compounds in cultured cells is mitotic failure. These agents are now being investigated in clinical studies and have demonstrated efficacy.

2.2.3.8 Aurora A–Specific Inhibitors

Second-generation Aurora kinase inhibitors inhibit each subtype of Aurora specifically. MLN8237 was developed from MLN8054 by Millennium Pharmaceuticals [90]. MLN8237 is the first orally available selective small-molecule inhibitor of Aurora A and is currently in early-phase clinical testing in patients with advanced solid tumors and acute myelogenous leukemia. MK5108 is also a potent Aurora A–specific inhibitor that was developed by Vertex Pharmaceuticals [91]. Although it also inhibits Aurora B and C, the IC_{50} values are 220- and 190-fold higher than for Aurora A, respectively [91]. MK5108 has been studied in a phase I clinical trial in patients with advanced solid tumors [92]. ENMD-2076, developed by EntreMed, also inhibits Aurora A specifically, although it has activity other kinases, such as FLT3, VEGFR2, Src, and FGFR1, at similar concentrations [93]. Phase II clinical trials of ENMD-2076 in patients with platinum-resistant ovarian cancer were reported.

2.2.3.9 Aurora B–Specific Inhibitors

The Aurora B–specific inhibitor AZD1152 (barasertib) was developed by AstraZeneca and its phase II clinical studies in patients with several types of leukemia have been reported [94] [95]. AZD1152 is a pyrazoloquinazoline derivative and has greater than 1000-fold selectivity for Aurora B over Aurora A in recombinant enzyme assays. AZD1152 is converted rapidly to the active drug AZD1152 hydroxy-QPA (AZD1152-HQPA) in human plasma [96]. AZD1152 inhibits the proliferation of AML cell lines (HL-60, MOLM13, NB4), an ALL cell line (PALL-2), biphenotypic leukemia (MV4–11), and acute eosinophilic leukemia (EOL-1). These cells had 4 N/8 N DNA content and underwent apoptosis, as measured by cell cycle analysis and annexin V staining, respectively. Notably, AZD1152 synergistically enhanced the antiproliferative effects of vincristine, a tubulin-depolymerizing agent, and daunorubicin, a topoisomerase II inhibitor, in MOLM13 and PALL-2 cells in vitro [97].

2.3 Cell Differentiation

The human body consists of approximately 6×10^{13} cells that are derived from a fertilized egg. A balance of proliferation and differentiation is coordinated during development. The cell must implement control mechanisms that decide whether it should differentiate or proliferate under specific environmental conditions, such as the nutritional state.

The molecular control of development has been studied extensively in fruit fly and nematodes. In mammals, cell cultures have served as surrogate models for entire organisms in identifying molecules that affect cell commitment to particular lineages.

The first edition of this book described mammalian differentiation systems, such as hematopoietic cells, neuron cells, skeletal cells, hepatocytes, and epidermal tissues, and small molecules. Fifteen years later, pluripotency is now known to be controlled, even in mammalian cells. Based on the findings of Yamanaka's group, pluripotent cells can be obtained from fully differentiated cells by gene transfection [98]. Recently, small molecules have been reported to induce pluripotent cells [99]. We will introduce these important findings in the following sections.

In addition to these conventional stem cells, the concept of cancer stem cells has advanced to the forefront of cancer research. Formal evidence that cancer cells can self-renew and differentiate has been obtained in animal transplant studies. Cancer stem cells and recent progress in cancer therapy that targets them are also discussed.

2.3.1 iPS Cells

Embryonic stem (ES) cells from the inner cell mass of mammalian blastocysts can grow indefinitely and maintain pluripotency and the ability to differentiate into cells of all three germ layers [100, 101]. Similarly, somatic cells can be reprogrammed by transferring their nuclear contents into oocytes [102] or by fusion with ES cells [103, 104]. These findings indicate that oocytes and ES cells contain factors that render somatic cells pluripotent.

Yamanaka et al. hypothesized that the factors that maintain ES cell identity also have essential functions in the induction of pluripotency in somatic cells. They examined whether several transcription factors that maintain pluripotency in ES cells and early embryonic cells, in addition to other gene products, and are specifically expressed in ES cells induce pluripotency in somatic cells. By combining four specific factors (Oct3/4, Sox2, c-Myc, and Klf4), they generated pluripotent cells, which they called induced pluripotent stem (iPS) cells, directly from mouse embryonic and adult fibroblast cultures [98].

2.3.2 iPS Cells Induced by Small-Molecule Compounds

As discussed, pluripotent stem cells can be induced from somatic cells and used for medical purposes. Because somatic cells can be obtained from patients as an unlimited resource, this approach is very useful strategy for regenerative medicine. However, there are several concerns with reprogramming by genetic manipulation, such as gene introduction, with regard to their clinical application.

Thus, being able to reprogram somatic cells with small-molecule compounds has several advantages. They can be cell permeable and have become more cost effective, nonimmunogenic, preserved, and more easily synthesized. Attempts to establish stem cells from somatic cells have been made by many groups. A group at Peking University identified a combination of seven small compounds that could drive the reprogramming of mouse somatic cells into pluripotent cells [99]. Initially, they developed a small-molecule combination, called VC6T [comprising valproate (an HDAC inhibitor), CHIR99021 (a GSK3beta inhibitor), 616452 (a TGFbetaRI inhibitor), and tranylcypromine (a monoamine oxidase inhibitor)], that reprograms cells with a single gene, *Oct4* [105].

After screening chemicals that substituted for Oct4, they identified three additional compounds, FTTZ [forskolin (a cAMP agonist), DZNep (an *S*-adenosylhomocysteine hydrolase inhibitor), and TTNPB (a retinoic acid receptor agonist)]. In total, these seven compounds successfully reprogramed mouse somatic cells into pluripotent stem cells, which the researchers named CiPSCs (chemically induced pluripotent stem cells). They also found that four of these compounds "C6FZ" induced CiPSCs from MEFs and MAFs, albeit with a tenfold lower efficiency than VC6TFZ. Although the complete chemical reprogramming approach needs to be improved to reprogram human somatic cells, these findings increase our understanding of the establishment of pluripotent cells.

2.3.3 Medical Purposes

The reprogramming of somatic cells toward pluripotency using small molecules indicates the capacity of differentiated cells to modulate molecular pathways that are nonspecific to pluripotency via small molecules. Thus, functionally desirable cell types might be able to be developed in regenerative medicine by reprogramming cell fate using specific small molecules. To this end, the complete chemical reprogramming approach might meet the needs of regenerative medicine.

2.3.4 Cancer Stem Cells

Like stem cells in normal tissues, which are essential for their maintenance through continuous growth and appropriate differentiation, cancer stem cells (CSCs) exist in tumors. CSCs have been proposed to be a subpopulation of cancer cells with high clonogenic capacity. Evidence of the existence of CSCs was first obtained in acute myeloid leukemia—a certain cell population in AML is self-maintaining and has the capacity for multilineage [106].

This concept has been widely accepted, and CSCs are now believed to exist not only in AML but also in many other tumors. Several CSC markers have been identified in a wide range of solid and hematopoietic malignancies. CSCs are considered to be the cause of resistance against chemotherapy, metastasis, and tumor relapse. Thus, much effort is under way to develop small molecules that target CSCs.

2.3.5 Small Molecules that Target CSCs

Based on analyses of the properties of CSCs, several small molecules have been developed to specifically target CSCs. One such target is the drug efflux pathway. Several members of the ATP binding cassette (ABC) transporter family, such as multidrug resistance protein 1 (MDR1/ABCB1), breast cancer resistant protein (BCRP/ABCG2), and multidrug resistance–associated protein 1 (MRP1/ABCC1), are up-regulated in CSCs [107]. Thus, CSCs can be selected as a population that effluxes Hoechst dye more frequently [108].

Small molecules that target these transporters are potential lead compounds that eliminate CSCs. Mithramycin A, a polyauroleic acid from *Streptomyces*, represses ABCG2 and inhibits stem cell signaling [109]. Salinomycin, a polyether ionophore, eliminates CSCs, likely due to interference with ABC transporters and Wnt pathways [110]. Because the ability to differentiate into various cell types is one of the chief properties of stem cells, inducing differentiation through loss of their stemness is an attractive strategy for eradicating CSCs.

In this regard, all-trans-retinoic acid (ATRA) induces the differentiation of stem-like glioma cells, rendering it a potential treatment for glioma [111]. Pathways in apoptotic signaling, DNA repair, and quiescence are also notable targets for the eradication of CSCs, for which small molecules are being extensively screened [112].

References

1. Ullrich A, Schlessinger J (1990) Signal transduction by receptors with tyrosine kinase activity. Cell 61(2):203–212
2. Schlessinger J (2002) Ligand-induced, receptor-mediated dimerization and activation of EGF receptor. Cell 110(6):669–672
3. Roberts PJ, Der CJ (2007) Targeting the Raf-MEK-ERK mitogen-activated protein kinase cascade for the treatment of cancer. Oncogene 26(22):3291–3310. doi:10.1038/sj.onc.1210422
4. Giancotti FG, Ruoslahti E (1999) Integrin signaling. Science 285(5430):1028–1032
5. Schlaepfer DD, Mitra SK (2004) Multiple connections link FAK to cell motility and invasion. Curr Opin Genet Dev 14(1):92–101. doi:10.1016/j.gde.2003.12.002
6. Rowley JD (1973) Letter: A new consistent chromosomal abnormality in chronic myelogenous leukaemia identified by quinacrine fluorescence and Giemsa staining. Nature 243 (5405):290–293
7. Groffen J, Stephenson JR, Heisterkamp N, de Klein A, Bartram CR, Grosveld G (1984) Philadelphia chromosomal breakpoints are clustered within a limited region, bcr, on chromosome 22. Cell 36(1):93–99
8. Hazlehurst LA, Bewry NN, Nair RR, Pinilla-Ibarz J (2009) Signaling networks associated with BCR-ABL-dependent transformation. Cancer Control 16(2):100–107
9. De Braekeleer E, Douet-Guilbert N, Rowe D, Bown N, Morel F, Berthou C, Ferec C, De Braekeleer M (2011) ABL1 fusion genes in hematological malignancies: a review. Eur J Haematol 86(5):361–371. doi:10.1111/j.1600-0609.2011.01586.x
10. Druker BJ, Tamura S, Buchdunger E, Ohno S, Segal GM, Fanning S, Zimmermann J, Lydon NB (1996) Effects of a selective inhibitor of the Abl tyrosine kinase on the growth of Bcr-Abl positive cells. Nat Med 2(5):561–566
11. Soda M, Choi YL, Enomoto M, Takada S, Yamashita Y, Ishikawa S, Fujiwara S, Watanabe H, Kurashina K, Hatanaka H, Bando M, Ohno S, Ishikawa Y, Aburatani H, Niki T, Sohara Y, Sugiyama Y, Mano H (2007) Identification of the transforming EML4-ALK fusion gene in non-small-cell lung cancer. Nature 448(7153):561–566. doi:10.1038/nature05945
12. Kwak EL, Bang YJ, Camidge DR, Shaw AT, Solomon B, Maki RG, Ou SH, Dezube BJ, Janne PA, Costa DB, Varella-Garcia M, Kim WH, Lynch TJ, Fidias P, Stubbs H, Engelman JA, Sequist LV, Tan W, Gandhi L, Mino-Kenudson M, Wei GC, Shreeve SM, Ratain MJ, Settleman J, Christensen JG, Haber DA, Wilner K, Salgia R, Shapiro GI, Clark JW, Iafrate AJ (2010) Anaplastic lymphoma kinase inhibition in non-small-cell lung cancer. N Engl J Med 363(18):1693–1703. doi:10.1056/NEJMoa1006448
13. Choi YL, Soda M, Yamashita Y, Ueno T, Takashima J, Nakajima T, Yatabe Y, Takeuchi K, Hamada T, Haruta H, Ishikawa Y, Kimura H, Mitsudomi T, Tanio Y, Mano H, Group ALKLCS (2010) EML4-ALK mutations in lung cancer that confer resistance to ALK inhibitors. N Engl J Med 363(18):1734–1739. doi:10.1056/NEJMoa1007478
14. Sasaki T, Okuda K, Zheng W, Butrynski J, Capelletti M, Wang L, Gray NS, Wilner K, Christensen JG, Demetri G, Shapiro GI, Rodig SJ, Eck MJ, Janne PA (2010) The neuroblastoma-associated F1174 L ALK mutation causes resistance to an ALK kinase inhibitor in ALK-translocated cancers. Cancer Res 70(24):10038–10043. doi:10.1158/0008-5472.CAN-10-2956
15. Kohno T, Ichikawa H, Totoki Y, Yasuda K, Hiramoto M, Nammo T, Sakamoto H, Tsuta K, Furuta K, Shimada Y, Iwakawa R, Ogiwara H, Oike T, Enari M, Schetter AJ, Okayama H, Haugen A, Skaug V, Chiku S, Yamanaka I, Arai Y, Watanabe S, Sekine I, Ogawa S, Harris CC, Tsuda H, Yoshida T, Yokota J, Shibata T (2012) KIF5B-RET fusions in lung adenocarcinoma. Nat Med 18(3):375–377. doi:10.1038/nm.2644
16. Rikova K, Guo A, Zeng Q, Possemato A, Yu J, Haack H, Nardone J, Lee K, Reeves C, Li Y, Hu Y, Tan Z, Stokes M, Sullivan L, Mitchell J, Wetzel R, Macneill J, Ren JM, Yuan J,

Bakalarski CE, Villen J, Kornhauser JM, Smith B, Li D, Zhou X, Gygi SP, Gu TL, Polakiewicz RD, Rush J, Comb MJ (2007) Global survey of phosphotyrosine signaling identifies oncogenic kinases in lung cancer. Cell 131(6):1190–1203. doi:10.1016/j.cell.2007.11.025
17. Takeuchi K, Soda M, Togashi Y, Suzuki R, Sakata S, Hatano S, Asaka R, Hamanaka W, Ninomiya H, Uehara H, Lim Choi Y, Satoh Y, Okumura S, Nakagawa K, Mano H, Ishikawa Y (2012) RET, ROS1 and ALK fusions in lung cancer. Nat Med 18(3):378–381. doi:10.1038/nm.2658
18. Bible KC, Kaufmann SH (1996) Flavopiridol: a cytotoxic flavone that induces cell death in noncycling A549 human lung carcinoma cells. Cancer Res 56(21):4856–4861
19. Parker BW, Kaur G, Nieves-Neira W, Taimi M, Kohlhagen G, Shimizu T, Losiewicz MD, Pommier Y, Sausville EA, Senderowicz AM (1998) Early induction of apoptosis in hematopoietic cell lines after exposure to flavopiridol. Blood 91(2):458–465
20. Lam LT, Pickeral OK, Peng AC, Rosenwald A, Hurt EM, Giltnane JM, Averett LM, Zhao H, Davis RE, Sathyamoorthy M, Wahl LM, Harris ED, Mikovits JA, Monks AP, Hollingshead MG, Sausville EA, Staudt LM (2001) Genomic-scale measurement of mRNA turnover and the mechanisms of action of the anti-cancer drug flavopiridol. Genome Biol 2(10): RESEARCH0041
21. Byrd JC, Peterson BL, Gabrilove J, Odenike OM, Grever MR, Rai K, Larson RA, Cancer, Leukemia Group B (2005) Treatment of relapsed chronic lymphocytic leukemia by 72-hour continuous infusion or 1-hour bolus infusion of flavopiridol: results from Cancer and Leukemia Group B study 19805. Clin Cancer Res 11(11):4176–4181. doi:10.1158/1078-0432.CCR-04-2276
22. Jones JA, Kraut EH, Deam D, Byrd JC, Grever MR (2012) Hematologic improvement after flavopiridol treatment of pentostatin and rituximab refractory hairy cell leukemia. Leuk Lymphoma 53(3):490–491. doi:10.3109/10428194.2011.600484
23. McClue SJ, Blake D, Clarke R, Cowan A, Cummings L, Fischer PM, MacKenzie M, Melville J, Stewart K, Wang S, Zhelev N, Zheleva D, Lane DP (2002) In vitro and in vivo antitumor properties of the cyclin dependent kinase inhibitor CYC202 (R-roscovitine). Int J Cancer 102(5):463–468. doi:10.1002/ijc.10738
24. Lacrima K, Rinaldi A, Vignati S, Martin V, Tibiletti MG, Gaidano G, Catapano CV, Bertoni F (2007) Cyclin-dependent kinase inhibitor seliciclib shows in vitro activity in diffuse large B-cell lymphomas. Leuk Lymphoma 48(1):158–167. doi:10.1080/10428190601026562
25. Parry D, Guzi T, Shanahan F, Davis N, Prabhavalkar D, Wiswell D, Seghezzi W, Paruch K, Dwyer MP, Doll R, Nomeir A, Windsor W, Fischmann T, Wang Y, Oft M, Chen T, Kirschmeier P, Lees EM (2010) Dinaciclib (SCH 727965), a novel and potent cyclin-dependent kinase inhibitor. Mol Cancer Ther 9(8):2344–2353. doi:10.1158/1535-7163.MCT-10-0324
26. Stephenson JJ, Nemunaitis J, Joy AA, Martin JC, Jou YM, Zhang D, Statkevich P, Yao SL, Zhu Y, Zhou H, Small K, Bannerji R, Edelman MJ (2014) Randomized phase 2 study of the cyclin-dependent kinase inhibitor dinaciclib (MK-7965) versus erlotinib in patients with non-small cell lung cancer. Lung Cancer 83(2):219–223. doi:10.1016/j.lungcan.2013.11.020
27. Fry DW, Harvey PJ, Keller PR, Elliott WL, Meade M, Trachet E, Albassam M, Zheng X, Leopold WR, Pryer NK, Toogood PL (2004) Specific inhibition of cyclin-dependent kinase 4/6 by PD 0332991 and associated antitumor activity in human tumor xenografts. Mol Cancer Ther 3(11):1427–1438
28. Rocca A, Farolfi A, Bravaccini S, Schirone A, Amadori D (2014) Palbociclib (PD 0332991): targeting the cell cycle machinery in breast cancer. Expert Opin Pharmacother 15 (3):407–420. doi:10.1517/14656566.2014.870555
29. Lens SM, Voest EE, Medema RH (2010) Shared and separate functions of polo-like kinases and aurora kinases in cancer. Nat Rev Cancer 10(12):825–841. doi:10.1038/nrc2964
30. Barr FA, Sillje HH, Nigg EA (2004) Polo-like kinases and the orchestration of cell division. Nat Rev Mol Cell Biol 5(6):429–440. doi:10.1038/nrm1401

31. van Vugt MA, Medema RH (2005) Getting in and out of mitosis with polo-like kinase-1. Oncogene 24(17):2844–2859. doi:10.1038/sj.onc.1208617
32. Lane HA, Nigg EA (1996) Antibody microinjection reveals an essential role for human polo-like kinase 1 (Plk1) in the functional maturation of mitotic centrosomes. J Cell Biol 135(6 Pt 2):1701–1713
33. Sumara I, Gimenez-Abian JF, Gerlich D, Hirota T, Kraft C, de la Torre C, Ellenberg J, Peters JM (2004) Roles of polo-like kinase 1 in the assembly of functional mitotic spindles. Curr Biol 14(19):1712–1722. doi:10.1016/j.cub.2004.09.049
34. van Vugt MA, van de Weerdt BC, Vader G, Janssen H, Calafat J, Klompmaker R, Wolthuis RM, Medema RH (2004) Polo-like kinase-1 is required for bipolar spindle formation but is dispensable for anaphase promoting complex/Cdc20 activation and initiation of cytokinesis. J Biol Chem 279(35):36841–36854. doi:10.1074/jbc.M313681200
35. Kumagai A, Dunphy WG (1996) Purification and molecular cloning of Plx1, a Cdc25-regulatory kinase from Xenopus egg extracts. Science 273(5280):1377–1380
36. Toyoshima-Morimoto F, Taniguchi E, Nishida E (2002) Plk1 promotes nuclear translocation of human Cdc25C during prophase. EMBO Rep 3(4):341–348. doi:10.1093/embo-reports/kvf069
37. Watanabe N, Arai H, Iwasaki J, Shiina M, Ogata K, Hunter T, Osada H (2005) Cyclin-dependent kinase (CDK) phosphorylation destabilizes somatic Wee1 via multiple pathways. Proc Natl Acad Sci U S A 102(33):11663–11668. doi:10.1073/pnas.0500410102
38. Watanabe N, Arai H, Nishihara Y, Taniguchi M, Watanabe N, Hunter T, Osada H (2004) M-phase kinases induce phospho-dependent ubiquitination of somatic Wee1 by SCFbeta-TrCP. Proc Natl Acad Sci U S A 101(13):4419–4424. doi:10.1073/pnas.0307700101
39. Elowe S, Hummer S, Uldschmid A, Li X, Nigg EA (2007) Tension-sensitive Plk1 phosphorylation on BubR1 regulates the stability of kinetochore microtubule interactions. Genes Dev 21(17):2205–2219. doi:10.1101/gad.436007
40. Ahonen LJ, Kallio MJ, Daum JR, Bolton M, Manke IA, Yaffe MB, Stukenberg PT, Gorbsky GJ (2005) Polo-like kinase 1 creates the tension-sensing 3F3/2 phosphoepitope and modulates the association of spindle-checkpoint proteins at kinetochores. Curr Biol 15 (12):1078–1089. doi:10.1016/j.cub.2005.05.026
41. Wong OK, Fang G (2005) Plx1 is the 3F3/2 kinase responsible for targeting spindle checkpoint proteins to kinetochores. J Cell Biol 170(5):709–719. doi:10.1083/jcb.200502163
42. Wong OK, Fang G (2007) Cdk1 phosphorylation of BubR1 controls spindle checkpoint arrest and Plk1-mediated formation of the 3F3/2 epitope. J Cell Biol 179(4):611–617. doi:10.1083/jcb.200708044
43. Nishino M, Kurasawa Y, Evans R, Lin SH, Brinkley BR, Yu-Lee LY (2006) NudC is required for Plk1 targeting to the kinetochore and chromosome congression. Curr Biol 16 (14):1414–1421. doi:10.1016/j.cub.2006.05.052
44. Losada A, Hirano M, Hirano T (2002) Cohesin release is required for sister chromatid resolution, but not for condensin-mediated compaction, at the onset of mitosis. Genes Dev 16(23):3004–3016. doi:10.1101/gad.249202
45. Sumara I, Vorlaufer E, Stukenberg PT, Kelm O, Redemann N, Nigg EA, Peters JM (2002) The dissociation of cohesin from chromosomes in prophase is regulated by polo-like kinase. Mol Cell 9(3):515–525
46. Howard S, Berdini V, Boulstridge JA, Carr MG, Cross DM, Curry J, Devine LA, Early TR, Fazal L, Gill AL, Heathcote M, Maman S, Matthews JE, McMenamin RL, Navarro EF, O'Brien MA, O'Reilly M, Rees DC, Reule M, Tisi D, Williams G, Vinkovic M, Wyatt PG (2009) Fragment-based discovery of the pyrazol-4-yl urea (AT9283), a multitargeted kinase inhibitor with potent aurora kinase activity. J Med Chem 52(2):379–388. doi:10.1021/jm800984v
47. Qi W, Tang Z, Yu H (2006) Phosphorylation- and polo-box-dependent binding of Plk1 to Bub1 is required for the kinetochore localization of Plk1. Mol Biol Cell 17(8):3705–3716. doi:10.1091/mbc.E06-03-0240

48. Lee KS, Yuan YL, Kuriyama R, Erikson RL (1995) Plk is an M-phase-specific protein kinase and interacts with a kinesin-like protein, CHO1/MKLP-1. Mol Cell Biol 15(12):7143–7151
49. Neef R, Preisinger C, Sutcliffe J, Kopajtich R, Nigg EA, Mayer TU, Barr FA (2003) Phosphorylation of mitotic kinesin-like protein 2 by polo-like kinase 1 is required for cytokinesis. J Cell Biol 162(5):863–875. doi:10.1083/jcb.200306009
50. Zhou T, Aumais JP, Liu X, Yu-Lee LY, Erikson RL (2003) A role for Plk1 phosphorylation of NudC in cytokinesis. Dev Cell 5(1):127–138
51. Hudson JW, Kozarova A, Cheung P, Macmillan JC, Swallow CJ, Cross JC, Dennis JW (2001) Late mitotic failure in mice lacking Sak, a polo-like kinase. Curr Biol 11(6):441–446
52. Seong YS, Kamijo K, Lee JS, Fernandez E, Kuriyama R, Miki T, Lee KS (2002) A spindle checkpoint arrest and a cytokinesis failure by the dominant-negative polo-box domain of Plk1 in U-2 OS cells. J Biol Chem 277(35):32282–32293. doi:10.1074/jbc.M202602200
53. Elia AE, Cantley LC, Yaffe MB (2003) Proteomic screen finds pSer/pThr-binding domain localizing Plk1 to mitotic substrates. Science 299(5610):1228–1231. doi:10.1126/science.1079079
54. Elia AE, Rellos P, Haire LF, Chao JW, Ivins FJ, Hoepker K, Mohammad D, Cantley LC, Smerdon SJ, Yaffe MB (2003) The molecular basis for phosphodependent substrate targeting and regulation of Plks by the polo-box domain. Cell 115(1):83–95
55. Liao JJ (2007) Molecular recognition of protein kinase binding pockets for design of potent and selective kinase inhibitors. J Med Chem 50(3):409–424. doi:10.1021/jm0608107
56. Medema RH, Lin CC, Yang JC (2011) Polo-like kinase 1 inhibitors and their potential role in anticancer therapy, with a focus on NSCLC. Clin Cancer Res 17(20):6459–6466. doi:10.1158/1078-0432.CCR-11-0541
57. Gumireddy K, Reddy MV, Cosenza SC, Boominathan R, Baker SJ, Papathi N, Jiang J, Holland J, Reddy EP (2005) ON01910, a non-ATP-competitive small molecule inhibitor of Plk1, is a potent anticancer agent. Cancer Cell 7(3):275–286. doi:10.1016/j.ccr.2005.02.009
58. Kothe M, Kohls D, Low S, Coli R, Rennie GR, Feru F, Kuhn C, Ding YH (2007) Selectivity-determining residues in Plk1. Chem Biol Drug Des 70(6):540–546. doi:10.1111/j.1747-0285.2007.00594.x
59. Steegmaier M, Hoffmann M, Baum A, Lenart P, Petronczki M, Krssak M, Gurtler U, Garin-Chesa P, Lieb S, Quant J, Grauert M, Adolf GR, Kraut N, Peters JM, Rettig WJ (2007) BI 2536, a potent and selective inhibitor of polo-like kinase 1, inhibits tumor growth in vivo. Curr Biol 17(4):316–322. doi:10.1016/j.cub.2006.12.037
60. Lenart P, Petronczki M, Steegmaier M, Di Fiore B, Lipp JJ, Hoffmann M, Rettig WJ, Kraut N, Peters JM (2007) The small-molecule inhibitor BI 2536 reveals novel insights into mitotic roles of polo-like kinase 1. Curr Biol 17(4):304–315. doi:10.1016/j.cub.2006.12.046
61. Jimeno A, Li J, Messersmith WA, Laheru D, Rudek MA, Maniar M, Hidalgo M, Baker SD, Donehower RC (2008) Phase I study of ON 01910.Na, a novel modulator of the polo-like kinase 1 pathway, in adult patients with solid tumors. J Clin Oncol 26(34):5504–5510. doi:10.1200/JCO.2008.17.9788
62. Mross K, Frost A, Steinbild S, Hedbom S, Rentschler J, Kaiser R, Rouyrre N, Trommeshauser D, Hoesl CE, Munzert G (2008) Phase I dose escalation and pharmacokinetic study of BI 2536, a novel polo-like kinase 1 inhibitor, in patients with advanced solid tumors. J Clin Oncol 26(34):5511–5517. doi:10.1200/JCO.2008.16.1547
63. Hofheinz RD, Al-Batran SE, Hochhaus A, Jager E, Reichardt VL, Fritsch H, Trommeshauser D, Munzert G (2010) An open-label, phase I study of the polo-like kinase-1 inhibitor, BI 2536, in patients with advanced solid tumors. Clin Cancer Res 16(18):4666–4674. doi:10.1158/1078-0432.CCR-10-0318
64. Sebastian M, Reck M, Waller CF, Kortsik C, Frickhofen N, Schuler M, Fritsch H, Gaschler-Markefski B, Hanft G, Munzert G, von Pawel J (2010) The efficacy and safety of BI 2536, a novel Plk-1 inhibitor, in patients with stage IIIB/IV non-small cell lung cancer who had

relapsed after, or failed, chemotherapy: results from an open-label, randomized phase II clinical trial. J Thorac Oncol 5(7):1060–1067. doi:10.1097/JTO.0b013e3181d95dd4
65. Gilmartin AG, Bleam MR, Richter MC, Erskine SG, Kruger RG, Madden L, Hassler DF, Smith GK, Gontarek RR, Courtney MP, Sutton D, Diamond MA, Jackson JR, Laquerre SG (2009) Distinct concentration-dependent effects of the polo-like kinase 1–specific inhibitor GSK461364A, including differential effect on apoptosis. Cancer Res 69(17):6969–6977. doi:10.1158/0008-5472.CAN-09-0945
66. Olmos D, Barker D, Sharma R, Brunetto AT, Yap TA, Taegtmeyer AB, Barriuso J, Medani H, Degenhardt YY, Allred AJ, Smith DA, Murray SC, Lampkin TA, Dar MM, Wilson R, de Bono JS, Blagden SP (2011) Phase I study of GSK461364, a specific and competitive polo-like kinase 1 inhibitor, in patients with advanced solid malignancies. Clin Cancer Res 17(10):3420–3430. doi:10.1158/1078-0432.CCR-10-2946
67. Rudolph D, Steegmaier M, Hoffmann M, Grauert M, Baum A, Quant J, Haslinger C, Garin-Chesa P, Adolf GR (2009) BI 6727, a polo-like kinase inhibitor with improved pharmacokinetic profile and broad antitumor activity. Clin Cancer Res 15(9):3094–3102. doi:10.1158/1078-0432.CCR-08-2445
68. Watanabe N, Sekine T, Takagi M, Iwasaki J, Imamoto N, Kawasaki H, Osada H (2009) Deficiency in chromosome congression by the inhibition of Plk1 polo box domain–dependent recognition. J Biol Chem 284(4):2344–2353. doi:10.1074/jbc.M805308200
69. Murugan RN, Park JE, Kim EH, Shin SY, Cheong C, Lee KS, Bang JK (2011) Plk1-targeted small molecule inhibitors: molecular basis for their potency and specificity. Mol Cells 32 (3):209–220. doi:10.1007/s10059-011-0126-3
70. Reindl W, Yuan J, Kramer A, Strebhardt K, Berg T (2008) Inhibition of polo-like kinase 1 by blocking polo-box domain–dependent protein-protein interactions. Chem Biol 15 (5):459–466. doi:10.1016/j.chembiol.2008.03.013
71. Liao C, Park JE, Bang JK, Nicklaus MC, Lee KS (2010) Exploring potential binding modes of small drug-like molecules to the polo-box domain of human polo-like kinase 1. ACS Med Chem Lett 1(3):110–114. doi:10.1021/ml100020e
72. Watanabe N, Osada H (2012) Phosphorylation-dependent protein-protein interaction modules as potential molecular targets for cancer therapy. Curr Drug Targets 13(13):1654–1658
73. Glover DM, Leibowitz MH, McLean DA, Parry H (1995) Mutations in Aurora prevent centrosome separation leading to the formation of monopolar spindles. Cell 81(1):95–105
74. Carmena M, Earnshaw WC (2003) The cellular geography of aurora kinases. Nat Rev Mol Cell Biol 4(11):842–854. doi:10.1038/nrm1245
75. Chen HL, Tang CJ, Chen CY, Tang TK (2005) Overexpression of an Aurora-C kinase-deficient mutant disrupts the Aurora-B/INCENP complex and induces polyploidy. J Biomed Sci 12(2):297–310. doi:10.1007/s11373-005-0980-0
76. Adams RR, Maiato H, Earnshaw WC, Carmena M (2001) Essential roles of *Drosophila* inner centromere protein (INCENP) and Aurora B in histone H3 phosphorylation, metaphase chromosome alignment, kinetochore disjunction, and chromosome segregation. J Cell Biol 153(4):865–880
77. Goto H, Yasui Y, Nigg EA, Inagaki M (2002) Aurora-B phosphorylates histone H3 at serine28 with regard to the mitotic chromosome condensation. Genes Cells 7(1):11–17
78. Yan X, Cao L, Li Q, Wu Y, Zhang H, Saiyin H, Liu X, Zhang X, Shi Q, Yu L (2005) Aurora C is directly associated with Survivin and required for cytokinesis. Genes Cells 10(6):617–626. doi:10.1111/j.1365-2443.2005.00863.x
79. Hauf S, Cole RW, LaTerra S, Zimmer C, Schnapp G, Walter R, Heckel A, van Meel J, Rieder CL, Peters JM (2003) The small molecule Hesperadin reveals a role for Aurora B in correcting kinetochore–microtubule attachment and in maintaining the spindle assembly checkpoint. J Cell Biol 161(2):281–294. doi:10.1083/jcb.200208092
80. Ditchfield C, Johnson VL, Tighe A, Ellston R, Haworth C, Johnson T, Mortlock A, Keen N, Taylor SS (2003) Aurora B couples chromosome alignment with anaphase by targeting

BubR1, Mad2, and Cenp-E to kinetochores. J Cell Biol 161(2):267–280. doi:10.1083/jcb.200208091
81. Bebbington D, Binch H, Charrier JD, Everitt S, Fraysse D, Golec J, Kay D, Knegtel R, Mak C, Mazzei F, Miller A, Mortimore M, O'Donnell M, Patel S, Pierard F, Pinder J, Pollard J, Ramaya S, Robinson D, Rutherford A, Studley J, Westcott J (2009) The discovery of the potent aurora inhibitor MK-0457 (VX-680). Bioorg Med Chem Lett 19 (13):3586–3592. doi:10.1016/j.bmcl.2009.04.136
82. Harrington EA, Bebbington D, Moore J, Rasmussen RK, Ajose-Adeogun AO, Nakayama T, Graham JA, Demur C, Hercend T, Diu-Hercend A, Su M, Golec JM, Miller KM (2004) VX-680, a potent and selective small-molecule inhibitor of the Aurora kinases, suppresses tumor growth in vivo. Nat Med 10(3):262–267. doi:10.1038/nm1003
83. Cheetham GM, Charlton PA, Golec JM, Pollard JR (2007) Structural basis for potent inhibition of the Aurora kinases and a T315I multi-drug resistant mutant form of Abl kinase by VX-680. Cancer Lett 251(2):323–329. doi:10.1016/j.canlet.2006.12.004
84. Fancelli D, Berta D, Bindi S, Cameron A, Cappella P, Carpinelli P, Catana C, Forte B, Giordano P, Giorgini ML, Mantegani S, Marsiglio A, Meroni M, Moll J, Pittala V, Roletto F, Severino D, Soncini C, Storici P, Tonani R, Varasi M, Vulpetti A, Vianello P (2005) Potent and selective Aurora inhibitors identified by the expansion of a novel scaffold for protein kinase inhibition. J Med Chem 48(8):3080–3084. doi:10.1021/jm049076m
85. Arbitrario JP, Belmont BJ, Evanchik MJ, Flanagan WM, Fucini RV, Hansen SK, Harris SO, Hashash A, Hoch U, Hogan JN, Howlett AR, Jacobs JW, Lam JW, Ritchie SC, Romanowski MJ, Silverman JA, Stockett DE, Teague JN, Zimmerman KM, Taverna P (2010) SNS-314, a pan-Aurora kinase inhibitor, shows potent anti-tumor activity and dosing flexibility in vivo. Cancer Chemother Pharmacol 65(4):707–717. doi:10.1007/s00280-009-1076-8
86. McLaughlin J, Markovtsov V, Li H, Wong S, Gelman M, Zhu Y, Franci C, Lang D, Pali E, Lasaga J, Low C, Zhao F, Chang B, Gururaja TL, Xu W, Baluom M, Sweeny D, Carroll D, Sran A, Thota S, Parmer M, Romane A, Clemens G, Grossbard E, Qu K, Jenkins Y, Kinoshita T, Taylor V, Holland SJ, Argade A, Singh R, Pine P, Payan DG, Hitoshi Y (2010) Preclinical characterization of Aurora kinase inhibitor R763/AS703569 identified through an image-based phenotypic screen. J Cancer Res Clin Oncol 136(1):99–113. doi:10.1007/s00432-009-0641-1
87. Hardwicke MA, Oleykowski CA, Plant R, Wang J, Liao Q, Moss K, Newlander K, Adams JL, Dhanak D, Yang J, Lai Z, Sutton D, Patrick D (2009) GSK1070916, a potent Aurora B/C kinase inhibitor with broad antitumor activity in tissue culture cells and human tumor xenograft models. Mol Cancer Ther 8(7):1808–1817. doi:10.1158/1535-7163.MCT-09-0041
88. Curry J, Angove H, Fazal L, Lyons J, Reule M, Thompson N, Wallis N (2009) Aurora B kinase inhibition in mitosis: strategies for optimising the use of aurora kinase inhibitors such as AT9283. Cell Cycle 8(12):1921–1929
89. Wang S, Midgley CA, Scaerou F, Grabarek JB, Griffiths G, Jackson W, Kontopidis G, McClue SJ, McInnes C, Meades C, Mezna M, Plater A, Stuart I, Thomas MP, Wood G, Clarke RG, Blake DG, Zheleva DI, Lane DP, Jackson RC, Glover DM, Fischer PM (2010) Discovery of *N*-phenyl-4-(thiazol-5-yl)pyrimidin-2-amine aurora kinase inhibitors. J Med Chem 53(11):4367–4378. doi:10.1021/jm901913s
90. Gorgun G, Calabrese E, Hideshima T, Ecsedy J, Perrone G, Mani M, Ikeda H, Bianchi G, Hu Y, Cirstea D, Santo L, Tai YT, Nahar S, Zheng M, Bandi M, Carrasco RD, Raje N, Munshi N, Richardson P, Anderson KC (2010) A novel Aurora-A kinase inhibitor MLN8237 induces cytotoxicity and cell-cycle arrest in multiple myeloma. Blood 115(25):5202–5213. doi:10.1182/blood-2009-12-259523
91. Shimomura T, Hasako S, Nakatsuru Y, Mita T, Ichikawa K, Kodera T, Sakai T, Nambu T, Miyamoto M, Takahashi I, Miki S, Kawanishi N, Ohkubo M, Kotani H, Iwasawa Y (2010) MK-5108, a highly selective Aurora-A kinase inhibitor, shows antitumor activity alone and in combination with docetaxel. Mol Cancer Ther 9(1):157–166. doi:10.1158/1535-7163.MCT-09-0609

92. Amin M, Minton SE, LoRusso PM, Krishnamurthi SS, Pickett CA, Lunceford J, Hille D, Mauro D, Stein MN, Wang-Gillam A, Trull L, Lockhart AC (2016) A phase I study of MK-5108, an oral aurora a kinase inhibitor, administered both as monotherapy and in combination with docetaxel, in patients with advanced or refractory solid tumors. Investig New Drugs 34(1):84–95. doi:10.1007/s10637-015-0306-7
93. Fletcher GC, Brokx RD, Denny TA, Hembrough TA, Plum SM, Fogler WE, Sidor CF, Bray MR (2011) ENMD-2076 is an orally active kinase inhibitor with antiangiogenic and antiproliferative mechanisms of action. Mol Cancer Ther 10(1):126–137. doi:10.1158/1535-7163.MCT-10-0574
94. Mortlock AA, Foote KM, Heron NM, Jung FH, Pasquet G, Lohmann JJ, Warin N, Renaud F, De Savi C, Roberts NJ, Johnson T, Dousson CB, Hill GB, Perkins D, Hatter G, Wilkinson RW, Wedge SR, Heaton SP, Odedra R, Keen NJ, Crafter C, Brown E, Thompson K, Brightwell S, Khatri L, Brady MC, Kearney S, McKillop D, Rhead S, Parry T, Green S (2007) Discovery, synthesis, and in vivo activity of a new class of pyrazoloquinazolines as selective inhibitors of aurora B kinase. J Med Chem 50(9):2213–2224. doi:10.1021/jm061335f
95. Collins GP, Eyre TA, Linton KM, Radford J, Vallance GD, Soilleux E, Hatton C (2015) A phase II trial of AZD1152 in relapsed/refractory diffuse large B-cell lymphoma. Br J Haematol 170(6):886–890. doi:10.1111/bjh.13333
96. Wilkinson RW, Odedra R, Heaton SP, Wedge SR, Keen NJ, Crafter C, Foster JR, Brady MC, Bigley A, Brown E, Byth KF, Barrass NC, Mundt KE, Foote KM, Heron NM, Jung FH, Mortlock AA, Boyle FT, Green S (2007) AZD1152, a selective inhibitor of Aurora B kinase, inhibits human tumor xenograft growth by inducing apoptosis. Clin Cancer Res 13 (12):3682–3688. doi:10.1158/1078-0432.CCR-06-2979
97. Yang J, Ikezoe T, Nishioka C, Tasaka T, Taniguchi A, Kuwayama Y, Komatsu N, Bandobashi K, Togitani K, Koeffler HP, Taguchi H, Yokoyama A (2007) AZD1152, a novel and selective aurora B kinase inhibitor, induces growth arrest, apoptosis, and sensitization for tubulin depolymerizing agent or topoisomerase II inhibitor in human acute leukemia cells in vitro and in vivo. Blood 110(6):2034–2040. doi:10.1182/blood-2007-02-073700
98. Takahashi K, Yamanaka S (2006) Induction of pluripotent stem cells from mouse embryonic and adult fibroblast cultures by defined factors. Cell 126(4):663–676. doi:10.1016/j.cell.2006.07.024
99. Hou P, Li Y, Zhang X, Liu C, Guan J, Li H, Zhao T, Ye J, Yang W, Liu K, Ge J, Xu J, Zhang Q, Zhao Y, Deng H (2013) Pluripotent stem cells induced from mouse somatic cells by small-molecule compounds. Science 341(6146):651–654. doi:10.1126/science.1239278
100. Evans MJ, Kaufman MH (1981) Establishment in culture of pluripotential cells from mouse embryos. Nature 292(5819):154–156
101. Martin GR (1981) Isolation of a pluripotent cell line from early mouse embryos cultured in medium conditioned by teratocarcinoma stem cells. Proc Natl Acad Sci U S A 78 (12):7634–7638
102. Wilmut I, Schnieke AE, McWhir J, Kind AJ, Campbell KH (1997) Viable offspring derived from fetal and adult mammalian cells. Nature 385(6619):810–813. doi:10.1038/385810a0
103. Cowan CA, Atienza J, Melton DA, Eggan K (2005) Nuclear reprogramming of somatic cells after fusion with human embryonic stem cells. Science 309(5739):1369–1373. doi:10.1126/science.1116447
104. Tada M, Takahama Y, Abe K, Nakatsuji N, Tada T (2001) Nuclear reprogramming of somatic cells by in vitro hybridization with ES cells. Curr Biol 11(19):1553–1558
105. Li Y, Zhang Q, Yin X, Yang W, Du Y, Hou P, Ge J, Liu C, Zhang W, Zhang X, Wu Y, Li H, Liu K, Wu C, Song Z, Zhao Y, Shi Y, Deng H (2011) Generation of iPSCs from mouse fibroblasts with a single gene, Oct4, and small molecules. Cell Res 21(1):196–204. doi:10.1038/cr.2010.142

106. Lapidot T, Sirard C, Vormoor J, Murdoch B, Hoang T, Caceres-Cortes J, Minden M, Paterson B, Caligiuri MA, Dick JE (1994) A cell initiating human acute myeloid leukaemia after transplantation into SCID mice. Nature 367(6464):645–648. doi:10.1038/367645a0
107. Gottesman MM, Fojo T, Bates SE (2002) Multidrug resistance in cancer: role of ATP-dependent transporters. Nat Rev Cancer 2(1):48–58. doi:10.1038/nrc706
108. Ho MM, Ng AV, Lam S, Hung JY (2007) Side population in human lung cancer cell lines and tumors is enriched with stem-like cancer cells. Cancer Res 67(10):4827–4833. doi:10.1158/0008-5472.CAN-06-3557
109. Zhang M, Mathur A, Zhang Y, Xi S, Atay S, Hong JA, Datrice N, Upham T, Kemp CD, Ripley RT, Wiegand G, Avital I, Fetsch P, Mani H, Zlott D, Robey R, Bates SE, Li X, Rao M, Schrump DS (2012) Mithramycin represses basal and cigarette smoke–induced expression of ABCG2 and inhibits stem cell signaling in lung and esophageal cancer cells. Cancer Res 72 (16):4178–4192. doi:10.1158/0008-5472.CAN-11-3983
110. Gupta PB, Onder TT, Jiang G, Tao K, Kuperwasser C, Weinberg RA, Lander ES (2009) Identification of selective inhibitors of cancer stem cells by high-throughput screening. Cell 138(4):645–659. doi:10.1016/j.cell.2009.06.034
111. Campos B, Wan F, Farhadi M, Ernst A, Zeppernick F, Tagscherer KE, Ahmadi R, Lohr J, Dictus C, Gdynia G, Combs SE, Goidts V, Helmke BM, Eckstein V, Roth W, Beckhove P, Lichter P, Unterberg A, Radlwimmer B, Herold-Mende C (2010) Differentiation therapy exerts antitumor effects on stem-like glioma cells. Clin Cancer Res 16(10):2715–2728. doi:10.1158/1078-0432.CCR-09-1800
112. Sotiropoulou PA, Christodoulou MS, Silvani A, Herold-Mende C, Passarella D (2014) Chemical approaches to targeting drug resistance in cancer stem cells. Drug Discov Today. doi:10.1016/j.drudis.2014.05.002

Chapter 3
Epigenetics

Bioprobes that Target Epigenetic Modifications

Akihiro Ito and Minoru Yoshida

Abstract Epigenetics refers to a heritable alteration in gene expression not associated with an alteration in the DNA sequence. In other words, epigenetics is a heritable system that changes the phenotype without changing the genotype. Although the cells that constitute the various organs contain identical genomic DNA sequences, they establish and maintain different terminal phenotypes. This nongenetic cellular memory is based on epigenetics. Epigenetic information is less stable than genetic information, and can be influenced by diverse factors such as age, environment, and stress. Aberrant epigenetic changes are associated with various diseases, including cancers. Epigenetic regulation of gene expression is mediated by histone modifications, such as acetylation and methylation, as well as DNA methylation. These modifications are reversibly modulated by specific enzymes. Therefore, chemical tools that target these epigenetic modulators could not only serve as useful tools for investigating the roles of epigenetics in biological systems and multiple human disorders, but would also have potential as drugs. Indeed, several small molecules that target histone acetylation and DNA methylation have been approved for treatment of cancers. This chapter describes our current knowledge of inhibitors of epigenetic modulators and their clinical development.

Keywords Cancer • DNA methylation • Epigenetics • Histone acetylation • Histone methylation • Inhibitor

A. Ito (✉) • M. Yoshida
Chemical Genetics Laboratory, RIKEN, 2-1 Hirosawa, Wako, Saitama 351-0198, Japan

Chemical Genomics Research Group, RIKEN Center for Sustainable Resource Science, 2-1 Hirosawa, Wako, Saitama 351-0198, Japan
e-mail: akihiro-i@riken.jp; yoshidam@riken.jp

© Springer Japan KK 2017
H. Osada (ed.), *Bioprobes*, DOI 10.1007/978-4-431-56529-1_3

3.1 Introduction

Over evolutionary history, organisms have acquired the ability to survive adverse environmental conditions including stresses related to nutrient availability, temperature, humidity, and osmotic pressure. Although an organism's offspring can survive by harboring a favorable mutation that improves tolerance of severe environmental changes, they can also adapt by altering the pattern of gene expression. Alterations in gene expression acquired in one cell may be conveyed to its daughter cells via a type of cell "memory" mediated by epigenetics. The term "epigenetics" refers collectively to systems that produce diverse gene expression patterns without changing the DNA sequence, i.e., systems that determine which genes are expressed at specific times and locations. Although all cells within an organism contain same DNA sequence, epigenetic factors organize their genome structures to ensure execution of the correct transcriptional program in each cell. In this manner, epigenetics are involved in establishment and maintenance of distinct terminal phenotypes in different cell types. Thus, epigenetic regulation of gene expression patterns determines cell fate and influences fundamental processes such as development, differentiation, and proliferation.

In general, epigenetic regulation is modulated by chemical modifications of histones, as well as by DNA methylation. The *N*-terminal amino acid residues of core histones undergo posttranslational modifications such as acetylation, methylation, phosphorylation, ADP-ribosylation, ubiquitination, and SUMOylation. These modifications, which are generally reversible, control gene expression in specific chromosomal regions via a kind of epigenetic code. These chemical modifications undergo "writing" by modifying enzymes, "erasing" by demodifying enzymes, and "reading" by proteins that interact with specific modifications (Fig. 3.1) [1, 2]. The factors involved in these three steps represent potential targets for development of small molecules that influence epigenetic regulation.

Recent studies revealed that dysregulation of epigenetic control is a common feature of diverse diseases, including cancers [3]. Accordingly, proteins that regulate histone modifications or DNA methylation have attracted attention as targets for cancer therapy [4]. Indeed, several drugs that target the epigenetic machinery have recently been approved for use in human patients. The first such drugs were azacitidine (5-azacytidine) and decitabine (5-aza-2′-deoxycytidine), which are irreversible inhibitors of the DNA methyltransferase (DNMT) enzymes; the approval of these compounds represented a milestone for epigenetics-based therapies. Both drugs are currently used as first-line treatments for myelodysplastic syndrome (MDS) [5, 6]. The next generation of epigenetic drugs consisted of histone deacetylase (HDAC) inhibitors. Four drugs targeting HDACs have been approved for use in humans: suberoylanilide hydroxamic acid (SAHA) (Vorinostat), FK228 (romidepsin or depsipeptide) and belinostat (PXD101) for the treatment of refractory cutaneous or peripheral T-cell lymphoma [7–9], and panobinostat (LBH-589) for the treatment of multiple myeloma [10]. Thus, small molecules that target epigenetics are not only useful for elucidating the functions of individual factors

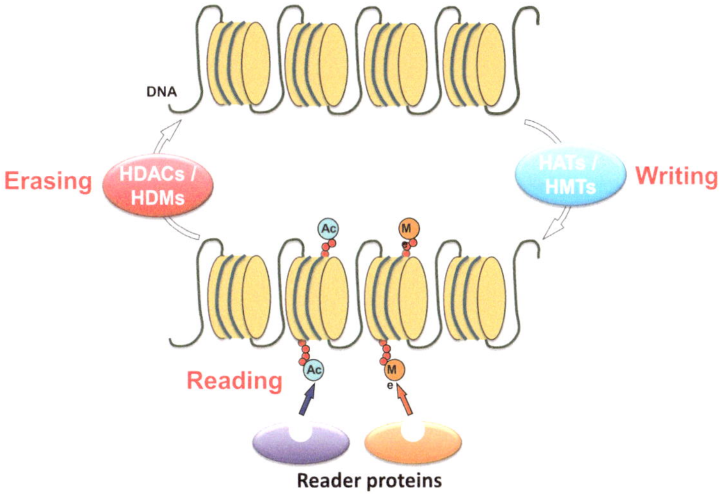

Fig. 3.1 Epigenetic regulation consists of three machineries: "writing" by histone-modifying enzymes including histone acetyltransferases (HATs) and histone methyltransferases (HMTs); "erasing" by histone-demodifying enzymes including histone deacetylases (HDACs) and histone demethylases (HDMs); and "reading" by proteins interacting with specific histone modifications. Reader proteins for recognizing an acetylated lysine include bromodomain-containing proteins, while those for recognizing a methylated lysine residues include proteins containing a chromodomain, plant homeodomain (PHD), tudor domain, or malignant brain tumor (MBT) domain

in epigenetic regulation and the biological processes regulated by epigenetics, but can also be used as therapeutic drugs against diseases such as cancers.

3.2 Histone Acetylation

Several lysine residues at the *N*-termini of core histones are posttranslationally acetylated at the ε-position (Fig. 3.2); these reversible modifications are generated by histone acetyltransferases (HATs) and removed by HDACs. Several HAT and HDAC isoforms have been identified to date, and a diverse array of DNA- or chromatin binding factors recruit them to specific chromatin region where they regulate histone acetylation levels. The epigenetic information carried in histone acetylation is interpreted by "reader" proteins containing bromodomains, which recognize specific acetylated lysine residues on target proteins. Proteins containing bromodomains include HATs, ATP-dependent chromatin remodeling factors, and general transcriptional factors.

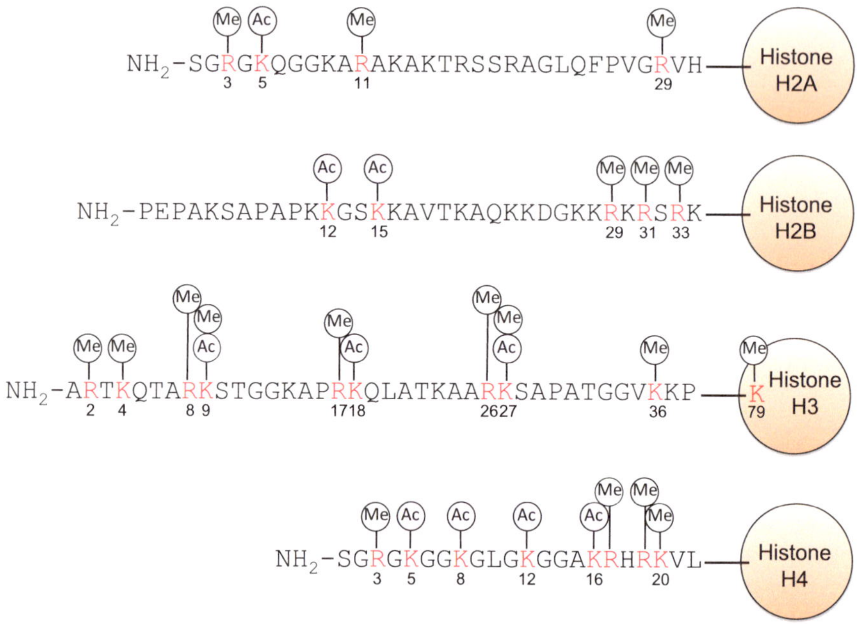

Fig. 3.2 Chemical modifications at lysine and arginine residues on the *N*-terminus of core histones. Acetylation occurs on several lysine residues at the *N*-termini of core histones including histones H2A, H2B, H3, and H4, while methylation occurs on both lysine and arginine residues on core histones. Some lysine residues including H3K9 and K27 undergo both acetylation and methylation

3.2.1 Histone Deacetylases

Eighteen subtypes of human HDACs have been identified to date. These proteins are classified into four groups according to their structural similarity with yeast homologs (Table 3.1) [11]. The class I HDACs, which comprise HDAC1, 2, 3, and 8, exhibit domain organization similar to that of yeast Rpd3. Members of this class are primarily localized in the nucleus. Except for HDAC8, all members of class I function as catalytic components in protein complexes [11]. For example, HDAC1 and HDAC2 interact with each other and act as the catalytic core of several protein complexes, including Sin3, Mi-2/NuRD, and CoREST.

Class II HDACs comprise HDAC4, 5, 6, 7, 9, and 10, and have catalytic domains similar to that of yeast Hda1. This class is further divided into two subclasses: class IIa, consisting of HDAC4, 5, 7, and 9; and class IIb, which includes HDAC6 and 10. Members of class IIa are localized in both the nucleus and cytoplasm and contain intrinsic nuclear import and export signals that facilitate dynamic nucleocytoplasmic trafficking [12–14]. The class IIb protein HDAC6 contains tandem catalytic domains and a conserved C-terminal zinc finger–containing domain, ZnF-UBP, that can bind ubiquitin; unlike class IIa proteins, it is primarily

Table 3.1 Human histone deacetylase (HDAC) family

Class	HDAC	Enzymatic activity	Major subcellular localization
I	HDAC1	Deacetylase	Nucleus
	HDAC2	Deacetylase	Nucleus
	HDAC3	Deacetylase	Nucleus
	HDAC8	Deacetylase	Nucleus
IIa	HDAC4	Deacetylase	Nucleus/cytoplasm
	HDAC5	Deacetylase	Nucleus/cytoplasm
	HDAC7	Deacetylase	Nucleus/cytoplasm
	HDAC9	Deacetylase	Nucleus/cytoplasm
IIb	HDAC6	Deacetylase	Cytoplasm
	HDAC10	Deacetylase	Nucleus/cytoplasm
III	SIRT1	Deacetylase	Nucleus/cytoplasm
	SIRT2	Deactylase/deacylase	Cytoplasm
	SIRT3	Deacetylase	Mitochondria
	SIRT4	ADP-ribosyltransferase/lipoamidase	Mitochondria
	SIRT5	Demalonylase/desuccinylase	Mitochondria
	SIRT6	Deacetylase/deacylase/ADP-ribosyltransferase	Nucleus
	SIRT7	Deacetylase	Nucleus (nucleolus)
IV	HDAC11	Deacetylase	Nucleus

localized in the cytoplasm [14–17] where it acts as an α-tubulin deacetylase [18, 19]. The other member of class IIb, HDAC10, has an *N*-terminal catalytic domain highly similar to the first deacetylase domain of HDAC6 and an incomplete C-terminal catalytic domain [20]. HDAC11 is homologous to both Rpd3 and Hda1 and is thus categorized as class IV. All class I, II, and IV HDACs require zinc for their enzymatic activity and appear to cleave the acetamide bond by activating a water molecule with a zinc atom coupled to a histidine–aspartate charge-relay system [21].

Class III HDACs are homologous to yeast Sir2 (silent information regulator 2), originally identified as a protein required for maintaining silent chromatin [22], and have no significant sequence similarity to the other classes of HDACs. Sir2 family proteins are referred to as sirtuins. Unlike the members of other classes, class III deacetylases/sirtuins possess nicotinamide adenine dinucleotide (NAD)-dependent deacetylase activity [23–25], which tightly couple the cleavage of NAD and protein deacetylation to yield nicotinamide, deacetylated product, and 2′-*O*-acetyl-ADP-ribose [26–29]. Accordingly, the mechanism of inhibition of this class and the chemical nature of the inhibitors differ from those of other classes of HDACs. Humans have seven sirtuin proteins (SIRT1–7) [30], which have different enzymatic activities and subcellular localizations [31]. SIRT1 is localized in both the nucleus and the cytoplasm, depending on cell type and/or environment conditions, whereas SIRT6 and SIRT7 are exclusively localized in the nucleus. SIRT2 is mainly located in the cytoplasm, and SIRT3, SIRT4, and SIRT5 reside in the mitochondria [32, 33]. In addition to their well-known deacetylation activity, SIRT2 and SIRT6 also possess deacylase activities; specifically, these enzymes catalyze long-chain deacylation [34–37]. SIRT4 and SIRT6 possess mono-ADP-ribosyltransferase activities [38], and a recent study demonstrated that SIRT4 is a lipoamidase [39, 40]. SIRT5 has both lysine demalonylase and desuccinylase activities [41]. Discovery of the new enzymatic activities of sirtuins suggest that these proteins regulate more diverse biological processes than previously thought.

3.2.2 Histone Deacetylase Inhibitors

As mentioned above, zinc-dependent class I/II/IV deacetylases and NAD-dependent class III deacetylases (sirtuins) catalyze lysine deacetylation via different enzymatic reactions. Accordingly, HDAC inhibitors can be broadly classified into two groups: zinc-dependent HDAC inhibitors and sirtuin inhibitors.

3.2.2.1 Zinc-Dependent Histone Deacetylase Inhibitors

Among the efforts to identify small molecules for use in epigenetic regulation, the development of zinc-dependent HDAC inhibitors is currently the furthest advanced. Accordingly, a considerable number of natural or synthetic HDAC inhibitors have been described in the literature. Zinc-dependent HDAC inhibitors generally consist of three substructures: a cap that interacts with the entrance surrounding the substrate binding pocket; a ligand that interacts with zinc ions in the active site, as a functional group; and a spacer that bridges between the cap and functional group (Fig. 3.3). Therefore, HDAC inhibitors can be classified according to the structure of the cap or functional group. Functional groups include carboxylates, hydroxamic acids, electrophilic ketones containing epoxyketones, anilides, and thiols.

HDACs not only influence epigenetic regulation of gene expression and other biological processes by deacetylating nonhistone substrates, but are also deeply implicated in diverse diseases such as cancer, neurodegeneration, and inflammatory disorders. Consequently, HDAC inhibitors have potential as therapeutics as well as research tools.

3.2.2.1.1 Carboxylates

The first small molecule shown to inhibit HDAC activity was butyrate (Fig. 3.3), a natural product generated in humans by both metabolism of fatty acids and bacterial fermentation in the colon [42]. In a variety of vertebrate cell lines, sodium butyrate causes hyperacetylation of core histones including histones H2A, H2B, H3, and H4.

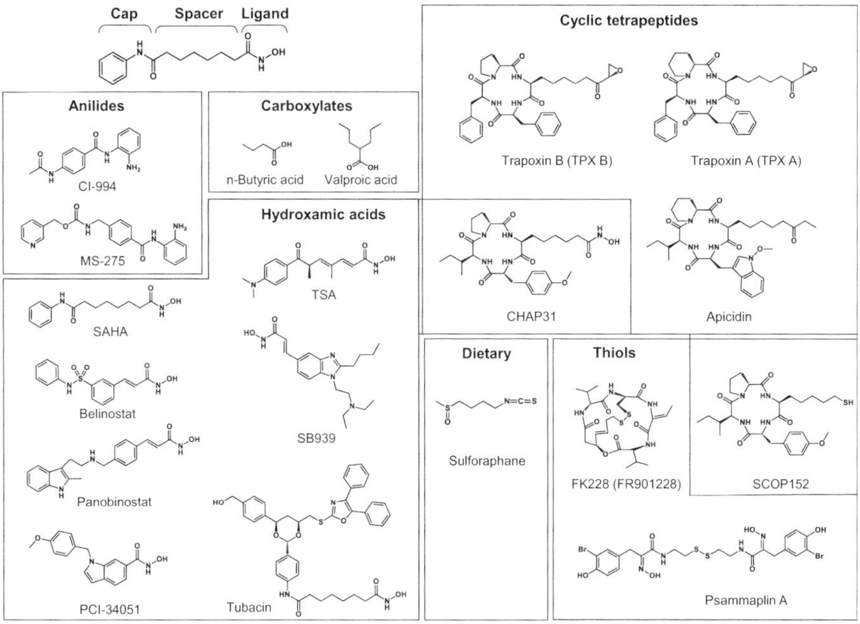

Fig. 3.3 Structure of zinc-dependent histone deacetylase (HDAC) inhibitors

Butyrate contains carboxylate as a functional group. Related compounds such as valproic acid [43, 44] (Fig. 3.3) were later also shown to be HDAC inhibitors. Members of this class of HDAC inhibitors generally have relatively poor inhibitory activity, possibly because the carboxylate coordinates zinc ions less strongly than other functional groups do. On the other hand, valproic acid has been used clinically as an antiepileptic or anxiolytic drug since the 1970s [45], and its safety has been established. Therefore, valproic acid could be repurposed for cancer treatment [46]. Indeed, several clinical trials have studied the efficacy of valproic acid against various types of cancers, including acute promyelocytic leukemia, adenoid cystic carcinoma, locally advanced rectal cancer, and glioblastoma [47–50].

3.2.2.1.2 Hydroxamic Acids

The first potent and specific HDAC inhibitor was the natural product trichostatin A (TSA), which contains hydroxamic acid as a functional group (Fig. 3.3). TSA was originally isolated as an antifungal antibiotic [51] and subsequently rediscovered as an inducer of differentiation in Friend cells [52]. TSA blocks the cell cycle in the G1 and G2 phases at very low concentrations [53]. During identification of its target, it was serendipitously discovered that TSA induces hyperacetylation of core histones in cells. Furthermore, TSA inhibits partially purified mouse HDAC activity at low nanomolar concentration in vitro, and a mutant cell line resistant to TSA does not accumulate acetylated histones in response to treatment with this compound [52]. These pieces of evidence clearly demonstrate that HDAC is the molecular target of TSA. X-ray crystallographic analysis of the bacterial histone deacetylase-like protein (HDLP) bound to TSA revealed the mechanism underlying the strong HDAC inhibition by this compound [21]: TSA mimics the substrate and binds by inserting its long aliphatic chain into a tube-like pocket, and inhibits enzyme activity by interacting with the zinc and active-site residues through the hydroxamic acid moiety at one end of the aliphatic chain. Hydrophobic amino acids surrounding the entrance of the substrate binding pocket in HDLP interact with an aromatic cap group on the other side of the aliphatic chain via hydrophobic interactions.

Another potent HDAC inhibitor is suberoylanilide hydroxamic acid (SAHA) (Fig. 3.3) [54]. This compound was created during a designed synthesis of hybrid polar compounds containing hydroxamic acid, and subsequently developed as an inducer of terminal differentiation and/or apoptosis in various transformed cells; it suppresses cell growth of various cancer cells both in vitro and in vivo [55–58]. SAHA was the first HDAC inhibitor to be approved as a drug for the treatment of refractory cutaneous T-cell lymphoma [59]. Two other members of this class of HDAC inhibitors, belinostat and panobinostat (Fig. 3.3), were recently approved for the treatment of refractory peripheral T-cell lymphoma and multiple myeloma, respectively [9, 10]. In addition to these approved compounds, a number of HDAC inhibitors containing hydroxamic acid, such as SB939 (Fig. 3.3), are currently being evaluated in clinical trials [60]. Although most hydroxamic acid–based HDAC inhibitors are nonselective and inhibit all zinc-dependent HDAC isoforms with various efficacies, several selective HDAC inhibitors have been developed. Tubacin (Fig. 3.3), the first specific HDAC6 inhibitor, was identified by a multidimensional chemical–genetic screen of small molecules [61]. Tubacin inhibits HDAC6-mediated α-tubulin deacetylation and cell migration without affecting the level of histone acetylation, gene expression patterns, or cell cycle progression. Furthermore, based on crystal structure analysis of HDAC8 bound to a compound [62, 63], several selective HDAC8 inhibitors have been designed and synthesized. One of these compounds, PCI-34051 (Fig. 3.3), contains a bulky aryl group in the linker group and inhibits HDAC8 with >200-fold selectivity over the other HDAC isoforms [64]. PCI-34051 and its derivatives induce cell death in T-cell lymphomas and neuroblastoma [64, 65].

3.2.2.1.3 Anilides

CI-994 (Fig. 3.3), originally shown to exert significant anticancer activity in a broad spectrum of tumor models [66–68], is the first member of the anilide class of HDAC inhibitors, but has poor inhibitory activity against HDACs in vitro. MS-275 (Fig. 3.3), which also contains anilide as its ligand, selectively inhibits class I HDACs [69] and inhibits growth of diverse solid tumor cells both in vitro and in vivo [70, 71]. The S3′-amino group of MS-275 is required for binding and inhibition of class I HDACs [70]. Kinetic analyses revealed that anilide derivatives behave as slow, tight binding inhibitors [72]. This observation was rationalized by analysis of crystal cocomplexes of HDAC2, which showed that anilide derivatives form stable complexes with the active site of HDAC2 [73]. Consistent with this compound's kinetic properties in vitro, analyses in which a fluorescent probe was used to visualize acetylation of histone H4K12 in living cells (Histac-K12) revealed that the effect of MS-275 lasts longer than those of hydroxamic acid derivatives such as TSA [74]. MS-275 is currently being evaluated as an antitumor drug in clinical trials [75–77].

3.2.2.1.4 Cyclic Tetrapeptides

Trapoxin (TPX) (Fig. 3.3), originally identified as an agent that induces morphological reversion from the transformed to the normal phenotype in v-*sis*-transformed NIH-3 T3 cells, also inhibits HDACs [78]. TPX is a cyclic tetrapeptide containing a unique amino acid, L-2-amino8-oxo-9,10-epoxydecanoic acid (Aoe). The epoxyketone group in Aoe appears to form a covalent bond between TPX and the active-site amino acid residues of HDACs. Taking advantage of the ability of TPX to form a covalent complex with HDACs, Taunton et al. used a synthesized TPX affinity probe to identify the first mammalian HDAC [79]. Among zinc-dependent mammalian HDACs, TPX only inhibits HDAC6 in vitro [80]. Taking advantage of this unique selectivity of TPX, Matsuyama et al. found that α-tubulin is a substrate of HDAC6 [19]. Although cyclic tetrapeptide derivatives containing Aoe exhibit very weak activity in animal models, probably due to the instability of the epoxyketone group in vivo, as a chemical probe TPX is one of the most valuable and useful HDAC inhibitors because it enabled identification of a mammalian HDAC and a substrate of HDAC6. CHAP (cyclic hydroxamic acid–containing peptide) 31 (Fig. 3.3), one the strongest CHAP derivatives, was designed as a hybrid of TSA and TPX [81]. CHAP31 is much more stable in the presence of cultured cells than TSA or TPX and exhibits antitumor activity in mice xenograft models.

Apicidin (Fig. 3.3), a fungal metabolite with potent and broad-spectrum antiprotozoal activity against apicomplexan parasites, exerts its biological effects by reversibly inhibiting HDAC activity [82]. Apicidin is a cyclic tetrapeptide that contains an ethyl ketone moiety in its side chain. The strong HDAC-inhibitory activity of apicidin further confirmed that cyclic tetrapeptides are useful scaffolds for designing potent HDAC inhibitors. Indeed, a number of cyclic tetrapeptides containing diverse functional groups have been synthesized, and many have strong HDAC-inhibitory activities [83–87].

3.2.2.1.5 Thiols

FK228 (also known as FR901228 and depsipeptide) (Fig. 3.3), originally isolated as a microbial metabolite, exhibits potent HDAC-inhibitory activity [88]. Unique among HDAC inhibitors, however, it does not contain a visible functional group that interacts with the zinc ion. Instead, FK228 is a cyclic depsipeptide with an intramolecular disulfide in its structure; reduction of the intramolecular disulfide bond activates FK228, and one of the sulfhydryl groups of the reduced form of FK228 (redFK) interacts with the active-site zinc, preventing access by the substrate [89, 90]. The disulfide bond in FK228 is rapidly reduced in cells by an activity involving glutathione. Thus, FK228 serves as a natural prodrug to inhibit HDACs and is activated by reduction after uptake into the cells. FK228 was the second HDAC inhibitor approved as a drug for the treatment of refractory cutaneous T-cell lymphoma [8].

Psammaplin A (Fig. 3.3), a marine natural product originally isolated from the sponge *Psammaplysilla,* exerts antibacterial activity against methicillin-resistant *Staphylococcus aureus* (MRSA) [91] as well as topoisomerase II inhibitory activity [92]. Psammaplin A is a structurally unique inhibitor of HDAC [93]; because it is a symmetrical bromotyrosine-derived disulfide, its inhibitory activity may be due to chelation of the zinc ion by a sulfhydryl in the active-site pocket. Studies of the mechanism of action of psammaplin A and its synthetic derivatives confirmed that the mechanism of inhibition involves cellular disulfide cleavage [94, 95]. Notably, certain derivatives of psammaplins isolated from the sponge *Pseudoceratina purpurea* exhibit DNA methyltransferase (DNMT) inhibitory activity [93].

On the basis of the corresponding hydroxamic acid (CHAP31) and FK228, a series of cyclic tetrapeptides containing thiol as the functional group, called sulfur-containing cyclic peptides (SCOPs) (Fig. 3.3), were synthesized [83]. SCOPs exhibit an intriguing pattern of target selectivity: they potently inhibit both HDAC1 and HDAC4, but have little effect on HDAC6 and HDAC8. Via disulfide formation with another copy of itself or another mercaptan, a SCOP can be converted to a "homodimer" or a "heterodimer," respectively, which (like FK228) acts as a prodrug. Once the homodimer is incorporated into the cells, it is reduced and activated by cellular enzymes, yielding two active molecules.

3.2.2.1.6 Dietary Inhibitors

Dietary isothiocyanates and allylsulfides possess HDAC-inhibitory activities [96]. Sulforaphane (Fig. 3.3), an isothiocyanate present in cruciferous vegetables such as broccoli and cabbage, inhibits HDACs in a dose-dependent manner [97]. Consumption of cruciferous vegetables is associated with a decrease in cancer risk, which may be mediated at least in part by sulforaphane-mediated HDAC inhibition [98].

3.2.2.2 Sirtuin Inhibitors

The first compounds shown to inhibit the in vitro deacetylase activity of sirtuins were nonhydrolyzable NAD analogs such as Carba-NAD, which compete for coenzyme binding in the active site [25]. However, such compounds are unsuitable for in vivo studies because they are not cell permeable and inhibit other NAD-dependent enzymes. To solve this problem, a wide variety of sirtuin inhibitors have recently been developed.

3.2.2.2.1 NAD Analogs

Based on the requirement of NAD for deacetylation by sirtuin, an uncleavable NAD analog was designed as a sirtuin inhibitor. This compound, Carba-NAD (Fig. 3.4) is a carbocyclic analog of NAD in which a 2,3-dihydroxycyclopentane methanol replaces the β-D-ribonucleotide ring of the nicotinamide riboside moiety [99]. Carba-NAD competitively inhibits the deacetylase activity of Hst2, a yeast homolog of Sir2 [25]. This observation is consistent with the requirement for NAD cleavage during the enzymatic reaction.

3.2.2.2.2 Nicotinamide and Its Analogs

Nicotinamide (Fig. 3.4), a by-product of the sirtuin-catalyzed deacetylase reaction, is a physiological inhibitor of sirtuins. Because nicotinamide inhibition is uniquely

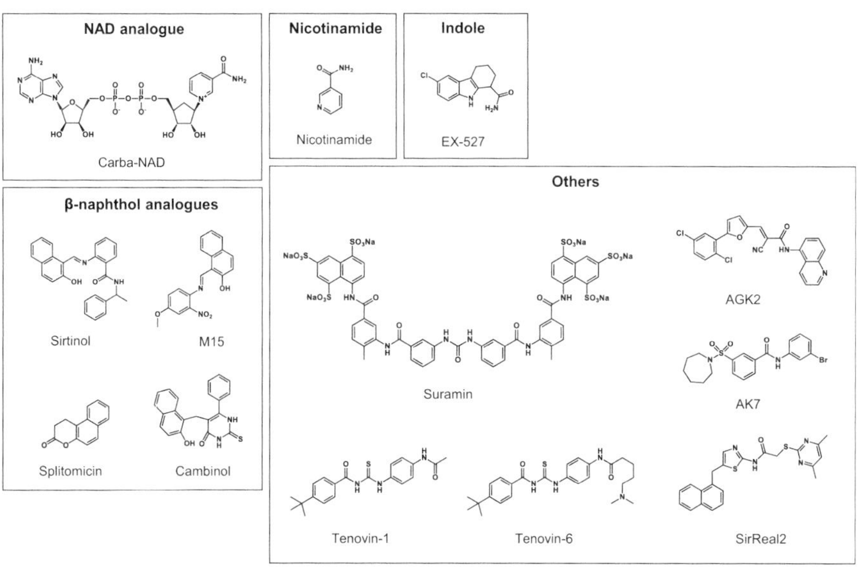

Fig. 3.4 Structure of sirtuin inhibitors

tied to the catalytic mechanism of Sir2 deacetylation [100], nicotinamide is likely to be a general inhibitor of all sirtuin family members. Indeed, the deacetylase activities of several yeast and human Sir2 family members are inhibited by nicotinamide. [25, 33, 101–103]. Moreover, nicotinamide strongly inhibits yeast silencing, increases rDNA recombination, and shortens replicative life-span, indicating that it phenocopies the *sir2* mutation in yeast [101]. Given that physiological concentrations of nicotinamide noncompetitively inhibit yeast Sir2 and human SIRT1 in vitro [101], nicotinamide may function as a physiologically relevant negative regulator of sirtuin enzymes [104].

3.2.2.2.3 β-Naphthol Analogs

A high-throughput cell-based phenotypic screen identified two small-molecule inhibitors of sirtuins, sirtinol and M15 (Fig. 3.4), which are structural analogs of α-substituted β-naphthol [105]. Sirtinol inhibits the in vivo silencing activity of yeast Sir2 and the in vitro deacetylase activities of yeast Sir2 and human SIRT2 but not HDAC1, indicating that it is a specific inhibitor of sirtuins. A chemical–genetic screen for *Arabidopsis* mutants resistant to sirtinol identified SIR1, a regulator of many auxin-inducible genes [106]. A series of sirtinol analogs were synthesized, of which two were 2- to 10-fold more potent than sirtinol against the human SIRT1 and SIRT2 enzymes [107].

In a yeast cell–based chemical screen for compounds that suppress Sir2-mediated silencing, Bedalov et al. identified splitomicin (Fig. 3.4), a structural relative of sirtinol, which phenocopies the *sir2* mutation in yeast and inhibits the in vitro deacetylase activity of yeast Sir2 [108]. Protein-based virtual screening resulted in identification of splitomicin derivatives as SIRT2 inhibitors [109]; some of these compounds exhibit antiproliferative properties and α-tubulin hyperacetylation in MCF7 breast cancer cells. Cambinol (Fig. 3.4), a chemically stable compound related to splitomicin, was identified as an inhibitor of SIRT1 and SIRT2 [110]. This compound exerts antitumor activity against BCL6-expressing Burkitt lymphoma in vitro and in mouse xenograft models.

3.2.2.2.4 Indoles

High-throughput screening using a fluorometric assay of human SIRT1 activity identified a series of indoles as potent selective SIRT1 inhibitors [111]. The most potent compound, EX-527 (Fig. 3.4), inhibited SIRT1 with an IC_{50} value of 98 nM. Kinetic analyses suggested that EX-527 binds after the release of nicotinamide from the enzyme and prevents the release of deacetylated peptide and *O*-acetyl-ADP-ribose, the products of sirtuin-catalyzed deacetylation. EX-527 was used to examine the role of SIRT1 in p53 acetylation and cell survival following DNA damage [112].

3.2.2.2.5 Other Sirtuin Inhibitors

Suramin (Fig. 3.4) is a symmetrical polyanionic naphthylurea originally used for treatment of trypanosomiasis and onchocerciasis. A screen for sirtuin activators revealed that suramin has inhibitory activity against SIRT1 [113]. Trapp et al. tested a diverse set of suramin analogs to elucidate the mechanism of SIRT1 and SIRT2 inhibition, and discovered selective inhibitors of human sirtuins with potency in the 10–100 nM range [114]. Crystal structures of SIRT5 in complex with suramin revealed that suramin binds at the NAD- and substrate binding site [115].

By searching for compounds that cause intracellular α-synuclein inclusions to grow from numerous small aggregates to larger bodies, Outeiro et al. identified AGK2 (Fig. 3.4) as a selective SIRT2 inhibitor with an IC_{50} of value of 3.5 μM [116]. In a cellular model of Parkinson's disease, AGK2 induces hyperacetylation of α-tubulin, rescues α-synuclein toxicity, and modifies inclusion morphology. In addition, AGK2 protects against dopaminergic cell death both in vitro and in a *Drosophila* model of Parkinson's disease. Genetic inhibition of SIRT2 via small interfering RNA similarly rescues α-synuclein toxicity, indicating a link between SIRT2 and Parkinson's disease. A role for SIRT2 in Parkinson's disease is further supported by the neuroprotective effect of AK7 (Fig. 3.4), a selective SIRT2 inhibitor structurally different from AGK2, in models of Parkinson's disease [117].

A cell-based screen aimed at discovering small molecules that activate p53 identified tenovin-1, along with its more water-soluble analog tenovin-6 (Fig. 3.4). These compounds inhibit SIRT1 and SIRT2 in the 1–10 μM range and prevent tumor growth in vivo by activating p53 [118]. Tenovin-6 suppresses the development of CML-like myeloproliferative disease in mice [119] and also induces apoptosis in acute promyelocytic leukemia (APL) [120] and acute lymphoblastic leukemia (ALL) cells [121]. These observations suggest that SIRT1 and/or SIRT2 are potential targets for antileukemia drug development.

By in vitro screening based on a fluorophore-labeled acetyl-lysine derivative for human SIRT1–3, a group of aminothiazoles called sirtuin-rearranging ligands (SirReals) were discovered as potent SIRT2-selective inhibitors. Of these, SirReal2 (Fig. 3.4) potently inhibits Sirt2 with an IC_{50} value of 140 nM but has very little effect on the activities of other SIRT isoforms [122]. Analyses of the crystal structure of the complex of SIRT2 and SirReal2 revealed that the compound's potency and unprecedented Sirt2 selectivity are based on ligand-induced structural rearrangement of the active site, revealing a previously unexploited binding pocket.

3.2.3 Histone Acetyltransferases

HATs catalyze the transfer of the acetyl group from acetyl-coenzyme A (acetyl-CoA) to the ε-amino group of a lysine residue on histone or nonhistone proteins. Human HATs are classified into four families based on their amino acid sequences, structures, and functions (Table 3.2). The CBP/p300 family has two members, p300 and CBP. These two proteins share extensive homology, but genetic and molecular analyses suggest that they possess both overlapping and unique functions [123]. The GCN5-related *N*-acetyltransferase (GNAT) family, which includes GCN5, PCAF, HAT1, and ELP3 [124], consists not only of protein lysine acetyltransferases (including HATs) but also acetyltransferases that catalyze *N*-terminal acetylation of proteins and acetylation of small molecules. Numerous genome sequencing and bioinformatics approaches suggest that this superfamily consists of some 10,000 members found in all domains of life. The MYST family, which comprises MOZ, TIP60, and HBO1, is defined by a distinct conserved histone acetyltransferase domain. The transcriptional factor–related HAT family includes TAF1 and TIFIIIC90. In addition, several steroid receptor coactivators such as SRC1, CLOCK, p160, and ACTR also possess HAT activities.

Table 3.2 Human histone acetyltransferase (HAT) family

Family	HAT	New name (KAT)	Function
CBP/p300	CBP	KAT3A	Transcriptional activation
	p300	KAT3B	Transcriptional activation
GNAT	HAT1	KAT1	Histone deposition, DNA repair
	GCN5	KAT2A	Transcriptional activation
	PCAF	KAT2B	Transcriptional activation
	ELP3	KAT9	Transcriptional elongation
MYST	TIP60	KAT5	Transcriptional activation, DNA repair
	MOZ	KAT6A	Transcriptional activation
	HBO1	KAT7	Transcription, DNA replication
Transcriptional factor–related/steroid receptor	TAF1	KAT4	Transcriptional activation
	TIFIIIC90	KAT12	Pol III transcription
	SRC1	KAT13A	Transcriptional activation
	ACTR	KAT13B	Transcriptional activation
	P160	KAT13C	Transcriptional activation
	CLOCK	KAT13D	Transcriptional activation

3.2.4 *Histone Acetyltransferases Inhibitors*

Several natural products or CoA derivatives act as HAT inhibitors, however; these compounds may be unsuited for cell-based studies because of poor specificity or cell permeability. Synthetic small-molecule inhibitors have recently been developed to address these problems.

3.2.4.1 Natural Product Inhibitors

Several dietary compounds such as curcumin (Fig. 3.5), a polyphenol present in turmeric spice [125], garcinol (Fig. 3.5), a polyisoprenylated benzophenone derivative from *Garcinia indica* fruit rind [126], and anacardic acid (Fig. 3.5), present in

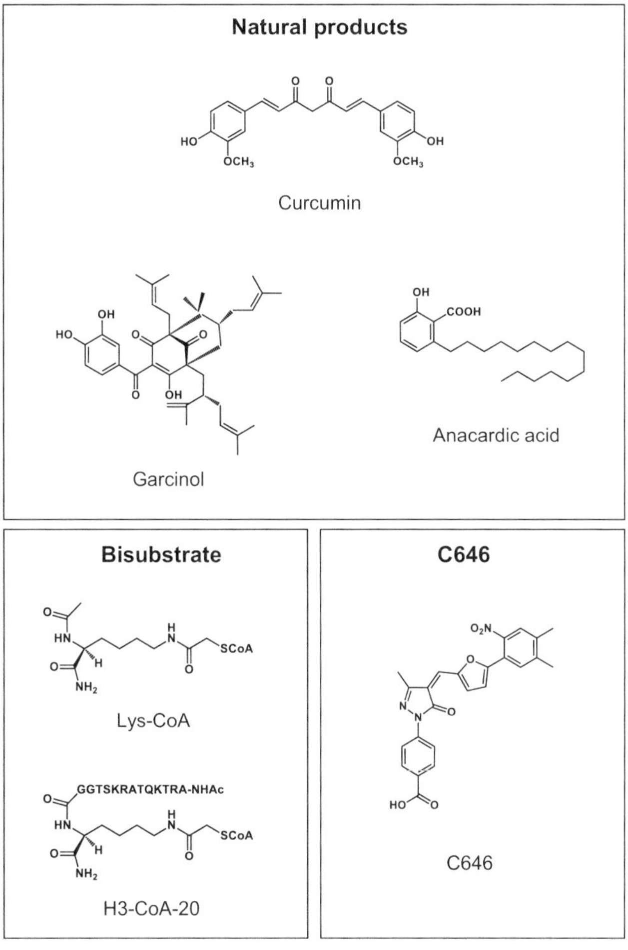

Fig. 3.5 Structure of histone acetyltransferase (HAT) inhibitors

cashew nut and gingko [125], inhibit HAT activity. A chemical screen using patient-specific induced pluripotent stem (iPS) cells revealed that anacardic acid can rescue the abnormal motor neuron phenotype in amyotrophic lateral sclerosis (ALS) [127]. However, these dietary compounds have diverse biological activities. For instance, anacardic acid also potently inhibits the SUMO E1 enzyme [128]. Therefore, the usefulness of these dietary HAT inhibitors as biological probes may be limited.

3.2.4.2 Bisubstrate Inhibitors

Lys-CoA (Fig. 3.5) and H3-CoA-20 (Fig. 3.5), designed and synthesized as bisubstrate HAT inhibitors that mimic the acetyl group donor acetyl-CoA and the acceptor lysine, selectively inhibit p300 and pCAF, respectively, with submicromolar IC_{50} values [129]. However, these bisubstrate inhibitors have poor pharmacokinetic properties because of their multiply charged phosphates. To address this issue, several truncated derivatives were designed and synthesized [130].

3.2.4.3 C646

In silico screening based on the crystal structure of the complex of p300 and Lys-CoA [131] identified a commercially available pyrazolone-containing small-molecule p300 HAT inhibitor, C646 (Fig. 3.5) [132]. C646 inhibits cellular proliferation, reduces colony formation, and induces apoptosis in AML1-ETO (AE) fusion protein–positive acute myeloid leukemia (AML) cell lines and primary blasts isolated from leukemic mice and AML patients [133]. In addition, C646 promotes sensitivity to DNA damage in melanoma [134] or non-small cell lung carcinoma cells [135]. These observations suggest that p300 is an attractive target for cancer therapy. To date, C646 is the most advanced commercially available chemical probe for investigating the cellular functions of p300.

3.2.5 Bromodomain and Extraterminal Protein Inhibitors

Epigenetic regulation by histone modifications and DNA methylation is mediated, at least in part, by reader proteins that physically interact with specific modifications. Thus, epigenetic "reader" as well as "writer" and "eraser" enzymes are attractive targets both for chemical probes for use in investigating epigenetic gene regulation and for antitumor drug development. Indeed, several potent bromodomain inhibitors of bromodomain and extraterminal (BET) proteins have been developed, including (+)-JQ1 (Fig. 3.6) and I-BET (Fig. 3.6), and these compounds exert potent antitumor activities [136]. These findings suggest that the bromodomains of BET family proteins are among the most promising

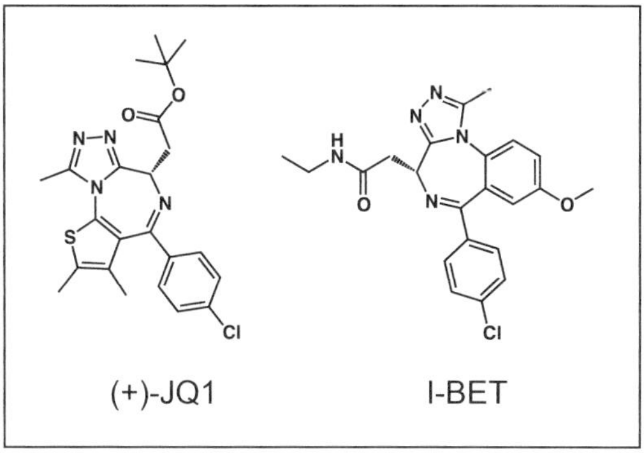

Fig. 3.6 Structure of bromodomain and extraterminal (BET) inhibitors

druggable epigenetic targets. In fact, some bromodomain inhibitors are currently being evaluated in clinical trials.

3.3 Histone Methylation

Histone methylation occurs predominantly on two highly abundant histone residues, Lys and Arg, on core histones including histone H3 and H4 (Fig. 3.2). There are three states of Lys methylation in nucleosomal histones: mono-, di-, and trimethylation; these reversible modifications are introduced by histone methyltransferases (HMTs) and removed by histone demethylases (HDMs). Unlike acetylation, methylation does not change the charge of histones, but instead serves as a binding site for histone "readers" such as proteins containing a chromodomain, plant homeodomain (PHD), tudor domain, or malignant brain tumor (MBT) domain [137, 138]. This modification is highly conserved across eukaryotic species [139] and is linked to either gene activation or repression depending on the target lysine residue. For instance, methylation of histones H3K4, H3K36, and H3K79 is involved in gene activation, whereas methylation of H3K9, H3K27, and H4K20 is generally associated with gene repression.

On the other hand, addition of methyl groups to Arg is catalyzed by protein arginine *N*-methyltransferases (PRMTs). These enzymes transfer the methyl group to the terminal guanidine nitrogens of Arg residues, and like HMTs generate three states of Arg methylation: mono-, symmetrical di-, and asymmetrical dimethylated arginine. Arginine methylation regulates signal transduction, mRNA splicing, transcriptional control, DNA repair, and protein translocation [140].

3.3.1 Histone Methyltransferases

Lys-specific HMTs have high site-specificity, which provides each methylation site with a distinct function. These enzymes can be mainly divided into two classes on the basis of their conserved catalytic domains (Table 3.3). One class shares a highly conserved catalytic domain called SET [Su(var)3–9, Enhancer-of-zeste, Trithorax] [141]. Members of the other class include yeast DOT1 (disruptor of telomeric silencing-1) and its eukaryotic homologs including human DOT1L (DOT1-like) [142], which contains a seven-β-strand core structure characteristic of the largest class of human *S*-adenosylmethionine (SAM)-dependent methyltransferases [143]. HMTs that contain a SET domain generally methylate lysine residues at the *N*-terminus of a core histone, whereas DOT1L methylates H3K79 within the histone globular core [144, 145].

PRMTs are classified into three groups based on the methylation pattern of the substrate [146]. Type 1 PRMTs, including PRMT1, PRMT2, PRMT3, PRMT4/

Table 3.3 Human lysine methyltransferase (KMT) family

Domain	KMT	New name	Function	Substrate specificity
SET	SUV39H1	KMT1A	Heterochromatin formation/ silencing	H3K9
	SUV39H2	KMT1B	Heterochromatin formation/ silencing	H3K9
	G9a	KMT1C	Heterochromatin formation/ silencing	H3K9
	GLP	KMT1D	Heterochromatin formation/ silencing	H3K9
	ESET/ SETDB1	KMT1E	Transcription repression	H3K9
	MLL1	KMT2A	Transcriptional activation	H3K4
	MLL2	KMT2B	Transcriptional activation	H3K4
	MLL3	KMT2C	Transcriptional activation	H3K4
	MLL4	KMT2D	Transcriptional activation	H3K4
	MLL5	KMT2E	Transcriptional activation	H3K4
	SET1A	KMT2F	Transcriptional activation	H3K4
	SET1B	KMT2G	Transcriptional activation	H3K4
	ASH1	KMT2H	Transcriptional activation	H3K36
	SET2	KMT3A	Pol III transcription	H3K36
	NSD1	KMT3B	Transcriptional activation	H3K36
	SYMD2	KMT3C	Transcriptional activation	H3K36
	PR-SET7/8	KMT5A	Transcription repression	H4K20
	SUV4-20H1	KMT5B	DNA-damage response	H4K20
	EZH2	KMT6	Silencing	H3K27
	SET7/9	KMT7	Transcription	H3K4
	RIZ1	KMT8	Transcription repression	H3K9
Seven-β-strand	DOT1L	KMT4	Transcriptional activation	H3K79

CARM1 (cofactor-associated arginine methyltransferase), PRMT6, and PRMT8, can catalyze mono- and asymmetrical dimethylation on arginines; type II, including PRMT5, mono- and symmetrically dimethylate their substrates; and type III, including PRMT7, only catalyze monomethylation on the guanidinium group [147]. The PRMTs catalyze methylation not only on arginine residues but also on nonhistone proteins [148].

3.3.2 Lys-Specific Histone Methyltransferase Inhibitors

Both groups of Lys-specific HMTs use SAM as the methyl group donor. Accordingly, *S*-adenosylhomocysteine (SAH) (Fig. 3.7), a product of the methyltransferase reaction, possesses inhibitory activity against methyltransferases, but its specificity is naturally poor. To date, several small molecules have been reported as enzyme-specific inhibitors. Here, we describe inhibitors of three HMTs including G9a, EZH2, and DOT1L, for which development is furthest advanced.

3.3.2.1 G9a Inhibitors

G9a, a histone methyltransferase that mainly catalyzes mono- and dimethylation of histone H3K9 [149], is involved in transcriptional gene silencing in euchromatic

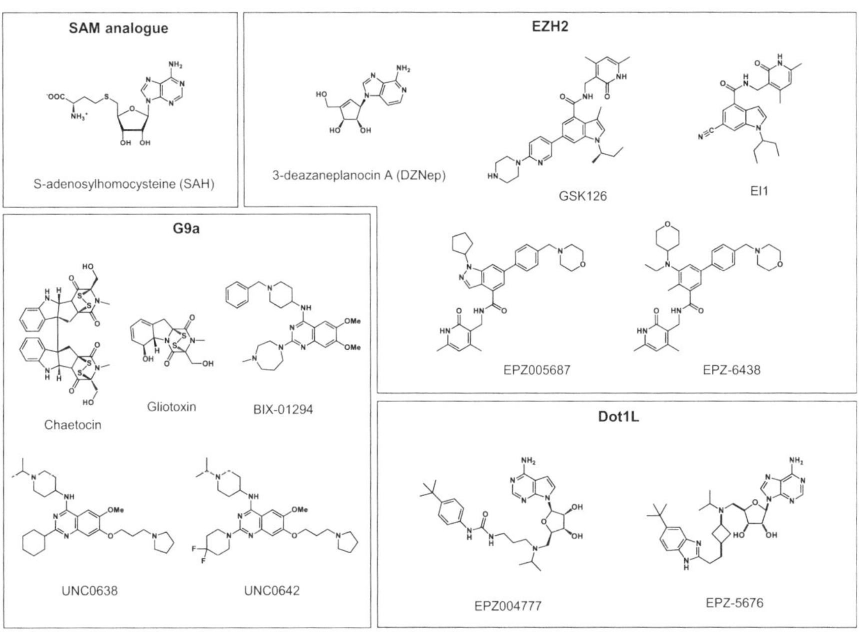

Fig. 3.7 Structure of Lys-specific histone methyltransferase (HMT) inhibitors

regions and is essential for embryonic development [150]. G9a represents an attractive target for cancer therapy because it is up-regulated in different types of cancer and its overexpression is associated with poor prognosis [151].

Chaetocin (Fig. 3.7), a 3–6 epi-dithio-diketopiperazines (ETP) originally isolated from the fermentation broth of *Chaetomium minutum,* was the first non-SAM analog G9a inhibitor to be identified. Its activity was revealed in a library screen for compounds that inhibit *Drosophila melanogaster* SU(VAR)3–9 protein, a SET domain–containing HMT responsible for histone H3K9 trimethylation [152]. Chaetocin is a competitive inhibitor of SAM and inhibits both and SU (VAR)3–9 and G9a, but not other HMTs such as E(z), PRSET7, or SET7/9, in vitro. Other EPTs such as gliotoxin (Fig. 3.7) also inhibit SU(VAR)3–9 and G9a with potency and selectivity similar to that of chaetocin [153, 154], suggesting that inhibition of HMTs may be a common pharmacological activity of ETPs. Although chaetocin was the first identified non-SAM analog small-molecule inhibitor of Lys-specific HMTs, its usefulness in cell studies may be limited because of its potent cytotoxicity. To overcome this problem, the total synthesis of chaetocin was accomplished [155] and the simple derivative with greatly reduced cytotoxicity was synthesized [156].

In a high-throughput screen that used rather high SAM concentrations to reduce the number of SAM-competitive compounds in the hit pool, BIX-01294 (Fig. 3.7), a diazepine–quinazoline–amine derivative, was identified as a G9a-specific inhibitor that does not compete with SAM [157]. BIX-01294 selectively inhibits the enzymatic activity of G9a with an IC_{50} of 1.7 μM and impairs the generation of H3K9me2 in cells. Based on x-ray cocrystal structures of G9a with BIX-01294 and its analogs, several BIX-01294 derivatives were designed and synthesized with the goal of improving in vitro and cellular potency. One of the resultant compounds, UNC0638 (Fig. 3.7), inhibits G9a activity with an IC_{50} of 15 nM and selectivity versus a wide range of epigenetic and nonepigenetic targets [158]. UNC0638 reduces global H3K9me2 levels in a variety of cell lines and reactivates G9a-silenced genes in a concentration-dependent manner. Thus, UNC0638 seems to be the first dependable chemical probe that can investigate cellular G9a functions. UNC0642 (Fig. 3.7), a UNC0638 analog with improved in vivo PK properties, was recently reported [159].

3.3.2.2 EZH2 Inhibitors

EZH2, the catalytic subunit of the polycomb repressive complex 2 (PRC2), catalyzes the methylation of histone H3K27, a repressive epigenetic mark. EZH2 is frequently amplified and/or overexpressed in several types of solid tumors, including prostate, breast, and lung cancers [160–162]. In addition, activating somatic mutations in the catalytic domain of EZH2 have been found in some leukemias, including diffuse large B-cell lymphoma (DLBCL) and follicular lymphoma (FL), leading to elevated trimethylation of histone H3K27 [163]. Accordingly, EZH2 has attracted a great deal of interest as an epigenetic target for cancer therapy, and multiple research groups and pharmaceutical companies have devoted effort to identifying potent and selective EZH2 inhibitors.

3-Deazaneplanocin A (DZNep) (Fig. 3.7) was the first reported EZH2 inhibitor [164], but this compound is not a direct inhibitor of EZH2. Instead, DZNep inhibits SAH hydrolase and depletes PRC2 subunits indirectly as a result of the ensuing increase in cellular SAH concentration. Consequently, DZNep decreases the level of methylation on multiple histone Lys residues, so its usefulness as a chemical probe and specific EZH2 inhibitor is limited. Research groups at pharmaceutical companies such as GSK, Epizyme, and Novartis, have recently reported small-molecule inhibitors of EZH2, including GSK-126, EI1, EPZ005687, and EPZ-6438 (Fig. 3.7), which exhibit high potency and selectivity against EZH2 [165–168]. These compounds decrease the level of cellular H3K27 methylation, activate the expression of EZH2 target genes, and selectively inhibit the growth of DLBCL cells harboring activating mutations in EZH2. Although they were independently identified by different groups, these EZH2 inhibitors all share a pyridone amide scaffold. EPZ-6438 is undergoing clinical trials against several types of lymphomas, including DLBCL and FL.

3.3.2.3 Dot1L Inhibitors

DOT1L catalyzes methylation at histone H3K79, which is associated with active gene transcription. DOT1L is a driver of leukemogenesis in mixed-lineage leukemia (MLL) [169]. In 2011, a research group at Epizyme reported EPZ004777 (Fig. 3.7) as a highly potent and selective inhibitor of DOT1L [169]. This compound selectively inhibits H3K79 methylation in MLL cells and blocks expression of leukemogenic genes. Exposure to EPZ004777 results in selective killing of leukemia cells bearing the MLL gene translocation and extends survival in a mouse MLL xenograft model. EPZ-5676 (Fig. 3.7), an analog of EPZ004777 with improved potency and drug-like properties [170], is undergoing clinical trials.

3.3.3 Histone Demethylases

Initially, histone lysine methylation was thought to be irreversible, but we now know it can be reversed by two different groups of histone lysine demethylases (KDMs): flavin adenine dinucleotide (FAD)-dependent amine oxidases, including lysine specific demethylase 1 (LSD1); and iron(II)-α-ketoglutarate (α-KG)–dependent dioxygenases of the Jumonji family, called Jumonji domain–containing histone demethylases (JHDM) (Table 3.4). Lysine demethylation by LSD1 requires FAD as a cofactor and produces hydrogen peroxide and formaldehyde. LSD1 can catalyze demethylation of secondary and tertiary but not quaternary amines, limiting its substrates to mono- and dimethylated lysines. By contrast, the JHDM family, which consists of over 20 members, catalyzes lysine demethylation using α-KG and

Table 3.4 Human histone lysine demethylase (KDM) family

Family	KDM	New name	Function	Substrate specificity
LSD1	LSD1	KDM1A	Transcription repression	H3K4
	LSD2	KDM1B	Transcription repression	H3K4
JHDM	JHDM1A	KDM2A	Maintenance of heterochromatin	H3K36
	JHDM1B	KDM2B	Maintenance of heterochromatin	H3K36
	JHDM2A	KDM3A	Transcriptional activation	H3K9
	JHDM2B	KDM3B	Transcriptional activation	H3K9
	JMJD2A/ JHDM3A	KDM4A	Transcriptional regulation	H3K9/K36
	JMJD2B	KDM4B	DNA-damage response	H3K9/K36
	JMJD2C/GASC1	KDM4C	Transcriptional regulation	H3K9/K36
	JMJD2D	KDM4D	Transcriptional activation	H3K9
	JARID1A	KDM5A	Transcriptional regulation	H3K4
	JARID1B	KDM5B	Transcriptional repression	H3K4
	JARID1C	KDM5C	Transcriptional repression	H3K4
	JARID1D	KDM5D	Transcriptional repression	H3K4
	UTX	KDM6A	Transcriptional activation	H3K27
	JMJD3	KDM6B	Transcriptional activation	H3K27
	JHDM1D	KDM7	Transcriptional activation	H3K9/K27

oxygen with production of succinic acid and formaldehyde as by-products. Unlike FAD-dependent demethylases, JHDMs can catalyze the demethylation of mono-, di-, and trimethylated lysines.

3.3.4 *Lysine Demethylases Inhibitors*

KDMs are divided into two families based on amino acid sequence conservation and catalytic mechanism. Accordingly, KDM inhibitors are classified into two groups: LSD1 and JHDM inhibitors. Mutations or changes in expression of particular KDMs are associated with several diseases, including cancers [171]. Therefore, small-molecule inhibitors of KDMs are in high demand, both for use in research to investigate KDM functions and for development of anticancer therapies.

3.3.4.1 Lysine Specific Demethylase 1 Inhibitors

LSD1 (KDM1A), which demethylates mono- and dimethylated histone H3K4, is involved in the regulation of a broad spectrum of biological processes including development, cellular differentiation, and embryonic pluripotency. LSD1 is also deeply implicated in cancer development. Consistent with this, LSD1 is highly expressed in multiple types of cancers including breast, prostate, bladder, and colorectal cancers [172–174]. LSD1 is essential for maintenance of AML cells harboring MLL translocations [175]. Thus, inhibition of LSD1 activity may have potential in cancer therapy.

LSD1 is structurally related to the monoamine oxidases MAO-A and MAO-B. Tranylcypromine (TCP) (Fig. 3.8), an inhibitor of MAO-A and MAO-B used as a drug for depression and Parkinson's disease, inhibits the reaction catalyzed by LSD1 [176]. Accordingly, compounds derived from TCP have been extensively investigated as LSD1 inhibitors [177, 178]. For instance, several TCP derivatives such as compound 2 and S1201 (Fig. 3.8), synthesized by structure-based design, are selective and potent LSD1 inhibitors [179, 180] and exert antileukemia activities [175]. Two TCP derivatives, ORY-1001 (Fig. 3.8) and GSK2879552 (Fig. 3.8), are currently being evaluated in phase I trials against acute leukemia and/or small cell lung carcinoma.

3.3.4.2 Jumonji Domain–Containing Histone Demethylase Inhibitors

N-oxalylglycine (NOG) (Fig. 3.8), the amide analog of α-KG, inhibits JHDM activity in vitro [181], but it naturally inhibits other α-KG-dependent enzymes such as prolyl hydroxylases and has poor cell permeability. Based on the crystal structure of the complex of JMJD2A (KDM3A), NOG, and histone trimethylated lysine peptide, Hamada et al. synthesized NOG derivatives that effectively inhibit JHDMs in cells [182]. Several isoform-specific JHDM inhibitors including daminozide (Fig. 3.8), a plant growth regulator, have been reported to date [183]. Among them, one of the most encouraging potential lead compounds is GSK-J1 (Fig. 3.8), an inhibitor of H3K27me3-specific demethylases JMJD3

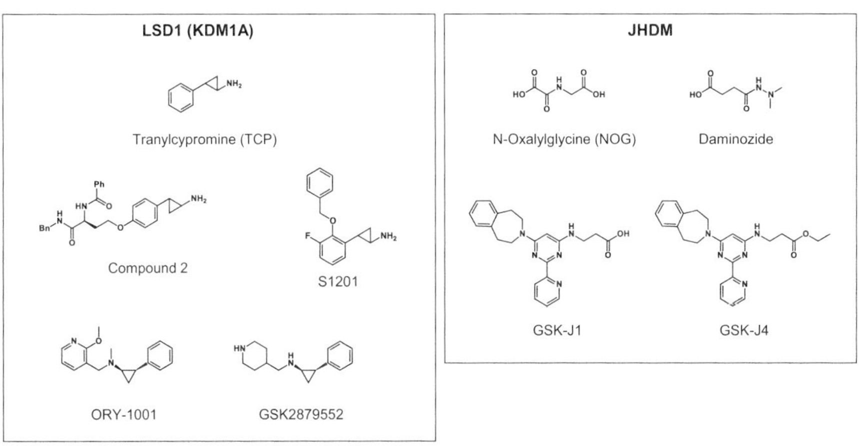

Fig. 3.8 Structure of histone demethylase (HDM) inhibitors

(KDM6B) and UTX (KDM6A) [184]. GSK-J1 inhibits KDM6B in vitro with an IC_{50} of 60 nM and is remarkably selective for the JMJD3 subfamily. However, the cellular effect of GSK-J1 is poor because the highly polar carboxylate group of GSK-J1 restricts its cell permeability. GSK-J4 (Fig. 3.8), the ethyl ester analog of GSK-J1, has improved cell penetrability and increases the level of histone H3K27 methylation in cells. GSK-J4 has shown effects in the inflammatory response and glioma xenografts [184, 185].

3.4 DNA Methylation

DNA methylation at position 5 of cytosine, 5-methylcytosine (5mC), is involved in transcriptional silencing of genes. In mammalian genomes, 5mC is present predominantly in a CpG dinucleotide context, and about 70–80% of CpGs are methylated. This covalent modification is catalyzed by DNA methyltransferases (DNMTs) using SAM as a methyl group donor. DNMTs are classified into two families [186]: DNMT1, which methylates newly synthesized DNA strands and is important for establishment and maintenance of tissue-specific patterns of DNA methylation; and DNMT3, consisting of DNMT3A and DNMT3B, which are responsible for de novo DNA methylation. Although DNA methylation has been recognized for a long time, the underlying mechanism remained unclear until the discovery that the ten–eleven translocation (TET) family enzymes are DNA demethylases [187–189]. The TET family has three members: TET1, TET2 and TET3, all of which can oxidize 5-methylcytosine to generate 5-hydroxymethylcytosine (5hmC). 5hmC, a key intermediate in active demethylation pathways, can either be passively depleted through DNA replication or actively reverted to cytosine via thymine DNA glycosylase (TDG)-mediated base excision repair. Because aberrant DNA methylation patterns are associated with abnormal gene expression, eventually leading to cancer [190], both DNMTs and TETs are attractive targets for cancer therapy.

3.4.1 DNA Methyltransferase Inhibitors

DNMT inhibitors are classified into two groups: nucleoside analogs and non-nucleoside analogs. The representative nucleoside analogs 5-azacytidine (Fig. 3.9) and decitabine (Fig. 3.9) have been approved for treatment of MDS [5, 6]. These inhibitors are incorporated into genomic DNA in place of deoxycytidine and covalently block DNMTs, thereby inducing reactivation of genes (such as tumor suppressor genes) silenced by aberrant DNA methylation. Despite their high efficiency and clinical use, these compounds have disadvantages including metabolic instability in the body, generation of cytotoxic metabolites, and undesired side effects such as dose-limiting neutropenia. Zebularine (Fig. 3.9),

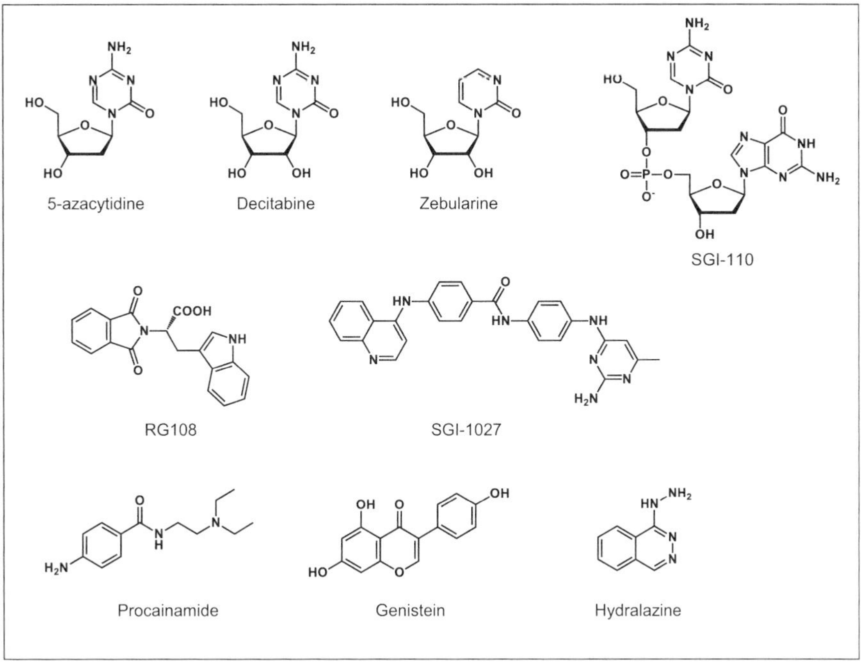

Fig. 3.9 Structure of DNA methyltransferase (DNMT) inhibitors

which was designed to solve these problems, is stable and less cytotoxic [191], but its cellular efficiency is poor [192]. To date, the most promising nucleoside analog DNMT inhibitor is SGI-110 (Fig. 3.9), an efficient prodrug of decitabine [193]. SGI-110 exhibits better aqueous solubility, cytotoxicity, cell permeability, and in vivo inhibition of DNA demethylation and tumor growth than decitabine [193, 194]. SGI-110 is currently undergoing clinical trials against multiple diseases, including MDS, AML, and ovarian and liver cancers.

Several nonnucleoside analogs that bind the catalytic sites of DNMTs, such as RG108 (Fig. 3.9), SGI-1027 (Fig. 3.9), procainamide (Fig. 3.9), genistein (Fig. 3.9), and hydralazine (Fig. 3.9), have been reported [195]. However, these compounds lack specificity for DNMTs and/or exhibit weak activity. To address these shortcomings, several analogs of SGI-1027 have been designed and synthesized [196].

3.5 Conclusions

Dysregulation of epigenetic events may be a key driving force in the development of cancers and other diseases. Thus, elucidation of the epigenetic regulation is one of the major tasks confronting biology following the completion of the Human

Genome Project. Efforts to develop potent and isoform-selective inhibitors of epigenetic machineries are necessary for the elucidation of epigenetic regulation and the development of new classes of drugs for human diseases including cancers.

References

1. Turner BM (2000) Histone acetylation and an epigenetic code. BioEssays 22(9):836–845. doi:10.1002/1521-1878(200009)22:9<836::AID-BIES9>3.0.CO;2-X
2. Jenuwein T, Allis CD (2001) Translating the histone code. Science 293(5532):1074–1080. doi:10.1126/science.1063127
3. Baylin SB, Jones PA (2011) A decade of exploring the cancer epigenome—biological and translational implications. Nat Rev Cancer 11(10):726–734. doi:10.1038/nrc3130
4. Ellis L, Atadja PW, Johnstone RW (2009) Epigenetics in cancer: targeting chromatin modifications. Mol Cancer Ther 8(6):1409–1420. doi:10.1158/1535-7163.MCT-08-0860
5. Issa JP, Kantarjian HM (2009) Targeting DNA methylation. Clin Cancer Res 15 (12):3938–3946. doi:10.1158/1078-0432.CCR-08-2783
6. Yang X, Lay F, Han H, Jones PA (2010) Targeting DNA methylation for epigenetic therapy. Trends Pharmacol Sci 31(11):536–546. doi:10.1016/j.tips.2010.08.001
7. Wagner JM, Hackanson B, Lubbert M, Jung M (2010) Histone deacetylase (HDAC) inhibitors in recent clinical trials for cancer therapy. Clin Epigenetics 1(3–4):117–136. doi:10.1007/s13148-010-0012-4
8. Khan O, La Thangue NB (2012) HDAC inhibitors in cancer biology: emerging mechanisms and clinical applications. Immunol Cell Biol 90(1):85–94. doi:10.1038/icb.2011.100
9. Rashidi A, Cashen AF (2015) Belinostat for the treatment of relapsed or refractory peripheral T-cell lymphoma. Future Oncol 11(11):1659–1664. doi:10.2217/fon.15.62
10. Laubach JP, Moreau P, San-Miguel JF, Richardson PG (2015) Panobinostat for the treatment of multiple myeloma. Clin Cancer Res 21(21):4767–4773. doi:10.1158/1078-0432.CCR-15-0530
11. Yang XJ, Seto E (2008) The Rpd3/Hda1 family of lysine deacetylases: from bacteria and yeast to mice and men. Nat Rev Mol Cell Biol 9(3):206–218. doi:10.1038/nrm2346
12. Khochbin S, Verdel A, Lemercier C, Seigneurin-Berny D (2001) Functional significance of histone deacetylase diversity. Curr Opin Genet Dev 11(2):162–166
13. Verdin E, Dequiedt F, Kasler HG (2003) Class II histone deacetylases: versatile regulators. Trends Genet 19(5):286–293. doi:10.1016/S0168-9525(03)00073-8
14. Yang XJ, Gregoire S (2005) Class II histone deacetylases: from sequence to function, regulation, and clinical implication. Mol Cell Biol 25(8):2873–2884. doi:10.1128/MCB.25.8.2873-2884.2005
15. Verdel A, Curtet S, Brocard MP, Rousseaux S, Lemercier C, Yoshida M, Khochbin S (2000) Active maintenance of mHDA2/mHDAC6 histone-deacetylase in the cytoplasm. Curr Biol 10(12):747–749
16. Seigneurin-Berny D, Verdel A, Curtet S, Lemercier C, Garin J, Rousseaux S, Khochbin S (2001) Identification of components of the murine histone deacetylase 6 complex: link between acetylation and ubiquitination signaling pathways. Mol Cell Biol 21(23):8035–8044
17. Boyault C, Sadoul K, Pabion M, Khochbin S (2007) HDAC6, at the crossroads between cytoskeleton and cell signaling by acetylation and ubiquitination. Oncogene 26 (37):5468–5476. doi:10.1128/MCB.21.23.8035-8044.2001
18. Hubbert C, Guardiola A, Shao R, Kawaguchi Y, Ito A, Nixon A, Yoshida M, Wang XF, Yao TP (2002) HDAC6 is a microtubule-associated deacetylase. Nature 417(6887):455–458. doi:10.1038/417455a

19. Matsuyama A, Shimazu T, Sumida Y, Saito A, Yoshimatsu Y, Seigneurin-Berny D, Osada H, Komatsu Y, Nishino N, Khochbin S, Horinouchi S, Yoshida M (2002) In vivo destabilization of dynamic microtubules by HDAC6-mediated deacetylation. EMBO J 21(24):6820–6831
20. Guardiola AR, Yao TP (2002) Molecular cloning and characterization of a novel histone deacetylase HDAC10. J Biol Chem 277(5):3350–3356. doi:10.1074/jbc.M109861200
21. Finnin MS, Donigian JR, Cohen A, Richon VM, Rifkind RA, Marks PA, Breslow R, Pavletich NP (1999) Structures of a histone deacetylase homologue bound to the TSA and SAHA inhibitors. Nature 401(6749):188–193. doi:10.1038/43710
22. Shore D, Squire M, Nasmyth KA (1984) Characterization of two genes required for the position-effect control of yeast mating-type genes. EMBO J 3(12):2817–2823
23. Imai S, Armstrong CM, Kaeberlein M, Guarente L (2000) Transcriptional silencing and longevity protein Sir2 is an NAD-dependent histone deacetylase. Nature 403(6771):795–800. doi:10.1038/35001622
24. Smith JS, Brachmann CB, Celic I, Kenna MA, Muhammad S, Starai VJ, Avalos JL, Escalante-Semerena JC, Grubmeyer C, Wolberger C, Boeke JD (2000) A phylogenetically conserved NAD^+–dependent protein deacetylase activity in the Sir2 protein family. Proc Natl Acad Sci U S A 97(12):6658–6663
25. Landry J, Slama JT, Sternglanz R (2000) Role of NAD(+) in the deacetylase activity of the SIR2-like proteins. Biochem Biophys Res Commun 278(3):685–690. doi:10.1006/bbrc.2000.3854
26. Tanner KG, Landry J, Sternglanz R, Denu JM (2000) Silent information regulator 2 family of NAD-dependent histone/protein deacetylases generates a unique product, 1-*O*-acetyl-ADP-ribose. Proc Natl Acad Sci U S A 97(26):14178–14182. doi:10.1073/pnas.250422697
27. Sauve AA, Celic I, Avalos J, Deng H, Boeke JD, Schramm VL (2001) Chemistry of gene silencing: the mechanism of NAD^+–dependent deacetylation reactions. Biochemistry 40 (51):15456–15463
28. Jackson MD, Denu JM (2002) Structural identification of 2′- and 3′-*O*-acetyl-ADP-ribose as novel metabolites derived from the Sir2 family of beta -NAD^+–dependent histone/protein deacetylases. J Biol Chem 277(21):18535–18544. doi:10.1074/jbc.M200671200
29. Chang JH, Kim HC, Hwang KY, Lee JW, Jackson SP, Bell SD, Cho Y (2002) Structural basis for the NAD-dependent deacetylase mechanism of Sir2. J Biol Chem 277(37):34489–34498. doi:10.1074/jbc.M205460200
30. Frye RA (2000) Phylogenetic classification of prokaryotic and eukaryotic Sir2-like proteins. Biochem Biophys Res Commun 273(2):793–798. doi:10.1006/bbrc.2000.3000
31. Michishita E, Park JY, Burneskis JM, Barrett JC, Horikawa I (2005) Evolutionarily conserved and nonconserved cellular localizations and functions of human SIRT proteins. Mol Biol Cell 16(10):4623–4635. doi:10.1091/mbc.E05-01-0033
32. Schwer B, North BJ, Frye RA, Ott M, Verdin E (2002) The human silent information regulator (Sir)2 homologue hSIRT3 is a mitochondrial nicotinamide adenine dinucleotide-dependent deacetylase. J Cell Biol 158(4):647–657. doi:10.1083/jcb.200205057
33. Onyango P, Celic I, McCaffery JM, Boeke JD, Feinberg AP (2002) SIRT3, a human SIR2 homologue, is an NAD-dependent deacetylase localized to mitochondria. Proc Natl Acad Sci U S A 99(21):13653–13658. doi:10.1073/pnas.222538099
34. Jiang H, Khan S, Wang Y, Charron G, He B, Sebastian C, Du J, Kim R, Ge E, Mostoslavsky R, Hang HC, Hao Q, Lin H (2013) SIRT6 regulates TNF-alpha secretion through hydrolysis of long-chain fatty acyl lysine. Nature 496(7443):110–113. doi:10.1038/nature12038
35. Feldman JL, Baeza J, Denu JM (2013) Activation of the protein deacetylase SIRT6 by long-chain fatty acids and widespread deacylation by mammalian sirtuins. J Biol Chem 288 (43):31350–31356. doi:10.1074/jbc.C113.511261
36. Teng YB, Jing H, Aramsangtienchai P, He B, Khan S, Hu J, Lin H, Hao Q (2015) Efficient demyristoylase activity of SIRT2 revealed by kinetic and structural studies. Sci Rep 5:8529. doi:10.1038/srep08529

37. Feldman JL, Dittenhafer-Reed KE, Kudo N, Thelen JN, Ito A, Yoshida M, Denu JM (2015) Kinetic and structural basis for acyl-group selectivity and NAD(+) dependence in sirtuin-catalyzed deacylation. Biochemistry 54(19):3037–3050. doi:10.1021/acs.biochem.5b00150
38. Haigis MC, Mostoslavsky R, Haigis KM, Fahie K, Christodoulou DC, Murphy AJ, Valenzuela DM, Yancopoulos GD, Karow M, Blander G, Wolberger C, Prolla TA, Weindruch R, Alt FW, Guarente L (2006) SIRT4 inhibits glutamate dehydrogenase and opposes the effects of calorie restriction in pancreatic beta cells. Cell 126(5):941–954. doi:10.1016/j.cell.2006.06.057
39. Mathias RA, Greco TM, Oberstein A, Budayeva HG, Chakrabarti R, Rowland EA, Kang Y, Shenk T, Cristea IM (2014) Sirtuin 4 is a lipoamidase regulating pyruvate dehydrogenase complex activity. Cell 159(7):1615–1625. doi:10.1016/j.cell.2014.11.046
40. Liszt G, Ford E, Kurtev M, Guarente L (2005) Mouse Sir2 homolog SIRT6 is a nuclear ADP-ribosyltransferase. J Biol Chem 280(22):21313–21320. doi:10.1074/jbc.M413296200
41. Du J, Zhou Y, Su X, Yu JJ, Khan S, Jiang H, Kim J, Woo J, Kim JH, Choi BH, He B, Chen W, Zhang S, Cerione RA, Auwerx J, Hao Q, Lin H (2011) Sirt5 is a NAD-dependent protein lysine demalonylase and desuccinylase. Science 334(6057):806–809. doi:10.1126/science.1207861
42. Candido EP, Reeves R, Davie JR (1978) Sodium butyrate inhibits histone deacetylation in cultured cells. Cell 14(1):105–113
43. Phiel CJ, Zhang F, Huang EY, Guenther MG, Lazar MA, Klein PS (2001) Histone deacetylase is a direct target of valproic acid, a potent anticonvulsant, mood stabilizer, and teratogen. J Biol Chem 276(39):36734–36741. doi:10.1074/jbc.M101287200
44. Göttlicher M, Minucci S, Zhu P, Krämer OH, Schimpf A, Giavara S, Sleeman JP, Lo Coco F, Nervi C, Pelicci PG, Heinzel T (2001) Valproic acid defines a novel class of HDAC inhibitors inducing differentiation of transformed cells. EMBO J 20(24):6969–6978. doi:10.1093/emboj/20.24.6969
45. Bruni J, Wilder BJ (1979) Valproic acid. Review of a new antiepileptic drug. Arch Neurol 36(7):393–398
46. Hrebackova J, Hrabeta J, Eckschlager T (2010) Valproic acid in the complex therapy of malignant tumors. Curr Drug Targets 11(3):361–379
47. Tassara M, Döhner K, Brossart P, Held G, Götze K, Horst HA, Ringhoffer M, Köhne CH, Kremers S, Raghavachar A, Wulf G, Kirchen H, Nachbaur D, Derigs HG, Wattad M, Koller E, Brugger W, Matzdorff A, Greil R, Heil G, Paschka P, Gaidzik VI, Göttlicher M, Döhner H, Schlenk RF (2014) Valproic acid in combination with all-trans retinoic acid and intensive therapy for acute myeloid leukemia in older patients. Blood 123(26):4027–4036. doi:10.1182/blood-2013-12-546283
48. Avallone A, Piccirillo MC, Delrio P, Pecori B, Di Gennaro E, Aloj L, Tatangelo F, D'Angelo V, Granata C, Cavalcanti E, Maurea N, Maiolino P, Bianco F, Montano M, Silvestro L, Terranova Barberio M, Roca MS, Di Maio M, Marone P, Botti G, Petrillo A, Daniele G, Lastoria S, Iaffaioli VR, Romano G, Caracò C, Muto P, Gallo C, Perrone F, Budillon A (2014) Phase 1/2 study of valproic acid and short-course radiotherapy plus capecitabine as preoperative treatment in low–moderate risk rectal cancer—V-shoRT-R3 (Valproic Acid–Short Radiotherapy–Rectum 3rd Trial). BMC Cancer 14:875. doi:10.1186/1471-2407-14-875
49. Bilen MA, Fu S, Falchook GS, Ng CS, Wheler JJ, Abdelrahim M, Erguvan-Dogan B, Hong DS, Tsimberidou AM, Kurzrock R, Naing A (2015) Phase I trial of valproic acid and lenalidomide in patients with advanced cancer. Cancer Chemother Pharmacol 75(4):869–874. doi:10.1007/s00280-015-2695-x
50. Krauze AV, Myrehaug SD, Chang MG, Holdford DJ, Smith S, Shih J, Tofilon PJ, Fine HA, Camphausen K (2015) A phase 2 study of concurrent radiation therapy, temozolomide, and the histone deacetylase inhibitor valproic acid for patients with glioblastoma. Int J Radiat Oncol Biol Phys 92(5):986–992. doi:10.1016/j.ijrobp.2015.04.038

51. Tsuji N, Kobayashi M, Nagashima K, Wakisaka Y, Koizumi K (1976) A new antifungal antibiotic, trichostatin. J Antibioti 29(1):1–6
52. Yoshida M, Kijima M, Akita M, Beppu T (1990) Potent and specific inhibition of mammalian histone deacetylase both in vivo and in vitro by trichostatin A. J Biol Chem 265 (28):17174–17179
53. Yoshida M, Beppu T (1988) Reversible arrest of proliferation of rat 3Y1 fibroblasts in both the G1 and G2 phases by trichostatin A. Exp Cell Res 177(1):122–131
54. Richon VM, Emiliani S, Verdin E, Webb Y, Breslow R, Rifkind RA, Marks PA (1998) A class of hybrid polar inducers of transformed cell differentiation inhibits histone deacetylases. Proc Natl Acad Sci U S A 95(6):3003–3007
55. Butler LM, Agus DB, Scher HI, Higgins B, Rose A, Cordon-Cardo C, Thaler HT, Rifkind RA, Marks PA, Richon VM (2000) Suberoylanilide hydroxamic acid, an inhibitor of histone deacetylase, suppresses the growth of prostate cancer cells in vitro and in vivo. Cancer Res 60 (18):5165–5170
56. Munster PN, Troso-Sandoval T, Rosen N, Rifkind R, Marks PA, Richon VM (2001) The histone deacetylase inhibitor suberoylanilide hydroxamic acid induces differentiation of human breast cancer cells. Cancer Res 61(23):8492–8497
57. Cohen LA, Amin S, Marks PA, Rifkind RA, Desai D, Richon VM (1999) Chemoprevention of carcinogen-induced mammary tumorigenesis by the hybrid polar cytodifferentiation agent, suberanilohydroxamic acid (SAHA). Anticancer Res 19(6B):4999–5005
58. He LZ, Tolentino T, Grayson P, Zhong S, Warrell RP Jr, Rifkind RA, Marks PA, Richon VM, Pandolfi PP (2001) Histone deacetylase inhibitors induce remission in transgenic models of therapy-resistant acute promyelocytic leukemia. J Clin Invest 108(9):1321–1330. doi:10.1172/JCI11537
59. Marks PA (2007) Discovery and development of SAHA as an anticancer agent. Oncogene 26 (9):1351–1356. doi:10.1038/sj.onc.1210204
60. Lakshmaiah KC, Jacob LA, Aparna S, Lokanatha D, Saldanha SC (2014) Epigenetic therapy of cancer with histone deacetylase inhibitors. J Cancer Res Ther 10(3):469–478. doi:10.4103/0973-1482.137937
61. Haggarty SJ, Koeller KM, Wong JC, Grozinger CM, Schreiber SL (2003) Domain-selective small-molecule inhibitor of histone deacetylase 6 (HDAC6)-mediated tubulin deacetylation. Proc Natl Acad Sci U S A 100(8):4389–4394. doi:10.1073/pnas.0430973100
62. Somoza JR, Skene RJ, Katz BA, Mol C, Ho JD, Jennings AJ, Luong C, Arvai A, Buggy JJ, Chi E, Tang J, Sang BC, Verner E, Wynands R, Leahy EM, Dougan DR, Snell G, Navre M, Knuth MW, Swanson RV, McRee DE, Tari LW (2004) Structural snapshots of human HDAC8 provide insights into the class I histone deacetylases. Structure 12(7):1325–1334. doi:10.1016/j.str.2004.04.012
63. Vannini A, Volpari C, Filocamo G, Casavola EC, Brunetti M, Renzoni D, Chakravarty P, Paolini C, De Francesco R, Gallinari P, Steinkühler C, Di Marco S (2004) Crystal structure of a eukaryotic zinc-dependent histone deacetylase, human HDAC8, complexed with a hydroxamic acid inhibitor. Proc Natl Acad Sci U S A 101(42):15064–15069. doi:10.1073/pnas.0404603101
64. Balasubramanian S, Ramos J, Luo W, Sirisawad M, Verner E, Buggy JJ (2008) A novel histone deacetylase 8 (HDAC8)–specific inhibitor PCI-34051 induces apoptosis in T-cell lymphomas. Leukemia 22(5):1026–1034. doi:10.1038/leu.2008.9
65. Rettig I, Koeneke E, Trippel F, Mueller WC, Burhenne J, Kopp-Schneider A, Fabian J, Schober A, Fernekorn U, von Deimling A, Deubzer HE, Milde T, Witt O, Oehme I (2015) Selective inhibition of HDAC8 decreases neuroblastoma growth in vitro and in vivo and enhances retinoic acid–mediated differentiation. Cell Death Dis 6:e1657. doi:10.1038/cddis.2015.24
66. el-Beltagi HM, Martens AC, Lelieveld P, Haroun EA, Hagenbeek A (1993) Acetyldinaline: a new oral cytostatic drug with impressive differential activity against leukemic cells and

normal stem cells—preclinical studies in a relevant rat model for human acute myelocytic leukemia. Cancer Res 53(13):3008–3014
67. LoRusso PM, Demchik L, Foster B, Knight J, Bissery MC, Polin LM, Leopold WR 3rd, Corbett TH (1996) Preclinical antitumor activity of CI-994. Investig New Drugs 14 (4):349–356
68. Seelig MH, Berger MR (1996) Efficacy of dinaline and its methyl and acetyl derivatives against colorectal cancer in vivo and in vitro. Eur J Cancer 32A(11):1968–1976
69. Suzuki T, Ando T, Tsuchiya K, Fukazawa N, Saito A, Mariko Y, Yamashita T, Nakanishi O (1999) Synthesis and histone deacetylase inhibitory activity of new benzamide derivatives. J Med Chem 42(15):3001–3003. doi:10.1021/jm980565u
70. Saito A, Yamashita T, Mariko Y, Nosaka Y, Tsuchiya K, Ando T, Suzuki T, Tsuruo T, Nakanishi O (1999) A synthetic inhibitor of histone deacetylase, MS-27-275, with marked in vivo antitumor activity against human tumors. Proc Natl Acad Sci U S A 96(8):4592–4597
71. Jaboin J, Wild J, Hamidi H, Khanna C, Kim CJ, Robey R, Bates SE, Thiele CJ (2002) MS-27-275, an inhibitor of histone deacetylase, has marked in vitro and in vivo antitumor activity against pediatric solid tumors. Cancer Res 62(21):6108–6115
72. Chou CJ, Herman D, Gottesfeld JM (2008) Pimelic diphenylamide 106 is a slow, tight-binding inhibitor of class I histone deacetylases. J Biol Chem 283(51):35402–35409. doi:10.1074/jbc.M807045200
73. Bressi JC, Jennings AJ, Skene R, Wu Y, Melkus R, De Jong R, O'Connell S, Grimshaw CE, Navre M, Gangloff AR (2010) Exploration of the HDAC2 foot pocket: synthesis and SAR of substituted *N*-(2-aminophenyl)benzamides. Bioorg Med Chem Lett 20(10):3142–3145. doi:10.1016/j.bmcl.2010.03.091
74. Ito T, Umehara T, Sasaki K, Nakamura Y, Nishino N, Terada T, Shirouzu M, Padmanabhan B, Yokoyama S, Ito A, Yoshida M (2011) Real-time imaging of histone H4K12-specific acetylation determines the modes of action of histone deacetylase and bromodomain inhibitors. Chem Biol 18(4):495–507. doi:10.1016/j.chembiol.2011.02.009
75. Ryan QC, Headlee D, Acharya M, Sparreboom A, Trepel JB, Ye J, Figg WD, Hwang K, Chung EJ, Murgo A, Melillo G, Elsayed Y, Monga M, Kalnitskiy M, Zwiebel J, Sausville EA (2005) Phase I and pharmacokinetic study of MS-275, a histone deacetylase inhibitor, in patients with advanced and refractory solid tumors or lymphoma. J Clin Oncol 23 (17):3912–3922. doi:10.1200/JCO.2005.02.188
76. Gore L, Rothenberg ML, O'Bryant CL, Schultz MK, Sandler AB, Coffin D, McCoy C, Schott A, Scholz C, Eckhardt SG (2008) A phase I and pharmacokinetic study of the oral histone deacetylase inhibitor, MS-275, in patients with refractory solid tumors and lymphomas. Clin Cancer Res 14(14):4517–4525. doi:10.1158/1078-0432.CCR-07-1461
77. Hauschild A, Trefzer U, Garbe C, Kaehler KC, Ugurel S, Kiecker F, Eigentler T, Krissel H, Schott A, Schadendorf D (2008) Multicenter phase II trial of the histone deacetylase inhibitor pyridylmethyl-*N*-{4-[(2-aminophenyl)-carbamoyl]-benzyl}-carbamate in pretreated metastatic melanoma. Melanoma Res 18(4):274–278. doi:10.1097/CMR.0b013e328307c248
78. Kijima M, Yoshida M, Sugita K, Horinouchi S, Beppu T (1993) Trapoxin, an antitumor cyclic tetrapeptide, is an irreversible inhibitor of mammalian histone deacetylase. J Biol Chem 268(30):22429–22435
79. Taunton J, Hassig CA, Schreiber SL (1996) A mammalian histone deacetylase related to the yeast transcriptional regulator Rpd3p. Science 272(5260):408–411
80. Furumai R, Komatsu Y, Nishino N, Khochbin S, Yoshida M, Horinouchi S (2001) Potent histone deacetylase inhibitors built from trichostatin A and cyclic tetrapeptide antibiotics including trapoxin. Proc Natl Acad Sci U S A 98(1):87–92. doi:10.1073/pnas.011405598
81. Komatsu Y, Tomizaki KY, Tsukamoto M, Kato T, Nishino N, Sato S, Yamori T, Tsuruo T, Furumai R, Yoshida M (2001) Cyclic hydroxamic-acid-containing peptide 31, a potent synthetic histone deacetylase inhibitor with antitumor activity. Cancer Res 61 (11):4459–4466

82. Darkin-Rattray SJ, Gurnett AM, Myers RW, Dulski PM, Crumley TM, Allocco JJ, Cannova C, Meinke PT, Colletti SL, Bednarek MA, Singh SB, Goetz MA, Dombrowski AW, Polishook JD, Schmatz DM (1996) Apicidin: a novel antiprotozoal agent that inhibits parasite histone deacetylase. Proc Natl Acad Sci U S A 93(23):13143–13147
83. Nishino N, Jose B, Okamura S, Ebisusaki S, Kato T, Sumida Y, Yoshida M (2003) Cyclic tetrapeptides bearing a sulfhydryl group potently inhibit histone deacetylases. Org Lett 5 (26):5079–5082. doi:10.1021/ol036098e
84. Nishino N, Yoshikawa D, Watanabe LA, Kato T, Jose B, Komatsu Y, Sumida Y, Yoshida M (2004) Synthesis and histone deacetylase inhibitory activity of cyclic tetrapeptides containing a retrohydroxamate as zinc ligand. Bioorg Med Chem Lett 14(10):2427–2431. doi:10.1016/j.bmcl.2004.03.018
85. Jose B, Oniki Y, Kato T, Nishino N, Sumida Y, Yoshida M (2004) Novel histone deacetylase inhibitors: cyclic tetrapeptide with trifluoromethyl and pentafluoroethyl ketones. Bioorg Med Chem Lett 14(21):5343–5346. doi:10.1016/j.bmcl.2004.08.016
86. Bhuiyan MP, Kato T, Okauchi T, Nishino N, Maeda S, Nishino TG, Yoshida M (2006) Chlamydocin analogs bearing carbonyl group as possible ligand toward zinc atom in histone deacetylases. Bioorg Med Chem 14(10):3438–3446. doi:10.1016/j.bmc.2005.12.063
87. Arai T, Ashraful Hoque M, Nishino N, Kim HJ, Ito A, Yoshida M (2013) Cyclic tetrapeptides with -SS- bridging between amino acid side chains for potent histone deacetylases' inhibition. Amino Acids 45(4):835–843. doi:10.1007/s00726-013-1527-8
88. Nakajima H, Kim YB, Terano H, Yoshida M, Horinouchi S (1998) FR901228, a potent antitumor antibiotic, is a novel histone deacetylase inhibitor. Exp Cell Res 241(1):126–133. doi:10.1006/excr.1998.4027
89. Furumai R, Matsuyama A, Kobashi N, Lee KH, Nishiyama M, Nakajima H, Tanaka A, Komatsu Y, Nishino N, Yoshida M (2002) FK228 (depsipeptide) as a natural prodrug that inhibits class I histone deacetylases. Cancer Res 62(17):4916–4921
90. Cole KE, Dowling DP, Boone MA, Phillips AJ, Christianson DW (2011) Structural basis of the antiproliferative activity of largazole, a depsipeptide inhibitor of the histone deacetylases. J Am Chem Soc 133(32):12474–12477. doi:10.1021/ja205972n
91. Kim D, Lee IS, Jung JH, Yang SI (1999) Psammaplin A, a natural bromotyrosine derivative from a sponge, possesses the antibacterial activity against methicillin-resistant *Staphylococcus aureus* and the DNA gyrase-inhibitory activity. Arch Pharm Res 22(1):25–29
92. Kim D, Lee IS, Jung JH, Lee CO, Choi SU (1999) Psammaplin A, a natural phenolic compound, has inhibitory effect on human topoisomerase II and is cytotoxic to cancer cells. Anticancer Res 19(5B):4085–4090
93. Pina IC, Gautschi JT, Wang GY, Sanders ML, Schmitz FJ, France D, Cornell-Kennon S, Sambucetti LC, Remiszewski SW, Perez LB, Bair KW, Crews P (2003) Psammaplins from the sponge *Pseudoceratina purpurea*: inhibition of both histone deacetylase and DNA methyltransferase. J Org Chem 68(10):3866–3873. doi:10.1021/jo034248t
94. Nicolaou KC, Hughes R, Pfefferkorn JA, Barluenga S (2001) Optimization and mechanistic studies of psammaplin A type antibacterial agents active against methicillin-resistant *Staphylococcus aureus* (MRSA). Chemistry 7(19):4296–4310
95. Nicolaou KC, Hughes R, Pfefferkorn JA, Barluenga S, Roecker AJ (2001) Combinatorial synthesis through disulfide exchange: discovery of potent psammaplin A type antibacterial agents active against methicillin-resistant *Staphylococcus aureus* (MRSA). Chemistry 7 (19):4280–4295
96. Nian H, Delage B, Ho E, Dashwood RH (2009) Modulation of histone deacetylase activity by dietary isothiocyanates and allyl sulfides: studies with sulforaphane and garlic organosulfur compounds. Environ Mol Mutagen 50(3):213–221. doi:10.1002/em.20454
97. Myzak MC, Karplus PA, Chung FL, Dashwood RH (2004) A novel mechanism of chemoprotection by sulforaphane: inhibition of histone deacetylase. Cancer Res 64 (16):5767–5774. doi:10.1158/0008-5472.CAN-04-1326

98. Tortorella SM, Royce SG, Licciardi PV, Karagiannis TC (2015) Dietary sulforaphane in cancer chemoprevention: the role of epigenetic regulation and HDAC inhibition. Antioxid Redox Signal 22(16):1382–1424. doi:10.1089/ars.2014.6097
99. Wall KA, Klis M, Kornet J, Coyle D, Ame JC, Jacobson MK, Slama JT (1998) Inhibition of the intrinsic NAD^+ glycohydrolase activity of CD38 by carbocyclic NAD analogues. Biochem J 335(Pt 3):631–636
100. Jackson MD, Schmidt MT, Oppenheimer NJ, Denu JM (2003) Mechanism of nicotinamide inhibition and transglycosidation by Sir2 histone/protein deacetylases. J Biol Chem 278 (51):50985–50998. doi:10.1074/jbc.M306552200
101. Bitterman KJ, Anderson RM, Cohen HY, Latorre-Esteves M, Sinclair DA (2002) Inhibition of silencing and accelerated aging by nicotinamide, a putative negative regulator of yeast sir2 and human SIRT1. J Biol Chem 277(47):45099–45107. doi:10.1074/jbc.M205670200
102. Luo J, Nikolaev AY, Imai S, Chen D, Su F, Shiloh A, Guarente L, Gu W (2001) Negative control of p53 by Sir2alpha promotes cell survival under stress. Cell 107(2):137–148
103. North BJ, Marshall BL, Borra MT, Denu JM, Verdin E (2003) The human Sir2 ortholog, SIRT2, is an NAD^+–dependent tubulin deacetylase. Mol Cell 11(2):437–444
104. Denu JM (2003) Linking chromatin function with metabolic networks: Sir2 family of NAD (+)-dependent deacetylases. Trends Biochem Sci 28(1):41–48
105. Grozinger CM, Chao ED, Blackwell HE, Moazed D, Schreiber SL (2001) Identification of a class of small molecule inhibitors of the sirtuin family of NAD-dependent deacetylases by phenotypic screening. J Biol Chem 276(42):38837–38843. doi:10.1074/jbc.M106779200
106. Zhao Y, Dai X, Blackwell HE, Schreiber SL, Chory J (2003) SIR1, an upstream component in auxin signaling identified by chemical genetics. Science 301(5636):1107–1110. doi:10.1126/science.1084161
107. Mai A, Massa S, Lavu S, Pezzi R, Simeoni S, Ragno R, Mariotti FR, Chiani F, Camilloni G, Sinclair DA (2005) Design, synthesis, and biological evaluation of sirtinol analogues as class III histone/protein deacetylase (sirtuin) inhibitors. J Med Chem 48(24):7789–7795. doi:10.1021/jm050100l
108. Bedalov A, Gatbonton T, Irvine WP, Gottschling DE, Simon JA (2001) Identification of a small molecule inhibitor of Sir2p. Proc Natl Acad Sci U S A 98(26):15113–15118. doi:10.1073/pnas.261574398
109. Neugebauer RC, Uchiechowska U, Meier R, Hruby H, Valkov V, Verdin E, Sippl W, Jung M (2008) Structure–activity studies on splitomicin derivatives as sirtuin inhibitors and computational prediction of binding mode. J Med Chem 51(5):1203–1213. doi:10.1021/jm700972e
110. Heltweg B, Gatbonton T, Schuler AD, Posakony J, Li H, Goehle S, Kollipara R, Depinho RA, Gu Y, Simon JA, Bedalov A (2006) Antitumor activity of a small-molecule inhibitor of human silent information regulator 2 enzymes. Cancer Res 66(8):4368–4377. doi:10.1158/0008-5472.CAN-05-3617
111. Napper AD, Hixon J, McDonagh T, Keavey K, Pons JF, Barker J, Yau WT, Amouzegh P, Flegg A, Hamelin E (2005) Discovery of indoles as potent and selective inhibitors of the deacetylase SIRT1. J Med Chem 48(25):8045–8054. doi:10.1021/jm050522v
112. Solomon JM, Pasupuleti R, Xu L, McDonagh T, Curtis R, DiStefano PS, Huber LJ (2006) Inhibition of SIRT1 catalytic activity increases p53 acetylation but does not alter cell survival following DNA damage. Mol Cell Biol 26(1):28–38. doi:10.1128/MCB.26.1.28-38.2006
113. Howitz KT, Bitterman KJ, Cohen HY, Lamming DW, Lavu S, Wood JG, Zipkin RE, Chung P, Kisielewski A, Zhang LL, Scherer B, Sinclair DA (2003) Small molecule activators of sirtuins extend *Saccharomyces cerevisiae* lifespan. Nature 425(6954):191–196. doi:10.1038/nature01960
114. Trapp J, Meier R, Hongwiset D, Kassack MU, Sippl W, Jung M (2007) Structure–activity studies on suramin analogues as inhibitors of NAD^+–dependent histone deacetylases (sirtuins). ChemMedChem 2(10):1419–1431. doi:10.1002/cmdc.200700003
115. Schuetz A, Min J, Antoshenko T, Wang CL, Allali-Hassani A, Dong A, Loppnau P, Vedadi M, Bochkarev A, Sternglanz R, Plotnikov AN (2007) Structural basis of inhibition

of the human NAD^+–dependent deacetylase SIRT5 by suramin. Structure 15(3):377–389. doi:10.1016/j.str.2007.02.002
116. Outeiro TF, Kontopoulos E, Altmann SM, Kufareva I, Strathearn KE, Amore AM, Volk CB, Maxwell MM, Rochet JC, McLean PJ, Young AB, Abagyan R, Feany MB, Hyman BT, Kazantsev AG (2007) Sirtuin 2 inhibitors rescue alpha-synuclein-mediated toxicity in models of Parkinson's disease. Science 317(5837):516–519. doi:10.1126/science.1143780
117. Chen X, Wales P, Quinti L, Zuo F, Moniot S, Herisson F, Rauf NA, Wang H, Silverman RB, Ayata C (2015) The sirtuin-2 inhibitor AK7 is neuroprotective in models of Parkinson's disease but not amyotrophic lateral sclerosis and cerebral ischemia. PLoS One 10(1): e0116919. doi:10.1371/journal.pone.0116919
118. Lain S, Hollick JJ, Campbell J, Staples OD, Higgins M, Aoubala M, McCarthy A, Appleyard V, Murray KE, Baker L, Thompson A, Mathers J, Holland SJ, Stark MJ, Pass G, Woods J, Lane DP, Westwood NJ (2008) Discovery, in vivo activity, and mechanism of action of a small-molecule p53 activator. Cancer Cell 13(5):454–463. doi:10.1016/j.ccr.2008.03.004
119. Yuan H, Wang Z, Li L, Zhang H, Modi H, Horne D, Bhatia R, Chen W (2012) Activation of stress response gene SIRT1 by BCR-ABL promotes leukemogenesis. Blood 119 (8):1904–1914. doi:10.1182/blood-2011-06-361691
120. Sunami Y, Araki M, Hironaka Y, Morishita S, Kobayashi M, Liew EL, Edahiro Y, Tsutsui M, Ohsaka A, Komatsu N (2013) Inhibition of the NAD-dependent protein deacetylase SIRT2 induces granulocytic differentiation in human leukemia cells. PLoS One 8(2):e57633. doi:10.1371/journal.pone.0057633
121. Jin Y, Cao Q, Chen C, Du X, Jin B, Pan J (2015) Tenovin-6-mediated inhibition of SIRT1/2 induces apoptosis in acute lymphoblastic leukemia (ALL) cells and eliminates ALL stem/progenitor cells. BMC Cancer 15:226. doi:10.1186/s12885-015-1282-1
122. Rumpf T, Schiedel M, Karaman B, Roessler C, North BJ, Lehotzky A, Olah J, Ladwein KI, Schmidtkunz K, Gajer M (2015) Selective Sirt2 inhibition by ligand-induced rearrangement of the active site. Nat Commun 6:6263. doi:10.1038/ncomms7263
123. Chan HM, La Thangue NB (2001) p300/CBP proteins: HATs for transcriptional bridges and scaffolds. J Cell Sci 114(Pt 13):2363–2373
124. Vetting MW, SdC LP, Yu M, Hegde SS, Magnet S, Roderick SL, Blanchard JS (2005) Structure and functions of the GNAT superfamily of acetyltransferases. Arch Biochem Biophys 433(1):212–226. doi:10.1016/j.abb.2004.09.003
125. Balasubramanyam K, Varier RA, Altaf M, Swaminathan V, Siddappa NB, Ranga U, Kundu TK (2004) Curcumin, a novel p300/CREB-binding protein-specific inhibitor of acetyltransferase, represses the acetylation of histone/nonhistone proteins and histone acetyltransferase-dependent chromatin transcription. J Biol Chem 279(49):51163–51171. doi:10.1074/jbc.M409024200
126. Balasubramanyam K, Altaf M, Varier RA, Swaminathan V, Ravindran A, Sadhale PP, Kundu TK (2004) Polyisoprenylated benzophenone, garcinol, a natural histone acetyltransferase inhibitor, represses chromatin transcription and alters global gene expression. J Biol Chem 279(32):33716–33726. doi:10.1074/jbc.M402839200
127. Egawa N, Kitaoka S, Tsukita K, Naitoh M, Takahashi K, Yamamoto T, Adachi F, Kondo T, Okita K, Asaka I, Aoi T, Watanabe A, Yamada Y, Morizane A, Takahashi J, Ayaki T, Ito H, Yoshikawa K, Yamawaki S, Suzuki S, Watanabe D, Hioki H, Kaneko T, Makioka K, Okamoto K, Takuma H, Tamaoka A, Hasegawa K, Nonaka T, Hasegawa M, Kawata A, Yoshida M, Nakahata T, Takahashi R, Marchetto MC, Gage FH, Yamanaka S, Inoue H (2012) Drug screening for ALS using patient-specific induced pluripotent stem cells. Sci Transl Med 4(145):145ra104. doi:10.1126/scitranslmed.3004052
128. Fukuda I, Ito A, Hirai G, Nishimura S, Kawasaki H, Saitoh H, Kimura K, Sodeoka M, Yoshida M (2009) Ginkgolic acid inhibits protein SUMOylation by blocking formation of the E1-SUMO intermediate. Chem Biol 16(2):133–140. doi:10.1016/j.chembiol.2009.01.009

129. Lau OD, Kundu TK, Soccio RE, Ait-Si-Ali S, Khalil EM, Vassilev A, Wolffe AP, Nakatani Y, Roeder RG, Cole PA (2000) HATs off: selective synthetic inhibitors of the histone acetyltransferases p300 and PCAF. Mol Cell 5(3):589–595
130. Cebrat M, Kim CM, Thompson PR, Daugherty M, Cole PA (2003) Synthesis and analysis of potential prodrugs of coenzyme A analogues for the inhibition of the histone acetyltransferase p300. Bioorg Med Chem 11(15):3307–3313
131. Liu X, Wang L, Zhao K, Thompson PR, Hwang Y, Marmorstein R, Cole PA (2008) The structural basis of protein acetylation by the p300/CBP transcriptional coactivator. Nature 451(7180):846–850. doi:10.1038/nature06546
132. Bowers EM, Yan G, Mukherjee C, Orry A, Wang L, Holbert MA, Crump NT, Hazzalin CA, Liszczak G, Yuan H (2010) Virtual ligand screening of the p300/CBP histone acetyltransferase: identification of a selective small molecule inhibitor. Chem Biol 17 (5):471–482. doi:10.1016/j.chembiol.2010.03.006
133. Gao XN, Lin J, Ning QY, Gao L, Yao YS, Zhou JH, Li YH, Wang LL, Yu L (2013) A histone acetyltransferase p300 inhibitor C646 induces cell cycle arrest and apoptosis selectively in AML1-ETO-positive AML cells. PLoS One 8(2):e55481. doi:10.1371/journal.pone.0055481
134. Yan G, Eller MS, Elm C, Larocca CA, Ryu B, Panova IP, Dancy BM, Bowers EM, Meyers D, Lareau L (2013) Selective inhibition of p300 HAT blocks cell cycle progression, induces cellular senescence, and inhibits the DNA damage response in melanoma cells. J Invest Dermatol 133(10):2444–2452. doi:10.1038/jid.2013.187
135. Oike T, Komachi M, Ogiwara H, Amornwichet N, Saitoh Y, Torikai K, Kubo N, Nakano T, Kohno T (2014) C646, a selective small molecule inhibitor of histone acetyltransferase p300, radiosensitizes lung cancer cells by enhancing mitotic catastrophe. Radiother Oncol 111 (2):222–227. doi:10.1016/j.radonc.2014.03.015
136. Gallenkamp D, Gelato KA, Haendler B, Weinmann H (2014) Bromodomains and their pharmacological inhibitors. ChemMedChem 9(3):438–464. doi:10.1002/cmdc.201300434
137. Adams-Cioaba MA, Min J (2009) Structure and function of histone methylation binding proteins. Biochem Cell Biol 87(1):93–105. doi:10.1139/O08-129
138. Eissenberg JC (2012) Structural biology of the chromodomain: form and function. Gene 496 (2):69–78. doi:10.1016/j.gene.2012.01.003
139. Garcia BA, Hake SB, Diaz RL, Kauer M, Morris SA, Recht J, Shabanowitz J, Mishra N, Strahl BD, Allis CD, Hunt DF (2007) Organismal differences in post-translational modifications in histones H3 and H4. J Biol Chem 282(10):7641–7655. doi:10.1074/jbc.M607900200
140. Bedford MT, Clarke SG (2009) Protein arginine methylation in mammals: who, what, and why. Mol Cell 33(1):1–13. doi:10.1016/j.molcel.2008.12.013
141. Jenuwein T, Laible G, Dorn R, Reuter G (1998) SET domain proteins modulate chromatin domains in eu- and heterochromatin. Cell Mol Life Sci 54(1):80–93
142. Singer MS, Kahana A, Wolf AJ, Meisinger LL, Peterson SE, Goggin C, Mahowald M, Gottschling DE (1998) Identification of high-copy disruptors of telomeric silencing in *Saccharomyces cerevisiae*. Genetics 150(2):613–632
143. Petrossian TC, Clarke SG (2011) Uncovering the human methyltransferasome. Mol Cell Proteomics 10(1):M110 000976. doi:10.1074/mcp.M110.000976
144. Dillon SC, Zhang X, Trievel RC, Cheng X (2005) The SET-domain protein superfamily: protein lysine methyltransferases. Genome Biol 6(8):227. doi:10.1186/gb-2005-6-8-227
145. Feng Q, Wang H, Ng HH, Erdjument-Bromage H, Tempst P, Struhl K, Zhang Y (2002) Methylation of H3-lysine 79 is mediated by a new family of HMTases without a SET domain. Curr Biol 12(12):1052–1058
146. McBride AE, Silver PA (2001) State of the arg: protein methylation at arginine comes of age. Cell 106(1):5–8
147. Zurita-Lopez CI, Sandberg T, Kelly R, Clarke SG (2012) Human protein arginine methyltransferase 7 (PRMT7) is a type III enzyme forming omega-NG-monomethylated arginine residues. J Biol Chem 287(11):7859–7870. doi:10.1074/jbc.M111.336271

148. Wei H, Mundade R, Lange KC, Lu T (2014) Protein arginine methylation of non-histone proteins and its role in diseases. Cell Cycle 13(1):32–41. doi:10.4161/cc.27353
149. Tachibana M, Sugimoto K, Fukushima T, Shinkai Y (2001) Set domain-containing protein, G9a, is a novel lysine-preferring mammalian histone methyltransferase with hyperactivity and specific selectivity to lysines 9 and 27 of histone H3. J Biol Chem 276(27):25309–25317. doi:10.1074/jbc.M101914200
150. Tachibana M, Sugimoto K, Nozaki M, Ueda J, Ohta T, Ohki M, Fukuda M, Takeda N, Niida H, Kato H, Shinkai Y (2002) G9a histone methyltransferase plays a dominant role in euchromatic histone H3 lysine 9 methylation and is essential for early embryogenesis. Genes Dev 16(14):1779–1791. doi:10.1101/gad.989402
151. Casciello F, Windloch K, Gannon F, Lee JS (2015) Functional role of G9a histone methyltransferase in cancer. Front Immunol 6:487. doi:10.3389/fimmu.2015.00487
152. Greiner D, Bonaldi T, Eskeland R, Roemer E, Imhof A (2005) Identification of a specific inhibitor of the histone methyltransferase SU(VAR)3-9. Nat Chem Biol 1(3):143–145. doi:10.1038/nchembio721
153. Sun Y, Takada K, Takemoto Y, Yoshida M, Nogi Y, Okada S, Matsunaga S (2012) (2012) Gliotoxin analogues from a marine-derived fungus, *Penicillium* sp., and their cytotoxic and histone methyltransferase inhibitory activities. J Nat Prod 75(1):111–114. doi:10.1021/np200740e
154. Takahashi M, Takemoto Y, Shimazu T, Kawasaki H, Tachibana M, Shinkai Y, Takagi M, Shin-ya K, Igarashi Y, Ito A, Yoshida M (2012) Inhibition of histone H3K9 methyltransferases by gliotoxin and related epipolythiodioxopiperazines. J Antibiot 65 (5):263–265. doi:10.1038/ja.2012.6
155. Iwasa E, Hamashima Y, Fujishiro S, Higuchi E, Ito A, Yoshida M, Sodeoka M (2010) Total synthesis of (+)-chaetocin and its analogues: their histone methyltransferase G9a inhibitory activity. J Am Chem Soc 132(12):4078–4079. doi:10.1021/ja101280p
156. Fujishiro S, Dodo K, Iwasa E, Teng Y, Sohtome Y, Hamashima Y, Ito A, Yoshida M, Sodeoka M (2013) Epidithiodiketopiperazine as a pharmacophore for protein lysine methyltransferase G9a inhibitors: reducing cytotoxicity by structural simplification. Bioorg Med Chem Lett 23(3):733–736. doi:10.1016/j.bmcl.2012.11.087
157. Kubicek S, O'Sullivan RJ, August EM, Hickey ER, Zhang Q, Teodoro ML, Rea S, Mechtler K, Kowalski JA, Homon CA, Kelly TA, Jenuwein T (2007) Reversal of H3K9me2 by a small-molecule inhibitor for the G9a histone methyltransferase. Mol Cell 25(3):473–481. doi:10.1016/j.molcel.2007.01.017
158. Liu F, Chen X, Allali-Hassani A, Quinn AM, Wigle TJ, Wasney GA, Dong A, Senisterra G, Chau I, Siarheyeva A, Norris JL, Kireev DB, Jadhav A, Herold JM, Janzen WP, Arrowsmith CH, Frye SV, Brown PJ, Simeonov A, Vedadi M, Jin J (2010) Protein lysine methyltransferase G9a inhibitors: design, synthesis, and structure activity relationships of 2,4-diamino-7-aminoalkoxy-quinazolines. J Med Chem 53(15):5844–5857. doi:10.1021/jm100478y
159. Liu F, Barsyte-Lovejoy D, Li F, Xiong Y, Korboukh V, Huang XP, Allali-Hassani A, Janzen WP, Roth BL, Frye SV (2013) Discovery of an in vivo chemical probe of the lysine methyltransferases G9a and GLP. J Med Chem 56(21):8931–8942. doi:10.1021/jm401480r
160. Varambally S, Dhanasekaran SM, Zhou M, Barrette TR, Kumar-Sinha C, Sanda MG, Ghosh D, Pienta KJ, Sewalt RG, Otte AP, Rubin MA, Chinnaiyan AM (2002) The polycomb group protein EZH2 is involved in progression of prostate cancer. Nature 419 (6907):624–629. doi:10.1038/nature01075
161. Kleer CG, Cao Q, Varambally S, Shen R, Ota I, Tomlins SA, Ghosh D, Sewalt RG, Otte AP, Hayes DF, Sabel MS, Livant D, Weiss SJ, Rubin MA, Chinnaiyan AM (2003) EZH2 is a marker of aggressive breast cancer and promotes neoplastic transformation of breast epithelial cells. Proc Natl Acad Sci U S A 100(20):11606–11611. doi:10.1073/pnas.1933744100
162. Takawa M, Masuda K, Kunizaki M, Daigo Y, Takagi K, Iwai Y, Cho HS, Toyokawa G, Yamane Y, Maejima K, Field HI, Kobayashi T, Akasu T, Sugiyama M, Tsuchiya E, Atomi Y,

Ponder BA, Nakamura Y, Hamamoto R (2011) Validation of the histone methyltransferase EZH2 as a therapeutic target for various types of human cancer and as a prognostic marker. Cancer Sci 102(7):1298–12305. doi:10.1111/j.1349-7006.2011.01958.x
163. Aumann S, Abdel-Wahab O (2014) Somatic alterations and dysregulation of epigenetic modifiers in cancers. Biochem Biophys Res Commun 455(1–2):24–34. doi:10.1016/j.bbrc.2014.08.004
164. Miranda TB, Cortez CC, Yoo CB, Liang G, Abe M, Kelly TK, Marquez VE, Jones PA (2009) DZNep is a global histone methylation inhibitor that reactivates developmental genes not silenced by DNA methylation. Mol Cancer Ther 28(6):1579–1588. doi:10.1158/1535-7163.MCT-09-0013
165. McCabe MT, Ott HM, Ganji G, Korenchuk S, Thompson C, Van Aller GS, Liu Y, Graves AP, Della Pietra A 3rd, Diaz E, LaFrance LV, Mellinger M, Duquenne C, Tian X, Kruger RG, McHugh CF, Brandt M, Miller WH, Dhanak D, Verma SK, Tummino PJ, Creasy CL (2012) EZH2 inhibition as a therapeutic strategy for lymphoma with EZH2-activating mutations. Nature 2492(7427):108–112. doi:10.1038/nature11606
166. Qi W, Chan H, Teng L, Li L, Chuai S, Zhang R, Zeng J, Li M, Fan H, Lin Y (2012) Selective inhibition of Ezh2 by a small molecule inhibitor blocks tumor cells proliferation. Proc Natl Acad Sci U S A 109(52):21360–21365. doi:10.1073/pnas.1210371110
167. Knutson SK, Wigle TJ, Warholic NM, Sneeringer CJ, Allain CJ, Klaus CR, Sacks JD, Raimondi A, Majer CR, Song J, Scott MP, Jin L, Smith JJ, Olhava EJ, Chesworth R, Moyer MP, Richon VM, Copeland RA, Keilhack H, Pollock RM, Kuntz KW (2012) A selective inhibitor of EZH2 blocks H3K27 methylation and kills mutant lymphoma cells. Nat Chem Biol 8(11):890–896. doi:10.1038/nchembio.1084
168. Nasveschuk CG, Gagnon A, Garapaty-Rao S, Balasubramanian S, Campbell R, Lee C, Zhao F, Bergeron L, Cummings R, Trojer P et al (2014) Discovery and optimization of tetramethylpiperidinyl benzamides as inhibitors of EZH2. ACS Med Chem Lett 5 (4):378–383. doi:10.1021/ml400494b
169. Daigle SR, Olhava EJ, Therkelsen CA, Majer CR, Sneeringer CJ, Song J, Johnston LD, Scott MP, Smith JJ, Xiao Y, Jin L, Kuntz KW, Chesworth R, Moyer MP, Bernt KM, Tseng JC, Kung AL, Armstrong SA, Copeland RA, Richon VM, Pollock RM (2011) Selective killing of mixed lineage leukemia cells by a potent small-molecule DOT1L inhibitor. Cancer Cell 20 (1):53–65. doi:10.1016/j.ccr.2011.06.009
170. Daigle SR, Olhava EJ, Therkelsen CA, Basavapathruni A, Jin L, Boriack-Sjodin PA, Allain CJ, Klaus CR, Raimondi A, Scott MP, Waters NJ, Chesworth R, Moyer MP, Copeland RA, Richon VM, Pollock RM (2013) Potent inhibition of DOT1L as treatment of MLL-fusion leukemia. Blood 122(6):1017–1025. doi:10.1182/blood-2013-04-497644
171. Hojfeldt JW, Agger K, Helin K (2013) Histone lysine demethylases as targets for anticancer therapy. Nat Rev Drug Discov 12(12):917–930. doi:10.1038/nrd4154
172. Kahl P, Gullotti L, Heukamp LC, Wolf S, Friedrichs N, Vorreuther R, Solleder G, Bastian PJ, Ellinger J, Metzger E, Schüle R, Buettner R (2006) Androgen receptor coactivators lysine-specific histone demethylase 1 and four and a half LIM domain protein 2 predict risk of prostate cancer recurrence. Cancer Res 66(23):11341–11347. doi:10.1158/0008-5472.CAN-06-1570
173. Kauffman EC, Robinson BD, Downes MJ, Powell LG, Lee MM, Scherr DS, Gudas LJ, Mongan NP (2011) Role of androgen receptor and associated lysine-demethylase coregulators, LSD1 and JMJD2A, in localized and advanced human bladder cancer. Mol Carcinog 50(12):931–944. doi:10.1002/mc.20758
174. Hayami S, Kelly JD, Cho HS, Yoshimatsu M, Unoki M, Tsunoda T, Field HI, Neal DE, Yamaue H, Ponder BA, Nakamura Y, Hamamoto R (2011) Overexpression of LSD1 contributes to human carcinogenesis through chromatin regulation in various cancers. Int J Cancer 128(3):574–586. doi:10.1002/ijc.25349
175. Harris WJ, Huang X, Lynch JT, Spencer GJ, Hitchin JR, Li Y, Ciceri F, Blaser JG, Greystoke BF, Jordan AM, Miller CJ, Ogilvie DJ, Somervaille TC (2012) The histone demethylase

KDM1A sustains the oncogenic potential of MLL-AF9 leukemia stem cells. Cancer Cell 21 (4):473–487. doi:10.1016/j.ccr.2012.03.014
176. Lee MG, Wynder C, Schmidt DM, McCafferty DG, Shiekhattar R (2006) Histone H3 lysine 4 demethylation is a target of nonselective antidepressive medications. Chem Biol 13 (6):563–567. doi:10.1016/j.chembiol.2006.05.004
177. Binda C, Valente S, Romanenghi M, Pilotto S, Cirilli R, Karytinos A, Ciossani G, Botrugno OA, Forneris F, Tardugno M, Edmondson DE, Minucci S, Mattevi A, Mai A (2010) Biochemical, structural, and biological evaluation of tranylcypromine derivatives as inhibitors of histone demethylases LSD1 and LSD2. J Am Chem Soc 132(19):6827–6833. doi:10.1021/ja101557k
178. Benelkebir H, Hodgkinson C, Duriez PJ, Hayden AL, Bulleid RA, Crabb SJ, Packham G, Ganesan A (2011) Enantioselective synthesis of tranylcypromine analogues as lysine demethylase (LSD1) inhibitors. Bioorg Med Chem 19(12):3709–3716. doi:10.1016/j.bmc.2011.02.017
179. Ueda R, Suzuki T, Mino K, Tsumoto H, Nakagawa H, Hasegawa M, Sasaki R, Mizukami T, Miyata N (2009) Identification of cell-active lysine specific demethylase 1-selective inhibitors. J Am Chem Soc 131(48):17536–17537. doi:10.1021/ja907055q
180. Mimasu S, Umezawa N, Sato S, Higuchi T, Umehara T, Yokoyama S (2010) Structurally designed trans-2-phenylcyclopropylamine derivatives potently inhibit histone demethylase LSD1/KDM1. Biochemistry 49(30):6494–64503. doi:10.1021/bi100299r
181. Cloos PA, Christensen J, Agger K, Maiolica A, Rappsilber J, Antal T, Hansen KH, Helin K (2006) The putative oncogene GASC1 demethylates tri- and dimethylated lysine 9 on histone H3. Nature 442(7100):307–311. doi:10.1038/nature04837
182. Hamada S, Kim TD, Suzuki T, Itoh Y, Tsumoto H, Nakagawa H, Janknecht R, Miyata N (2009) Synthesis and activity of *N*-oxalylglycine and its derivatives as Jumonji C-domain-containing histone lysine demethylase inhibitors. Bioorg Med Chem Lett 19(10):2852–2855. doi:10.1016/j.bmcl.2009.03.098
183. Maes T, Carceller E, Salas J, Ortega A, Buesa C (2015) Advances in the development of histone lysine demethylase inhibitors. Curr Opin Pharmacol 23:52–60. doi:10.1016/j.coph.2015.05.009
184. Kruidenier L, Chung CW, Cheng Z, Liddle J, Che K, Joberty G, Bantscheff M, Bountra C, Bridges A, Diallo H, Eberhard D, Hutchinson S, Jones E, Katso R, Leveridge M, Mander PK, Mosley J, Ramirez-Molina C, Rowland P, Schofield CJ, Sheppard RJ, Smith JE, Swales C, Tanner R, Thomas P, Tumber A, Drewes G, Oppermann U, Patel DJ, Lee K, Wilson DM (2012) A selective Jumonji H3K27 demethylase inhibitor modulates the proinflammatory macrophage response. Nature 488(7411):404–408. doi:10.1038/nature11262
185. Hashizume R, Andor N, Ihara Y, Lerner R, Gan H, Chen X, Fang D, Huang X, Tom MW, Ngo V, Solomon D, Mueller S, Paris PL, Zhang Z, Petritsch C, Gupta N, Waldman TA, James CD (2014) Pharmacologic inhibition of histone demethylation as a therapy for pediatric brainstem glioma. Nat Med 20(12):1394–1396. doi:10.1038/nm.3716
186. Goll MG, Bestor TH (2005) Eukaryotic cytosine methyltransferases. Annu Rev Biochem 74:481–514. doi:10.1146/annurev.biochem.74.010904.153721
187. Tahiliani M, Koh KP, Shen Y, Pastor WA, Bandukwala H, Brudno Y, Agarwal S, Iyer LM, Liu DR, Aravind L, Rao A (2009) Conversion of 5-methylcytosine to 5-hydroxymethylcytosine in mammalian DNA by MLL partner TET1. Science 324 (5929):930–935. doi:10.1126/science.1170116
188. Kriaucionis S, Heintz N (2009) The nuclear DNA base 5-hydroxymethylcytosine is present in Purkinje neurons and the brain. Science 324(5929):930–935. doi:10.1126/science.1170116
189. Ito S, D'Alessio AC, Taranova OV, Hong K, Sowers LC, Zhang Y (2010) Role of Tet proteins in 5mC to 5hmC conversion, ES-cell self-renewal and inner cell mass specification. Nature 466(7310):1129–1133. doi:10.1038/nature09303
190. Sharma S, Kelly TK, Jones PA (2010) Epigenetics in cancer. Carcinogenesis 31(1):27–36. doi:10.1093/carcin/bgp220

191. Cheng JC, Matsen CB, Gonzales FA, Ye W, Greer S, Marquez VE, Jones PA, Selker EU (2003) Inhibition of DNA methylation and reactivation of silenced genes by zebularine. J Natl Cancer Inst 95(5):399–409
192. Flotho C, Claus R, Batz C, Schneider M, Sandrock I, Ihde S, Plass C, Niemeyer CM, Lubbert M (2009) The DNA methyltransferase inhibitors azacitidine, decitabine and zebularine exert differential effects on cancer gene expression in acute myeloid leukemia cells. Leukemia 23 (6):1019–1028. doi:10.1038/leu.2008.397
193. Yoo CB, Jeong S, Egger G, Liang G, Phiasivongsa P, Tang C, Redkar S, Jones PA (2007) Delivery of 5-aza-2′-deoxycytidine to cells using oligodeoxynucleotides. Cancer Res 67 (13):6400–6408. doi:10.1158/0008-5472.CAN-07-0251
194. Chuang JC, Warner SL, Vollmer D, Vankayalapati H, Redkar S, Bearss DJ, Qiu X, Yoo CB, Jones PA (2010) S110, a 5-aza-2′-deoxycytidine-containing dinucleotide, is an effective DNA methylation inhibitor in vivo and can reduce tumor growth. Mol Cancer Ther 9 (5):1443–1450. doi:10.1158/1535-7163.MCT-09-1048
195. Erdmann A, Halby L, Fahy J, Arimondo PB (2015) Targeting DNA methylation with small molecules: what's next? J Med Chem 58(6):2569–2583. doi:10.1021/jm500843d
196. Rilova E, Erdmann A, Gros C, Masson V, Aussagues Y, Poughon-Cassabois V, Rajavelu A, Jeltsch A, Menon Y, Novosad N, Gregoire JM, Vispé S, Schambel P, Ausseil F, Sautel F, Arimondo PB, Cantagrel F (2014) Design, synthesis and biological evaluation of 4-amino-*N*-(4-aminophenyl)benzamide analogues of quinoline-based SGI-1027 as inhibitors of DNA methylation. ChemMedChem 9(3):590–601. doi:10.1002/cmdc.201300420

Chapter 4
Apoptosis and Autophagy

Etsu Tashiro, Mitsuhiro Kitagawa, and Masaya Imoto

Abstract Apoptosis and autophagy are highly coordinated mechanisms to maintain cellular homeostasis against various intrinsic and/or extrinsic stresses in eukaryotic cells. Since the early 1990s, our knowledge of the molecular mechanisms and the physiological functions of apoptosis have increased substantially. Dysfunction of apoptosis leads to various diseases, including cancers and degenerative diseases. Therefore, small molecules that induce apoptosis or synthetic lethality with mutated genes have been examined for the development of cancer-specific apoptosis-inducing agents; conversely, small molecules that suppress apoptosis have been identified for protective agents against neuronal cell death.

On the other hand, autophagy is an evolutionarily conserved pathway involved in the degradation of intracellular components and is critical for the maintenance of cellular homeostasis. The mechanisms and functions of autophagy have been revealed by genetic studies in yeast, which identified a series of autophagy-related genes. Many small molecules that have been discovered to induce or inhibit autophagy also provide insight into the mechanisms and functions of the autophagic process. In this chapter, we introduce several small molecules identified by synthetic lethality screening, apoptosis inhibitors, and autophagy modulators.

Keywords Apoptosis • Synthetic lethality • Oncogene • Cancer • Neurodegenerative disease • Neuroprotection • Autophagy

4.1 Introduction

Eukaryotic cells have evolved highly coordinated mechanisms to maintain cellular homeostasis against various intrinsic and/or extrinsic stresses. These highly coordinated mechanisms include apoptosis and autophagy.

Since the early 1990s, our understanding of the molecular mechanisms and physiological functions of apoptosis has increased substantially. Apoptosis can be

E. Tashiro • M. Kitagawa • M. Imoto (✉)
Department of Biosciences and Informatics, Faculty of Science and Technology, Keio University, 3-14-1 Hiyoshi, Kohoku-ku, Yokohama, Japan, 223-8522
e-mail: imoto@bio.keio.ac.jp

© Springer Japan KK 2017

H. Osada (ed.), *Bioprobes*, DOI 10.1007/978-4-431-56529-1_4

triggered by a variety of stimuli, including viral infection [1], growth factor withdrawal [2–7], DNA damage resulting from radiation [8–10], and chemotherapeutic drugs [11]. During apoptosis, a family of cysteine proteases, caspases, is activated. In response to various apoptotic stimuli, "initiator" caspases (e.g., caspase-2, -8, -9, or -10) are activated, which in turn cleave and activate the zymogenic forms of "executioner" caspases (e.g., caspase-3 or -7), resulting in proteolytic cleavage of specific cellular substrates and subsequent cell death (Fig. 4.1). At present, two apoptotic pathways are known, namely, extrinsic and intrinsic apoptosis pathways. In the extrinsic pathway, initiator caspases-8 and -10 are activated through the formation of a death-inducing signal complex (DISC) upon binding of extracellular ligands such as Fas, tumor necrosis factor (TNF), and TNF-related apoptosis-inducing ligand (TRAIL) to their respective cell-surface receptors. In the intrinsic pathway, a decrease in mitochondrial outer-membrane permeabilization triggers the release of various pro-apoptotic proteins, including cytochrome *c* and second mitochondria-derived activator of caspases (SMAC), from the mitochondrial intermembrane space into the cytoplasm. In the cytoplasm, cytochrome *c* binds to the adaptor protein apoptotic protease-activating factor 1 (APAF-1) to form a caspase-9-activating complex, called apoptosome. SMAC enhances cytochrome *c*-induced caspase activation by binding and neutralizing X-linked inhibitor of apoptosis protein (XIAP), an inhibitor of caspases-3, -7, and -9. On the other hand, the B-cell lymphoma (BCL)-2 family of proteins also plays important roles in the regulation of apoptosis, and they are classified based on the presence of BCL-2 homology (BH) domains. Upon cytotoxic or genotoxic stress, BH3-only proteins such as BCL-2-interacting mediator of cell death (BIM), BH3-interacting domain death agonist (BID), and BCL-2-associated agonist of cell death (BAD) induce activation and promote oligomerization of multi-BH (BH1-BH3) domain proteins (e.g., BCL-2-associated protein X (BAX) and BCL-2 antagonist/killer (BAK)), resulting in the decrease in permeabilization of the mitochondrial outer membrane. In contrast, prosurvival members such as BCL-2, B-cell lymphoma–extra large (BCL-xL), and myeloid cell leukemia sequence 1 (MCL-1), which contains all four BH domains, counteract this process by sequestering pro-apoptotic family members. Importantly, the ratio between pro- and anti-apoptotic BCL-2 family proteins determines the susceptibility of cells to apoptotic stimuli.

As dysfunction of apoptosis leads to various diseases, small molecules that induce or suppress apoptosis may be chemotherapeutic candidates for cancer or neurodegenerative diseases, respectively. Indeed, most anticancer drugs used today were discovered based on their ability to induce apoptosis in rapidly dividing cancer cells in vitro. However, these drugs affect targets shared between normal cells and cancer cells, including enzymes involved in fundamental processes such as DNA replication and cell division. Therefore, the development of safe and effective anticancer drugs that selectively induce apoptosis in cancer cells is an urgent issue for cancer chemotherapy. Molecularly targeted therapeutics represent a promising new approach to anticancer drug discovery [12]. This approach involves searching for genotype-selective anticancer agents that induce apoptosis in cancer

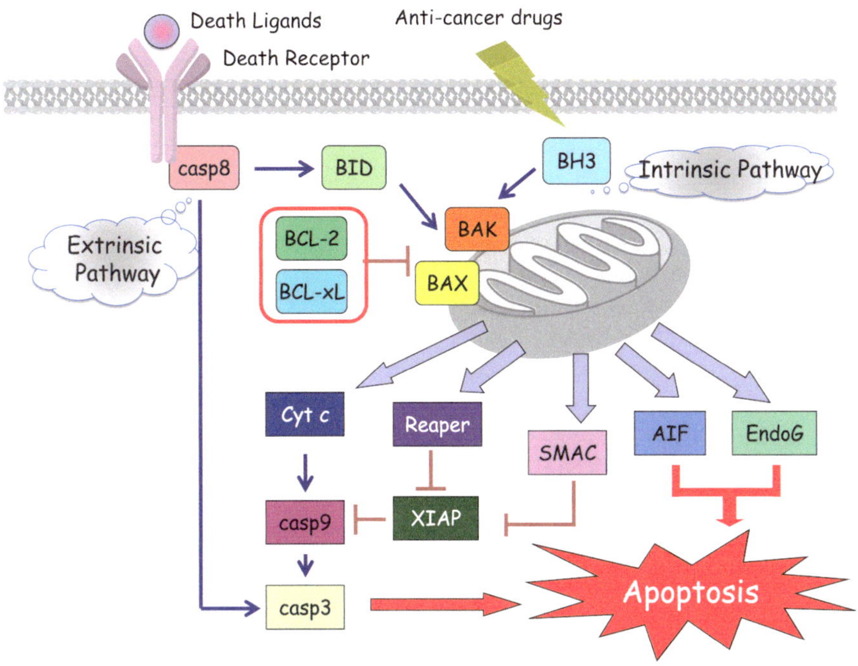

Fig. 4.1 Extrinsic and intrinsic apoptosis pathway. During apoptosis, a family of cysteine proteases, caspases, is activated. In response to various apoptotic stimuli, "initiator" caspases (e.g., caspase-2, -8, -9, or -10) become activated, which in turn cleave and activate the zymogenic forms of "executioner" caspases (e.g., caspase-3 or -7), resulting in proteolytic cleavage of specific cellular substrates and subsequent cell death. At present, two apoptotic pathways are known, namely, extrinsic and intrinsic apoptosis. Extrinsic apoptosis pathway: stimulation of death receptors, such as APO-1/Fas (CD95) or TNF-related apoptosis-inducing ligand (TRAIL) receptors, by CD95 ligand or TRAIL results in receptor aggregation and recruitment of the adaptor molecule Fas-associated death domain (FADD) and caspase-8 to form DISC. Then, caspase-8 is activated and the activated caspase-8 initiates apoptosis by direct cleavage of downstream effector caspases. Intrinsic apoptosis pathway: the intrinsic apoptosis pathway is initiated by several stress signals through the release of apoptogenic factors such as cytochrome *c*, apoptosis-inducing factor (AIF), Smac, or EndoG from mitochondrial intermembrane space. The release of cytochrome *c* into the cytosol triggers caspase-3 activation through formation of the cytochrome *c*/Apaf-1/caspase-9-containing apoptosome complex. The Smac promotes caspase activation through neutralizing the inhibitory effects to XIAP, whereas AIF or EndoG causes DNA condensation

cells only in the presence of specific oncoproteins or in the absence of specific tumor suppressors. An example of a successful molecularly targeted anticancer therapeutics is Gleevec, an inhibitor of the breakpoint cluster region–Abelson (Bcr-Abl) kinase oncoprotein found in Philadelphia chromosome–positive chronic myelogenous leukemia (CML) [13]. Another example is lapatinib, an inhibitor of tyrosine kinase that is available for human epidermal growth factor receptor 2 (HER2)–positive breast cancer [14]. Despite these successes, developing drugs that selectively induce apoptosis in cancer cells without harming normal cells remains a considerable challenge because of the following problems. First, some cancers can escape a given state of oncogene addiction probably because of the frequent genomic and epigenetic abnormalities of cancers. Oncogene addiction is a phenomenon in which some cancers rely on a single dominant oncogene for growth and survival [15]. Because of this and tumor heterogeneity, anticancer agents that target one dominant oncogene are insufficient for cancer chemotherapeutics. Second, many of the genetic alterations frequently found in cancers are loss-of-function mutations in tumor suppressor genes; drugs that restore the function of a missing or altered protein have not been developed. To resolve this problem, the "synthetic lethality" concept, first observed in the fruit fly *Drosophila melanogaster* almost a century ago, was proposed. Malignant transformation driven by gain-of-function mutations in oncogenes and loss-of-function mutations in tumor suppressor genes results in cell deregulation that is frequently associated with enhanced cellular stressors. Adaptation to this stress phenotype is required for cancer cells to survive, thus, cancer cells may become dependent upon non-oncogenes that do not ordinarily perform such a vital function in normal cells [16]. Therefore, synthetic lethality provides a rationale for molecular targeted therapy.

In contrast to cancer, neurodegenerative diseases such as Parkinson's disease (PD), Huntington's disease (HD), and Alzheimer's disease (AD) are caused by apoptosis induction of neuronal cells. Increasing evidence suggests that intracellular deposition of misfolded proteins is a common neuropathological hallmark of these neurodegenerative diseases. For instances, Lewy bodies (LBs) are the cardinal histological hallmark in PD, which contain abnormal filamentous α-synuclein (αSYN) aggregates. Secreted monomeric and oligomeric forms of αSYN re-enter neighboring cells, resulting in various cytotoxic effects such as production of reactive oxygen species (ROS), glial-cell inflammatory responses [17, 18], and synaptic malfunction [19]. Another example is HD. HD is an autosomal dominant neurodegenerative disorder caused by a CAG trinucleotide repeat expansion (>35 repeats), which encodes an abnormally long polyglutamine (polyQ) segment in the *N*-terminus of the huntingtin protein [20, 21]. Mutant huntingtin toxicity is thought to arise after cleavage of the protein, which forms *N*-terminal fragments comprising the first 100–150 residues within the expanded polyQ tract. These fragments are also toxic species found in aggregates and inclusions [21]. Another example is AD. AD is characterized by extracellular amyloid beta (Aβ) plaques and intracellular neurofibrillary tangles (NFTs) [22]. NFTs consist of paired helical filaments of microtubule-associated tau protein that is hyperphosphorylated [23]. The density of tau tangles correlates with regional and global aspects of AD-associated cognitive dysfunction. Furthermore, studies on tau toxicity in certain genetic forms of frontotemporal dementia [24] strongly suggest that tau aggregation may result in a toxic gain of function leading to AD-associated neurodegeneration. Because of these observations, several small molecules that suppress deposition of misfolded proteins and neuronal apoptotic cell death have been identified.

Autophagy is an evolutionarily conserved pathway involved in the degradation of intracellular components, including organelles, and is critical for the maintenance of cellular homeostasis. Initially, cytoplasmic components are sequestered by a unique membrane called an "isolated membrane." Dynamic membrane organization from small membrane particles to autophagosomes is activated by recruitment of autophagy-related genes (ATG) and microtubule-associated protein 1 light chain 3 (LC3) [25]. Then, the autophagosomes fuse with lysosomes to form autolysosome. The inner membrane of the autophagosomes and the cytoplasm-derived materials contained within the autophagosomes are then degraded by lysosome hydrolases [26]. Amino acids produced by protein degradation in autolysosomes are then returned by lysosomal membrane permeases to the cytoplasm for reuse. Autophagy occurs in all cells at low basal levels under normal conditions to maintain homeostasis, but it is also associated with physiological and pathological processes such as development, differentiation, neurodegenerative diseases, stress, infection, and cancer [27].

In this chapter, we introduce various small molecules that were identified through synthetic lethality screening as cancer chemotherapeutics, apoptosis inhibitors for neurodegenerative diseases, and autophagy modulators.

4.2 Apoptosis Modulators

4.2.1 Apoptosis Inducers and Inhibitors

In the first edition of *Bioprobes*, we introduced various small molecules that were reported to induce or inhibit apoptosis through different modes of action. Thus, in this chapter, we update small molecules that induce or inhibit the apoptotic pathway.

Binding between pro- and anti-apoptotic BCL-2 family proteins is mediated through a hydrophobic groove on anti-apoptotic proteins and the BH3 domain. Thus, small molecules that specifically bind to the hydrophobic pocket and thereby antagonize the function of anti-apoptotic BCL-2 family proteins have been reported. ABT-737 (Fig. 4.2) was identified by examining the structure–activity relationship in a screening strategy combining nuclear magnetic resonance (NMR), structure-based design, and combinatory chemical synthesis. ABT-737 mimics the BH3 domain of BAD and binds selectively to BCL-2, BCL-xL, and BCL-w [28]. Another BH3 mimetic small molecule is gossypol (Fig. 4.2), a natural polyphenolic compound derived from cottonseed [29]. Gossypol acts as a pan-BCL-2 inhibitor and can inactivate BCL-2, BCL-xL, MCL-1, and BCL-w; therefore, gossypol was demonstrated to show pro-apoptotic effects in various in vivo and in vitro models [30]. Obatoclax (GX15-070) (Fig. 4.2), another pan-BCL-2 inhibitor, has been found to bind to all anti-apoptotic BCL-2 family proteins, including BCL-2, BCL-xL, and MCL-1, with high affinity. Obatoclax was proposed to competitively inhibit binding of pro-apoptotic proteins to the hydrophobic groove of anti-apoptotic proteins [31]. Several MCL-1 inhibitors have been developed. Marinopyrrole A (Fig. 4.2) was originally isolated from marine *Streptomyces* as an antibiotic compound against methicillin-resistant *Staphylococcus*

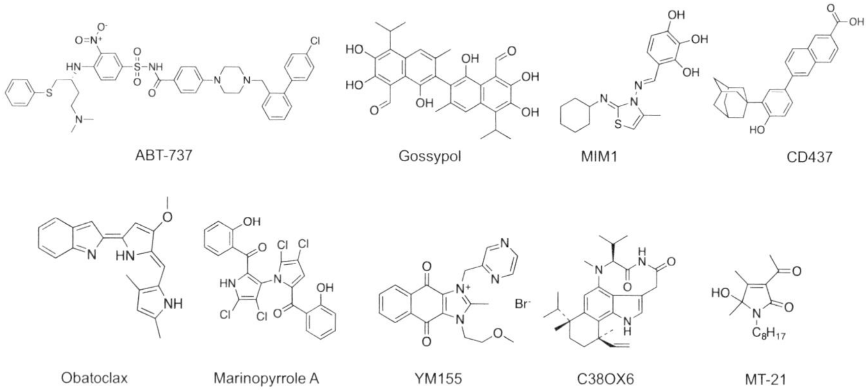

Fig. 4.2 Structures of apoptosis inducers

aureus (MRSA) [32], and later, it was shown to selectively bind to MCL-1 but not to BCL-xL, which disrupted the interaction between MCL-1 and BIM [33]. Moreover, MIM1 (MCL-1 Inhibitor Molecule 1) (Fig. 4.2) was shown to bind selectively to MCL-1 and induce BAX/BAK-dependent apoptosis in leukemia cells [34]. Because the IC_{50} values of these small molecules against MCL-1 are modest (micromolar range), more potent MCL-1 inhibitors are needed for clinical applications.

Overexpression of inhibitor of apoptosis (IAP) family proteins, has been observed in many cancer cells [35]. Survivin is the smallest member of the IAP family proteins, and is expressed at high levels during fetal development, while it is rarely expressed in healthy adult normal tissues. However, survivin is up-regulated in the majority of cancer cells; therefore, survivin is considered to be a good molecular target for development of anticancer chemotherapy. YM155 (Fig. 4.2) is the first identified small molecule targeting survivin. YM155 was obtained by high-throughput screening using a survivin promoter–luciferase assay that resulted in diminished survivin expression [36]. XIAP, another member of the IAP family, is an endogenous direct inhibitor of caspases-3, -7, and -9. Therefore, we screened for XIAP inhibitors from chemically modified microbial metabolites called "unnatural natural products". Eventually, we found that C38OX6 (Fig. 4.2) is an inhibitor of XIAP and that it restored XIAP-suppressed enzymatic activity of caspase-3 by inhibition of XIAP–caspase-3 interactions. Moreover, C38OX6 sensitized cancer cells to anticancer drugs [37].

The permeability transition pore complex (PTPC) is a highly dynamic supramolecular structure that comprises the voltage-dependent anion channel (VDAC), adenine nucleotide translocase (ANT), hexokinase, and cyclophilin D. Sustained opening of the PTPC causes loss of mitochondrial membrane potential, leading to osmotic imbalance and swelling of the mitochondrial matrix, a phenomenon called mitochondrial permeability transition [38]. Therefore, small molecules targeting the PTPC were considered to induce apoptosis. A synthetic compound, MT-21 (Fig. 4.2), was shown to specifically inhibit the molecular interaction between ANT and cyclophilin D [39], and it induced apoptosis in human promyelotic leukemia HL-60 cells through induction of cytochrome *c* release from the mitochondria [40]. 6-[3-(1-Adamantyl)-4-hydroxyphenyl]-2-naphtalenecarboxylic acid (CD437) (Fig. 4.2), a retinoid-related compound, triggers ANT-dependent mitochondrial permeability transition and subsequently induces apoptosis that is independent of its ability to bind to the nuclear receptor [41, 42].

4.2.2 Compounds Inducing Synthetic Lethality

Synthetic lethality occurs when a combination of mutations in two (*X* and *Y*) or more genes leads to cell death, whereas a mutation in only one of those genes does not [43, 44]. Therefore, chemical inhibitors of *Y* should selectively induce apoptosis in cancer cells with mutant *X*. In this section, we introduce various small molecules that were identified by synthetic lethality screening for cancer chemotherapeutics (Fig. 4.3).

4.2.2.1 PARP Inhibitor in BRCA Mutant Cancer Cells

DNA continually becomes damaged by environmental stressors and endogenous activities, such as DNA replication and cellular free-radical generation, which cause diverse changes, including base modifications, double-strand breaks (DSBs), single-strand breaks (SSBs), and intrastrand and interstrand cross-links [45]. These aberrations are repaired by distinct DNA repair pathways, which are coordinated to maintain the stability and integrity of the genome. Cancer cells frequently harbor defects in these DNA repair pathways, leading to genomic instability and tumorigenesis. Poly (adenosine diphosphate-ribose) polymerases (PARPs) are a large family of multifunctional enzymes, the most abundant of which is PARP1. PARPs play an important role in the repair of DNA SSBs by repairing base excision [45]. Inhibition of PARPs leads to the accumulation of DNA SSBs, which can lead to DNA DSBs at replication forks. Normally, these breaks are repaired by means of the error-free homologous-recombination double-stranded DNA repair pathway, the key components of which are the tumor suppressor proteins BRCA1 and BRCA2 [46]. BRCA1 and BRCA2 are important for DNA-DSB repair by homologous recombination [46]. Therefore, it was predicted that inhibition of PARP1 in a BRCA1- or BRCA2-defective background might result in chromosomal instability, cell-cycle arrest, and subsequent apoptosis. In 2005, Farmer [47] and Bryant [48] clearly proved the hypothesis that PARP inhibitors selectively induce cell death in BRCA1- or BRCA2-deficient cancer cells. Knockdown of PARP1 by small interfering RNA (siRNA) caused a reduction in the clonogenic survival of BRCA1- or BRCA2-deficient embryonic stem (ES) cells compared with that of wild-type ES cells. Consistent with the knockdown experiments using siRNA, BRCA1- or BRCA2-deficient ES cells were about 100-fold more sensitive to the PARP inhibitors KU0058684 and KU0058948 (Fig. 4.4) than were heterozygous mutant or wild-

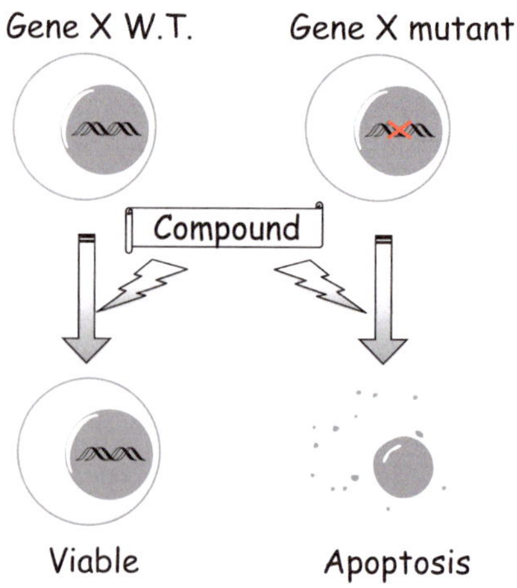

gene	compound
BRCA	PARP inhibitors [KU0058684, KU0058948 DHQ, NU1025, AG14361, olaparib]
Ras	SC-D, erastin, oncrasin1 Lanperisone
Neu	F16
Mad2	cantharidin
Akt	auranofin
VHL	STF-31

Fig. 4.3 Strategy for screening of selective apoptosis-inducing agents. Gene X and chemical inhibitor's target gene are synthetic lethal only when their simultaneous inactivation results in cell death. At present, various small molecules are identified by synthetic lethality screening for cancer chemotherapeutics. Among them, olaparib was approved by the European Medicines Agency (EMA) and the Food and Drug Administration (FDA) for patients with BRCA-deficient ovarian cancer in December 2014

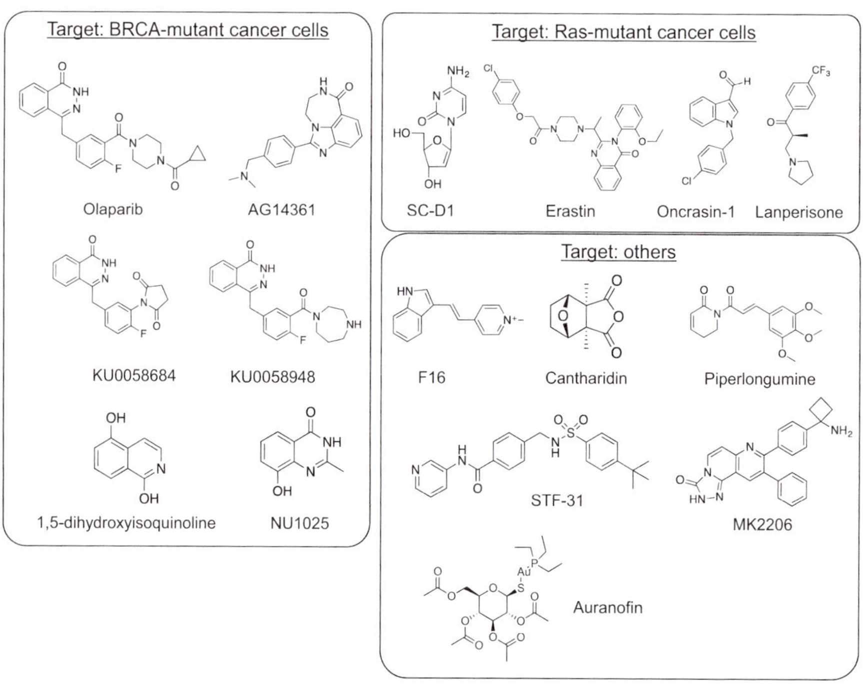

Fig. 4.4 Structures of synthetic lethality–inducing compounds

type cells. Similarly, knockdown of BRCA1 in MCF7 human breast cancer cells by siRNA induced sensitivity to KU0058684 and KU0058948 [47]. Moreover, other PARP inhibitors 1,5-dihydroxyisoquinoline [49], NU1025 [50], and AG14361 [51] (Fig. 4.4) profoundly reduced the survival of MCF7 and MDA-MB-231, another human breast cancer cell line, but only when siRNA knocked down BRCA2 [48]. Both KU0058684 and AG14361 selectively blocked the growth of BRCA2-deficient tumors in xenografts. Since mutations in BRCA1 and BRCA2 predispose individuals to breast and other cancers [52], inhibition of PARP1 is a potent strategy for synthetic lethal therapeutics against cancers in carriers of BRCA1 or BRCA2 mutations. The profound sensitivity of BRCA mutant cells to PARP inhibition has led to the development of a number of clinical trials for testing the efficiency of this approach. Indeed, the potent PARP inhibitor olaparib (AZD2281, previously known as KU0059436; AstraZeneca [53]) (Fig. 4.4) was shown to induce selective killing of BRCA1- or BRCA2-deficient cancer cells. Olaparib was approved by the European Medicines Agency (EMA) and the Food and Drug Administration (FDA) in December 2014.

4.2.2.2 Compounds Inducing Synthetic Lethality in *Ras* Mutant Cancer Cells

Activating mutations in the Ras oncogene (*H-*, *N-*, and *K-Ras*) are found in ~20% of all human tumors. These mutations occur most frequently in *K-Ras* and are predominantly missense mutations that compromise the guanosine-5′-triphosphatase (GTPase) activity of Ras. Canonical Ras signaling pathways include the Raf/mitogen-activated protein kinase kinase (MEK)/extracellular-signal-regulated kinase (ERK) pathway and the phosphatidylinositol-4,5-bisphosphate-3-kinase (PI3K)/AKT pathway. Many compounds targeting canonical Ras signaling pathways have been developed. However, targeting these pathways has become complicated because of the essential role of these pathways in the proliferation and survival of normal cell types. Therefore, synthetic lethal screenings targeting oncogenic Ras have been performed by several research groups.

4.2.2.2.1 SC-D

Kinzler and colleagues described a strategy for drug screening based on isogenic human cancer cell lines in which key tumorigenic genes were deleted by targeted homologous recombination. At first, they established two cell lines: a yellow fluorescent-protein (YFP) expression vector was introduced into the colon cancer cell line DLD-1 (K-Ras mutant cell line), and a blue fluorescent-protein (BFP) expression vector was introduced into an isogenic derivative in which the mutant *K-Ras* allele had been deleted. Coculture of both cell lines allowed facile screening for compounds with selective toxicity toward the mutant *Ras* genotype. After screening 30,000 compounds from the ChemBridge and National Cancer Institute

(NCI) libraries, they identified triphenyl tetrazolium (TPT) and SC-D (a novel sulfinyl cytidine analog) (Fig. 4.4) and found that these compounds displayed selective toxicity for the mutant K-Ras-containing cell line (DLD-1) compared with the K-Ras-knockout. SC-D but not TPT selectively inhibited the growth of rat kidney epithelial (RK3E) cells stably transformed with K-Ras^{G12V}. SC-D also inhibited tumor xenografts of two colon cell lines, DLD-1 and HCT116, both of which harbor a single G13D point mutation in the *K-Ras* gene. Although SC-D and TPT are potent synthetic lethality–inducing agents against cancer cell lines with K-Ras mutations, the mechanisms underlying the selective toxicity of SC-D and TPT are unclear. The literature provided us with a broadly applicable approach for mining therapeutic agents targeted to specific genetic alterations responsible for cancer development [54].

4.2.2.2.2 Erastin

In 2003, for synthetic lethality screening, Stockwell and colleagues established experimentally transformed human cells expressing the human catalytic subunit of the enzyme telomerase (*hTERT*), a genomic construct encoding simian virus 40 (SV40), as well as large T (LT) and small T (ST) oncoproteins. In addition, they introduced an oncogenic allele of *H-Ras* (*Ras*^{V12}) into primary human foreskin fibroblast BJ cells. They screened 23,550 compounds comprising 20,000 compounds from a combinatorial library, 1990 compounds from the NCI diversity collection, and 1540 biologically active compounds. As a result, they found a novel compound called erastin (Fig. 4.4) from the combinatorial library that showed greater cytotoxicity against BJ-TERT/LT/ST/Ras^{V12} cells compared with that against primary BJ, BJ-TERT, BJ-TERT/LT/Ras^{V12}, or BJ-TERT/LT/ST cells. The data indicates that erastin-induced cell death requires the presence of both ST and Ras^{V12}. Moreover, erastin did not induce morphological alterations such as nuclear karyorrhexis and chromatin margination, DNA fragmentation, and caspase-3 cleavage, suggesting that cell death induced by erastin is non-apoptotic. These findings together suggest that erastin-induced non-apoptotic cell death is selective to ST- and Ras^{V12}–expressing cells [55]. They further addressed the molecular mechanism of erastin. They synthesized erastin analogs that could be linked to a solid-phase resin for biochemical purification of potential targets, and obtained erastin A6, which retained the ability to kill BJ-TERT/LT/ST/Ras^{V12} cells but not BJ-TERT cells, as well as erastin B2, which lacked activity. Erastins A6 and B2 were immobilized on solid-phase resin, and proteins that interacted with erastin A6 but not B2 were sought. Using BJ-TERT/LT/ST/Ras^{V12} cell lysates, they isolated two isoforms of human mitochondrial VDACs (VDAC2 and VDAC3), as VDACs bound more efficiently to resin with erastin A6 than to resin with erastin B2. Knockdown of VDAC2 or VDAC3 caused resistance to erastin, implying that these two VDAC isoforms are involved in the mechanism of action of erastin. Using purified mitochondria expressing a single VDAC isoform, they also found that erastin altered the permeability of the outer mitochondrial membrane.

Therefore, erastin binds to VDAC2 and VDAC3, resulting in selective induction of non-apoptotic cell death in certain tumor cells harboring activating mutations in the Ras pathway [56].

4.2.2.2.3 Oncrasin-1

Another group, Fan and colleagues, also performed synthetic lethality screening for compounds that selectively kill tumor cells by establishing immortalized human ovarian surface epithelial cells (designated T29) and their tumorigenic derivatives transformed with mutant *K-Ras* (T29Kt1). Among 10,000 compounds from a chemical library purchased from the ChemBridge Corporation, 1-[(4-chlorophenyl)methyl]-1*H*-indole-3-carboxaldehyde, designated as oncogenic Ras tumor-inhibiting compound 1 (oncrasin-1) (Fig. 4.4), was shown to have highly selective cytotoxicity toward T29Kt1 cells compared to T29 cells. Examining the effects of oncrasin-1 on native human cancer cell lines revealed that it effectively killed *K-Ras*-mutant lung cancer cells. However, it had minimal effects not only on lung cancer cells harboring wild-type *Ras* genes, but also on lung cancer cells harboring mutant *N-Ras* genes and human bronchial epithelial cells. These results suggest that oncrasin-1 is effective against various lung cancer cells with *K-Ras* mutations. Furthermore, K-Ras knockdown by siRNA inhibited oncrasin-1-mediated apoptosis, suggesting that K-Ras is required for effective apoptosis induction by oncrasin-1. Furthermore, treatment with oncrasin-1 was found to lead to formation of large foci of protein kinase C iota (PKCι) in the nuclei of K-Ras-mutated T29Kt1 and H460 cells but not in oncrasin-1-resistant T29 and H1299 cells. Knockdown of PKCι resulted in almost complete resistance to oncrasin-1, suggesting that oncrasin-1 has synthetic lethality for K-Ras and PKCι [57]. However, the mechanism behind abnormal PKCι aggregation by oncrasin-1-induced apoptosis remains largely unclear.

4.2.2.2.4 Lanperisone

In addition to the abovementioned studies, Jacks and colleagues also demonstrated synthetic lethality using mouse embryonic fibroblast (MEFs) derived from a mouse model harboring a conditional oncogenic *K-Ras*^{G12D} [58]. Among over 50,000 compounds screened, tolperisone (piperidine analog) was identified with potent and selective activity against *K-Ras* mutant cells compared with wild-type control cells. Tolperisone is an orally available, centrally acting muscle relaxant used to treat painful muscle spasms [59]. Among tolperisone derivatives synthesized or obtained from commercial sources, lanperisone (Fig. 4.4) [60] was found to have the most potent and selective activity. To explore the molecular mechanism by which lanperisone selectively kills *K-Ras* mutant cells, gene set enrichment analysis (GSEA) [61] was performed. This analysis compares signatures associated with lanperisone treatment against a curated database of gene expression changes associated with various biological pathways. GSEA showed that the gene expression signatures for lanperisone were similar to those associated with oxidative stress pathways. In addition, the connectivity map [62] revealed that oxidative stress-inducing agents induced similar changes in gene expression to those induced by lanperisone. These results suggested that lanperisone induces oxidative stress. Indeed, lanperisone significantly increased intracellular ROS levels in K-Ras mutant MEFs compared with those in wild-type MEFs. Moreover, ROS scavengers such as deferoxamine, butylated hydroxyanisole, and the vitamin E analog, trolox, completely abolished cell killing by lanperisone. Together, these results demonstrate that selective killing of K-Ras mutant cells by lanperisone is mediated by the induction of ROS activity [58]. At present, the molecular target of lanperisone is still unknown. However, the potential mechanism of action of lanperisone seems to resemble most closely that of erastin, which also induces oxidative cell death in cell lines expressing oncogenic K-Ras through interaction with VDAC [55, 56]. Identification of the molecular target of lanperisone would provide new insight into how oxidative stress contributes to synthetic lethality in K-Ras mutant tumor cells.

4.2.2.3 Nononcogene-Perturbing Compounds

Tumor progression results in deregulation of cell proliferation and differentiation that are frequently associated with enhanced cellular stress. For survival, tumor cells acquire adaptation to cellular stress. As results, tumor cells may become dependent upon nononcogenes that do not ordinarily perform such vital functions in normal cells [16]. Thus, targeting these nononcogene dependencies may result in synthetic lethal interactions and selective death of tumor cells.

4.2.2.3.1 F16

The HER-2/ERBB-2/Neu proto-oncogene is a member of the epidermal growth factor (EGF) receptor family. This family of receptor tyrosine kinases plays an essential role during growth and differentiation of many tissues, and their overexpression is associated with several types of human cancers. Thus, anticancer agents that directly affect the activity of the EGF receptor family would interfere with tumor cell growth. Accordingly, many researchers attempted to develop small molecules or antibody targeting EGF receptor family as anticancer agents. Herceptin (Genentech, Inc.) is the humanized recombinant version of the neutralizing mouse monoclonal antibody directed against the extracellular domain of HER-2. Herceptin causes HER-2 dephosphorylation and down-regulation in HER-2-overexpressing cell lines. However, Herceptin does not always lead to growth inhibition. Moreover, experiments on Herceptin also demonstrated that activation of both EGF receptor family members, HER-1/EGF receptor and HER-3/ERBB-3, and downstream signaling cascades could influence the overall response of cells to HER-2 inactivation by monoclonal antibodies [63, 64]. These observations suggest that activation of parallel or downstream pathways that aid HER-2 in inducing cellular transformation may override cell growth control independently of HER-2 status. Therefore, to find novel approaches to identify poorly understood events that assist HER-2 in promoting cellular transformation, Leder and colleagues screened a chemical library for compounds that selectively inhibit proliferation of Neu (HER-2/ERBB-2)-overexpressing EpH4 cells (mouse mammary epithelial cells). Among 16,000 compounds, F16 (Fig. 4.4) was found to selectively inhibit cell proliferation of Neu-overexpressing EpH4 cells. F16 showed marked sensitivity against mouse tumor cell lines derived from *Neu*- and *v-H-Ras*-transgenic mice, whereas mouse fibrosarcoma cell lines derived from *v-H-Ras*-transgenic mice did not show any changes in growth. In addition, β-catenin-initiated mammary tumor cell lines had moderate responses to F16, whereas *Myc*-initiated mouse tumor cells were F16 resistant. F16 produced a lower-energy absorption band with a maximum at 420 nm in the visible region. Conformational rigidity of the molecule favors relaxation of its excited state during fluorescence, resulting in an intense emission band at 520 nm. Therefore, F16 could be visualized in living cells. Fluorescence microscopy revealed that F16 produced an intracellular staining pattern of punctated green fluorescence that overlaps with mitochondria, suggesting that F16 selectively accumulated in mitochondria. Consistent with the results of the antiproliferative effects of F16 against tumor cells, intracellular accumulation of F16 was observed in F16-sensitive Neu-overexpressing EpH4, *v-H-Ras*-, *Neu*-, and β-catenin-initiated tumor cell lines, but not in immortalized and nontransformed mouse mammary epithelial cell lines. Moreover, F16 decreased adenosine triphosphate (ATP) levels and increased production of ROS selectively in F16-sensitive tumor cell lines. F16 caused opening of the permeability transition pore in isolated mitochondria, suggesting that the mitochondrion is probably the primary target of F16-mediated toxicity. Thus, although the precise mechanism by which F16 inhibited mitochondrial function leads to selective killing of cancer cells is unclear, the broad action spectrum of F16 may be explained by the commonality of mitochondrial hyperpolarization among many tumor cell lines [65].

4.2.2.3.2 Piperlongumine

Lee and colleagues established cell-based small-molecule screening using a luciferase reporter gene fused with the cell death involved p53 target promoter (*CDIP*) to identify compounds acting through novel pro-apoptotic mechanisms [16]. The compound with the highest luciferase activity was piperlongumine (Fig. 4.4), a natural product isolated from the plant species, *Piper longum* L [62]. Piperlongumine treatment markedly induced cell death in cancer cells with wild-type as well as mutant p53 and *ERBB2* and/or in H-Ras-overexpressing breast epithelial cells. However, piperlongumine caused little reduction in cell viability of normal cells and of nontransformed immortalized cells. Moreover, significant antitumor effects were observed in piperlongumine-treated mouse xenograft tumor models, with no apparent toxicity in normal mice. Therefore, piperlongumine possesses killing properties selective toward cancer cells. To investigate the molecular mechanism by which piperlongumine induces death of cancer cells, quantitative proteomics was analyzed by synthesizing piperlongumine linked to *p*-*t*Bu-polyethylene glycol for piperlongumine target identification. As a result, glutathione *S*-transferase pi 1 (GSTP1) and carbonyl reductase 1 (CBR1) were identified as piperlongumine binding proteins. Piperlongumine was shown to interact directly with purified GSTP1. Since GSTP1 and CBR1 are part of a common complex [66, 67], CBR1 may be an indirect partner. Moreover, GSTP1 and CBR1 are known to participate in the cellular response leading to oxidative stress caused by elevated levels of ROS. Indeed, piperlongumine caused a marked increase in ROS levels in cancer cells. In contrast to the results for cancer cells, piperlongumine did not cause an increase in ROS levels in normal cells. Moreover, overexpression of CRB1 or GSTP1 markedly reduced piperlongumine-induced ROS levels and partially rescued piperlongumine-induced apoptosis. In a complementary study, knockdown of GSTP1 or CBR1 did not affect piperlongumine-induced ROS levels. These results indicate that piperlongumine induces apoptosis by interfering with redox and ROS homeostatic regulators such as GSTP1 and CBR1. Perturbing redox and ROS homeostasis is, therefore, a promising strategy for cancer treatment [16].

4.2.2.3.3 Other Compounds

The spindle checkpoint is essential for ensuring proper chromosome segregation and, therefore, for maintaining genomic stability. Mitotic arrest deficiency 2 (MAD2) is a critical component of the spindle checkpoint [68] and is overexpressed in many cancer cells [69]. To investigate the genes inducing synthetic lethality in MAD2-overexpressing cells, analysis of a synthetic genetic array in yeast was performed. This study revealed that MAD2 overexpression induced lethality in 13 gene deletions [70]. Among the human homologs of the 13 candidate genes in yeast, knockdown of protein phosphatase 2 regulatory subunit A alpha (*PPP2R1A*), a gene encoding a constant regulatory subunit of protein phosphatase 2 (PP2A), significantly inhibited the growth of MAD2-overexpressing cancer cells. Furthermore, the PP2A inhibitor cantharidin (Fig. 4.4), which was originally isolated from *Lytta vesicatoria*, also inhibited the growth of MAD2-overexpressing cancer cells [71]. Cantharidin displayed high levels of cytotoxic activity against a broad range of tumor cell lines. These results suggest that PP2A is a good therapeutic target in MAD2-overexpressing tumors and might explain the mechanism of cytotoxic activity of cantharidin [70].

The PI3K/AKT pathway plays a critical role in cancer cell growth and survival. Therefore, inhibitors against this pathway have been developed and tested for cancer treatment. MK2206 (Merck, Inc.) (Fig. 4.4), a small-molecule inhibitor that targets AKT, is currently being tested in phase I and II clinical trials [72]. However, some cancer cell lines do not respond to MK2206. Therefore, identification and inhibition of additional targets are needed to improve the efficacy of MK2206. To identify targets or networks that can induce synthetic lethality with MK2206, genome-wide siRNA library screening was performed. As a result, thioredoxin reductase-1 (TXNRD1) was found to sensitize lung cancer cells to MK2206. Moreover, a TXNRD1 inhibitor, auranofin (Fig. 4.4), also induced synthetic lethality with MK2206 [73]. Auranofin is an oral medicine based on a gold compound used for antirheumatoid arthritis. To examine the mechanism of synthetic lethality interaction between AKT and TXNRD1, cellular ROS production was measured because TXNRD1 is one of the major regulators of cellular redox processes. The combination of auranofin and MK2206 induced more ROS production than MK2206 alone. The ROS scavenger, mordihydroguaiaretic acid, suppressed cell growth inhibition due to combination treatment with auranofin and MK2206. Furthermore, MK2206 was shown to inhibit another major cellular antioxidant pathway, the Kelch-like ECH-associated protein 1 (KEAP1)–nuclear factor (erythroid-derived 2)-like 2 (NRF2) pathway. Therefore, combination of auranofin and MK2206 inhibited two major cellular antioxidant systems, TXNRD1 and NRF2, thereby inducing ROS production and apoptosis [73].

The von Hippel–Lindau (VHL) tumor suppressor gene is inactivated in about 80% of renal cell carcinomas (RCCs). To develop a compound that shows selective toxicity to cells that lost functional VHL, synthetic lethality screening was performed. Among 64,000 compounds, STF-31 (Fig. 4.4) was found to selectively kill RCCs by binding glucose transporter 1 (GLUT1) directly and impeding glucose uptake in vivo without causing toxicity to normal tissues [74].

4.2.3 Compounds that Protect Against Neuronal Cell Death

As mentioned above, neurodegenerative diseases such as AD, PD, and HD possess common pathological mechanisms, raising the possibility of shared therapeutic targets. All approaches that are in clinical use today merely treat the symptoms, so there is an urgent need for development of neurorestorative therapies that target pathogenic mechanisms. In this section, we introduce various small molecules that show neuroprotective effects in neurodegenerative disease models.

4.2.3.1 Geldanamycin

Geldanamycin (GA) (Fig. 4.5) [75, 76], a naturally occurring benzoquinone ansamycin that specifically binds to and inhibits the molecular chaperone heat shock protein (HSP) 90 [77], is known as a potent antitumor agent. Moreover, GA exhibited neuroprotective abilities in a variety of in vivo neurodegeneration models [78, 79]. GA was also shown to prevent αSYN-induced toxicity in human H4 neuroglioma cells [80] and tau aggregation [81]. Therefore, GA is considered a promising candidate for chemotherapy against neurodegenerative diseases such as

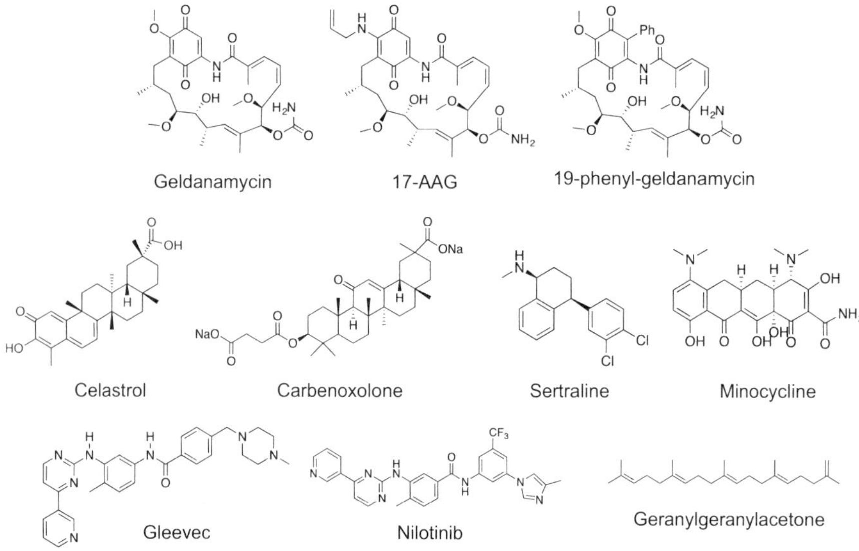

Fig. 4.5 Structures of inhibitors of neuronal cell death

PD and AD. However, GA itself cannot cross the blood–brain barrier [82] and has considerable toxicity. On the other hand, 17-(allylamino)-17-demethoxygeldanamycin (17-AAG) (Fig. 4.5), one of GA derivatives that shares its important biological activities but shows less toxicity, passes through the blood–brain barrier [82]. However, preclinical studies showed that 17-AAG has limited oral availability and causes varying degrees of hepatotoxicity, possibly through reactions with biological nucleophiles with the reactive 19-position of its quinone ring [83]. Therefore, GA derivatives other than 17-AAG have been explored. In 2013, Moody and colleagues synthesized 19-substituted GA derivatives (Fig. 4.5) on the hypothesis that blocking the reactive 19-position might suppress the reaction with biological nucleophiles, ameliorate toxicity, and result in conformational changes arising from amide *trans* to *cis* isomerism. As expected, 19-phenyl GA markedly reduced benzoquinone ansamycin toxicity to human umbilical vein endothelial cells (HUVECs) and to retinal pigmented epithelial cells. Moreover, 19-phenyl GA was more potent than 17-AAG in the induction of Hsp70 in SH-SY5Y neuroblastoma cells [84].

Expression of normal and mutant αSYN in *Drosophila melanogaster* causes selective loss of dopaminergic neurons [85], which can be fully prevented by raising HSP70 levels through transgenic expression [86]. This finding suggests that dopaminergic neurons may be sensitive to HSP70 and supports the hypothesis that pharmacological enhancement of HSP70 activity might be an effective therapeutic approach. A number of small molecules have been reported to up-regulate HSP70 expression with different mechanisms of action (for more details see ref. [87]). For example, celastrol (Fig. 4.5), a quinone methide triterpene, was first identified as a potent inhibitor of lipid peroxidation in mitochondria [88] and was subsequently reported to induce HSP70 up-regulation [89]. Geranylgeranylacetone (Fig. 4.5), an antiulcer drug [90], and carbenoxolone (Fig. 4.5), a glycyrrhizic acid derivative widely used in the treatment of peptic ulcers [91], were also shown to up-regulate HSP70 [92, 93]. Celastrol was reported to protect rotenone-induced cell death in SH-SY5Y cells, an experimental cellular model of PD [94]. Carbenoxolone was also demonstrated to lower αSYN aggregation and prevent αSYN-induced cytotoxicity in human neuroglioma H4 cells [95]. The above studies therefore suggest that GA derivatives and Hsp70-modulating agents may be excellent chemotherapeutic candidates for neurodegenerative diseases.

4.2.3.2 Minocycline

Minocycline (Fig. 4.5) is a second-generation broad-spectrum tetracycline antibiotic that effectively crosses the blood–brain barrier [96]. Recently, minocycline was found to delay mortality in R6/2 transgenic mice expressing exon-1 of huntingtin with an expanded polyQ repeat. Minocycline significantly delayed the characteristic decline in rotarod performance and extended survival [97]. It also showed neuroprotective effects in an in vitro AD model. It was found to attenuate neuronal cell death induced by Aβ peptide$_{1\text{-}42}$ exposure of PC12 cells differentiated by using nerve growth factor [98]. The molecular mechanism underlying its neuroprotective properties remains elusive. However, in 2011, Perez-Paya and colleagues identified a minocycline binding protein as APAF-1 by evaluation using two complementary ligand-based NMR techniques that analyze the effects of ligand binding on the NMR signal. APAF-1 is a key protein in the formation of the apoptosome, a multiprotein complex involved in caspase activation. Minocycline was shown to inhibit apoptosome activity possibly by binding to multiple WD40 domain repeats at the C-terminus of APAF-1 that binds to cytochrome *c* [99]. Such binding may explain the protective effect of minocycline against neuronal cell death.

4.2.3.3 Gleevec and Nilotinib

c-ABL, a nonreceptor tyrosine kinase distributed in the nucleus and cytosol, is involved in a wide range of functions that includes cell-cycle regulation and apoptosis induction. In the brain, c-ABL is involved in neuronal plasticity, neurite outgrowth, and neurogenesis [100]. Recently, c-ABL was reported to play a role in neurologic disorders such as PD [101], AD [102], and Niemann–Pick type-2 disease [103] through aberrant activation. When hippocampal neurons were treated with Aβ peptide fibrils, phosphorylation of c-ABL was observed. c-ABL activates p73, a structural and functional homolog of p53, through phosphorylation. The phosphorylation of p73 by c-ABL results in p73 translocation to the nucleus and in transcriptional up-regulation of p53-responsive genes. Consistent with previous reports, translocation of p73 into the nucleus was observed in hippocampal neurons exposed to Aβ fibrils. Moreover, neuronal cell death induced by Aβ fibrils was suppressed by the ABL inhibitor Gleevec or by siRNA-mediated ABL knockdown. Therefore, phosphorylation of c-ABL may be involved in neuronal cell death induced by Aβ fibrils [102]. On the other hand, Dawson and colleagues revealed that 1-methyl-4-phenylpyridinium (MPP^+) (Fig. 4.8) and 1-methyl-4-phenyl-1,2,3,6-tetrahydropyridine (MPTP) activate c-ABL in vitro and in vivo, respectively. The activated c-ABL leads to inactivation of ubiquitin E3 ligase activity of PARKIN through Tyr143 phosphorylation and cell death. Furthermore, Gleevec suppressed PARKIN phosphorylation and MPP^+-induced cell death. Since knockdown of PARKIN by small hairpin RNA prevented Gleevec protection, the protective effects of Gleevec are dependent on PARKIN. Thus, tyrosine phosphorylation of PARKIN by c-ABL is a major post-translational modification that inhibits PARKIN function, possibly contributing to the pathogenesis of PD [101].

According to the above observations, Gleevec (Fig. 4.5) may be a promising chemotherapeutic candidate for neurodegenerative diseases. However, one of the limitations of Gleevec as a treatment for neurodegenerative diseases is the limited penetration of the drug into the central nervous system [104]. On the other hand, nilotinib (Fig. 4.5) was suggested to enter the brain. This drug is a second-generation ABL inhibitor that was approved by the FDA in 2007 for the treatment of adult patients with chronic and accelerated-phase Philadelphia chromosome–positive CML. Administration of nilotinib by intraperitoneal injection decreased αSYN levels in 7- to 8-month-old transgenic αSYN mice that harbored the A53T mutation of αSYN [105]. Moreover, a phase I clinical trial of nilotinib in a patient with PD was done in 2015. Therefore, nilotinib is expected to be approved in the near future.

4.2.3.4 SSRIs

Depression, anxiety disorders, and various personality disorders are considered to be caused by lack of serotonin in the synaptic cleft. Since selective serotonin reuptake inhibitors (SSRIs) may increase the levels of the serotonin in the synaptic cleft by inhibiting its uptake into presynaptic cells, the increased amount of serotonin with SSRIs treatment could bind to postsynaptic receptors. Thus, SSRIs are typically available for the treatment of these disorders. On the other hand, several SSRIs were reported to inhibit dynamin GTPase. Among them, sertraline (Zoloft®) (Fig. 4.5), a second-generation SSRI, exhibited the strongest effect against dynamin [106, 107]. Recently, dynamin 1 has been revealed to contribute to αSYN incorporation into neuronal cells. As K44A dynamin 1 is a dominant-negative mutant that blocks formation of endocytic vesicles, K44A dynamin 1 reduces αSYN aggregate incorporation into neuronal cells. In addition, inhibition of dynamin 1 with siRNA also resulted in a decrease in αSYN aggregate incorporation into SH-SY5Y cells [108–111]. Therefore, as expected, inhibition of dynamin GTPase activity by sertraline treatment prevented the uptake of αSYN aggregates into SH-SY5Y cells [111]. There is an increasing interest in dynamin GTPase as a new target for PD.

4.3 Autophagy Modulators

4.3.1 Mechanism and Function of Autophagy

In the 1990s, genetic studies on yeast identified a series of ATG genes [112]. These studies elucidated the mechanism and functions of autophagy. The critical regulator of autophagy induction is the mammalian target of rapamycin (mTOR) complex 1 (mTORC1), which was identified when rapamycin treatment induced autophagy in yeast and mammalian cells [113, 114]. Rapamycin stabilizes the raptor–mTOR association by forming a complex with a 12 kDa FK506 binding protein (FKBP12), thereby specifically inhibiting the kinase activity of mTORC1 [115]. mTORC1 suppresses the unc-51-like autophagy-activating kinase 1 (ULK1; ATG1 homolog) complex under nutrient-rich conditions. Upon autophagy induction, the ULK1 complex (which includes ULK1, ATG13, FIP200, and ATG101) activates and translocates to the endoplasmic reticulum (ER). Afterward, the ULK1 complex regulates the class III PI3K complex, which contains Vps34 (a class III PI3K, also known as PIK3C3), Beclin 1 (a mammalian homolog of yeast ATG6), ATG14-like protein, Vps15, and activating molecule in Beclin-1-regulated autophagy (AMBRA1), thereby generating phosphatidylinositol 3-phosphate (PI(3)P). Localized production of PI(3)P leads to recruitment of effector proteins containing FYVE or PX domains that control membrane docking and fusion during formation of internal vesicles. The ATG12–ATG5–ATG16L1 complex and LC3–phosphatidylethanolamine conjugate play important roles in the elongation and closure of the isolated membrane during formation of autophagosomes. The origin of the isolated membrane and the site of its nucleation have been investigated. It is proposed that the isolated membrane might come from pre-existing cellular membranes, such as the ER, the Golgi, the endosome, the mitochondria, and the plasma membrane. Moreover, it was recently demonstrated that the ER–mitochondria contact site is important in autophagosome formation [116]. Finally, maturation occurs when the autophagosome fuses with the endolysosomal compartment, which causes autolysosome formation. Final degradation of cargos occurs in autolysosomes. The molecular mechanism of autophagosome formation has finally been elucidated, but that of autolysosome formation has largely remained unclear. (For a more detailed review of the molecular mechanisms of autophagy, see ref. [27]).

Many small molecules that induce or inhibit autophagy have been discovered (Fig. 4.6). These autophagy modulators may be divided into two categories, depending on whether they act on mTOR.

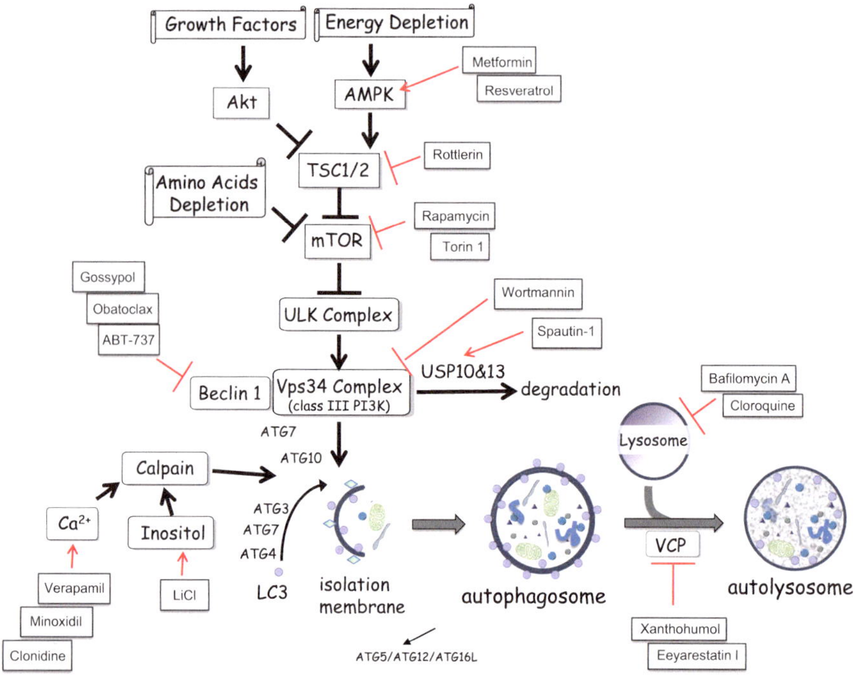

Fig. 4.6 Autophagy modulating agents. Autophagy is an evolutionarily conserved pathway involved in the degradation of intracellular components, including organelles, and is critical for the maintenance of cellular homeostasis. Energy depletion, amino acid depletion, or growth factor withdrawal inactivate mammalian target of rapamycin (mTOR), which results in activation of class III phosphatidylinositol-4,5-bisphosphate-3-kinase (PI3K) (Vps34) complex. Then, autophagosome formation and subsequent autolysosome formation occurs. On the other hand, intracellular Ca^{2+} levels and inositol levels also control autophagy through calpain. Many small molecules that induce or inhibit autophagy have been discovered

4.3.2 mTOR-Dependent Autophagy Regulation

4.3.2.1 mTOR Inhibitors

mTOR is present in two distinct complexes, mTORC1 and mTOR complex 2 (mTORC2). mTORC1 phosphorylates S6 kinases, eukaryotic translation initiation factor 4E binding proteins, and proline-rich AKT substrate of 40 kDa. On the other hand, mTORC2 catalyzes the phosphorylation of AKT and of serum/glucocorticoid-induced kinase 1 (SGK1). Various signals such as growth factors, amino acids, glucose and energy status regulate the activity of the autophagic machinery, and these signals are integrated by the kinase activity of mTOR. Torin 1 (Fig. 4.7) is a selective ATP-competitive small-molecule mTOR inhibitor which directly inhibits both mTORC1 and mTORC2 complexes. The effect of torin 1 on autophagy induction was found to be independent of mTORC2 but dependent on mTORC1 [117]. Roberge and his group screened for small molecules that modulate

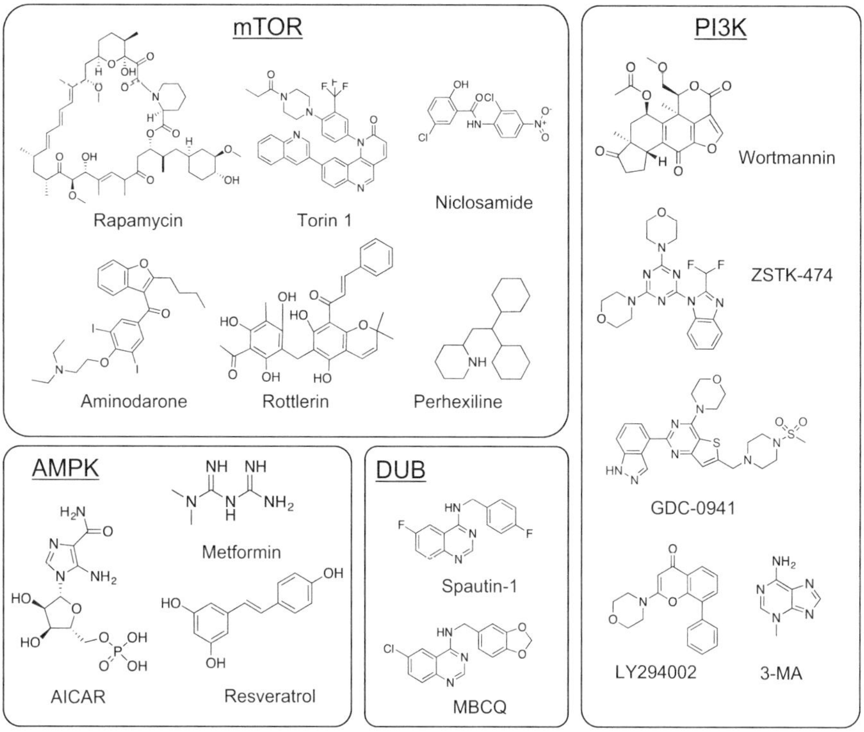

Fig. 4.7 Structures of autophagy modulators (mammalian target of rapamycin (mTOR)-dependent)

autophagy by monitoring the number of autophagosomes in MCF-7 cells expressing enhanced green fluorescent-protein (EGFP)-LC3. A collection of 3584 drugs and pharmacologically active chemicals was tested, and four inhibitors (perhexiline, niclosamide, amiodarone, and rottlerin (Fig. 4.7)) were shown to rapidly increase autophagosome levels. Further biochemical assays revealed that rottlerin inhibited mTORC1 signaling through tuberous sclerosis complex 2 (TSC2), a negative regulator of mTORC1, whereas other inhibitors suppressed mTORC1 signaling in a TSC2-independent manner. Furthermore, these four inhibitors did not inhibit mTORC2 signaling. Perhexiline, niclosamide, and rottlerin do not show toxicity under sufficient levels of nutrients and growth factors but they selectively kill starved cells. These results suggest that small molecules that inhibit mTORC1 and stimulate autophagy may be valuable tools for evaluating the therapeutic potential of manipulating mTORC1 and autophagy in disease [118].

4.3.2.2 PI3K Inhibitors

PI3K signaling plays an important role in the regulation of autophagy. The PI3K family may be divided into three classes. Class I PI3K inhibits the activity of mTOR; therefore, class I PI3K negatively regulates autophagy. Indeed, the class I PI3K–specific inhibitors GDC-0941 [119] and ZSTK-474 [120] (Fig. 4.7) induce autophagy [121, 122]. Class II PI3K activity is thought to have no relevance to autophagic activity. In contrast to the negative role of class I PI3K, class III PI3K complex accelerates the autophagic process. The essential role of Vps34 in autophagy has been established in many studies, largely by using small-molecule inhibitors of PI3K such as 3-methyladenine, wortmannin, and 2-(4-morpholinyl)-8-phenylchromone (LY294002) [123, 124] (Fig. 4.7) to suppress autophagy. However, the precise role of Vps34 in autophagy remains largely unclear, because these PI3K inhibitors lack specificity. Therefore, there is an urgent need to develop small molecules that inhibit Vps34-controlled autophagy. Using GFP-LC3-expressing cells, Yuan and his group found 4-[[3,4-(methylenedioxy)benzyl]amino]-6-chloroquinazoline (MBCQ) (Fig. 4.7) to be an inhibitor against rapamycin-induced autophagy [125]. MBCQ has been known as an inhibitor of phosphodiesterase type 5 (PDE5), an enzyme that degrades cyclic guanosine monophosphate (cGMP) by hydrolysis [126]. However, other PDE5 inhibitors, including MY-5445, dipyridamole, IBMX, and sildenafil (Viagra), failed to inhibit rapamycin-induced autophagy. These results indicate that the inhibitory activity of MBCQ against rapamycin-induced autophagy is not related to its PDE5 inhibitory activity. In order to separate the autophagy inhibitory activity of MBCQ from its inhibition of PDE5, Yuan's group developed the MBCQ derivative spautin-1 (Fig. 4.7). Spautin-1 had the same autophagy inhibitory activity as that of MBCQ, but possessed lower PDE5 inhibitory activity compared with MBCQ. Furthermore, spautin-1 was shown to reduce the expression levels of Vps34 complexes by promoting the degradation of Vps34 complexes through the proteasomal pathway, raising the possibility that spautin-1 targets deubiquitinating enzymes (DUBs). DUBs normally participate in negative regulation of ubiquitination of Vps34 complexes, thereby leading to inhibition of autophagy. To identify the DUB for Vps34 complexes, a collection of 127 siRNAs targeting human DUBs from the Dharmacon library SMART Pools was screened for inhibition of autophagy. According to a report, knockdown of ubiquitin-specific protease (USP)10 or USP13 has been found to produce a consistent phenotype of reduced expression levels of Vps34 complexes. Moreover, spautin-1 reduced the expression levels of USP10 and USP13, which resulted in inhibition of the deubiquitinating activities of USP10 and USP13 on ubiquitinated Beclin 1 in vitro. Thus, inhibition of USP10 and USP13 by spautin-1 promoted the ubiquitination and degradation of Vps34 complexes, which in turn led to inhibition of autophagy [125].

4.3.2.3 AMPK Activators

5′-Adenosine monophosphate (AMP)-activated protein kinase (AMPK) is a key energy sensor that regulates cellular metabolism to maintain energy homeostasis

[127]. There is evidence that supports the role of AMPK in autophagy induction in response to various cellular stresses, including glucose starvation. Under nutrient-sufficient conditions, high mTOR activity disrupts ULK1–AMPK interaction by phosphorylating ULK1 Ser757, which results in inactivation of ULK1 and consequently inhibits autophagy. Conversely, under glucose starvation, the AMP/ATP ratio increases, resulting in AMPK activation. Activated AMPK directly activates ULK1 through phosphorylation of Ser317 and Ser777 [128]. Therefore, small molecules that activate AMPK may induce autophagy through inhibition of mTOR-dependent signaling. Indeed, metformin (Fig. 4.7), the most widely used drug for the treatment of type 2 diabetes, was reported to activate AMPK in vitro and in vivo [129–131] and was shown to induce autophagy [132]. However, its mode of action is not completely understood. Moreover, resveratrol (*trans*-3,4′,5-trihydroxystilbene) (Fig. 4.7), found in abundance in red wine, activates AMPK and thereby inhibits mTOR, resulting in induction of autophagy. Resveratrol-induced autophagy involves lysosomal degradation of Aβ [133], which partly explains the lower incidence of dementia and AD among populations that moderately consume red wine [134]. The AMP analog 5-aminoimidazole-4-carboxamide ribonucleoside (AICAR) (Fig. 4.7), a well-known AMPK activator, also induces autophagy and subsequent lysosomal degradation of Aβ [133].

4.3.3 mTOR-Independent Autophagy Regulation

4.3.3.1 Intracellular Ca^{2+} and Inositol Modulators

HD is associated with intraneuronal polyQ aggregation, which results in neuronal toxicity and cell death. Since autophagy is a key pathway for the clearance of aggregation-prone cytosolic proteins, small molecules that induce autophagy may be excellent chemotherapeutic candidates for this neurodegenerative disorder. In 2002, Rubinsztein and colleagues found that lithium chloride (LiCl) protected against neuronal cell death and inclusion formation induced by EGFP-tagged HD exon 1 fragment with 74 glutamines (EGFP-HDQ74) [135]. Furthermore, LiCl induces autophagy and enhances the clearance of known autophagy substrates such as EGFP-HDQ74 and A53T mutant αSYN aggregates [136]. On the other hand, LiCl was reported to inhibit a number of enzymes that include glycogen synthase kinase-3β (GSK-3β) and inositol monophosphatase (IMPase) [137, 138]. L-690,330 (Fig. 4.8), a specific inhibitor of IMPase [139], reduced aggregation and cell death caused by EGFP-HDQ74 in SK-N-SH cells, whereas SB216763, a specific inhibitor of GSK-3β [140], increased EGFP-HDQ74 aggregates but reduced cell death. Furthermore, *myo*-inositol significantly reversed the protective effect of LiCl on EGFP-HDQ74-induced aggregation and cell death. Similar to LiCl, carbamazepine and valproic acid, both of which deplete intracellular inositol levels [141], reduced EGFP-HDQ74 aggregates and cell death. Accumulated data indicate that depletion

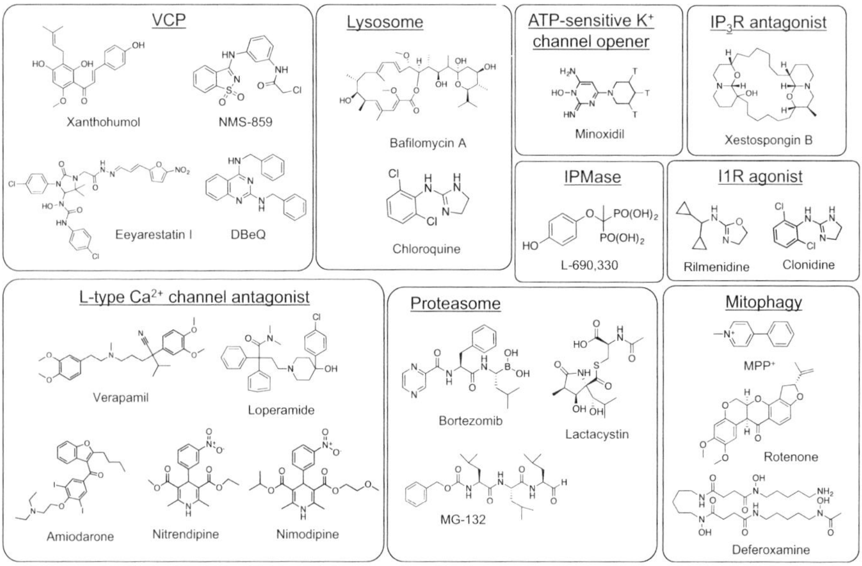

Fig. 4.8 Structures of autophagy modulators (mammalian target of rapamycin (mTOR)-independent)

of intracellular inositol induces autophagy and subsequent clearance of aggregate proteins [136].

In addition, Rubinsztein and colleagues also screened FDA-approved drugs to identify compounds that induce mTOR-independent autophagy and protect against EGFP-HDQ74 and mutant αSYN aggregates and against neuronal cell death [142]. As a result, L-type Ca^{2+} channel antagonists (verapamil, loperamide, nimodipine, nitrendipine, and amiodarone), an ATP-sensitive K^+ channel opener (minoxidil), and imidazoline-1 receptor (I1R) agonists (clonidine and rilmenidine) (Fig. 4.8) were discovered. Among them, clonidine and rilmenidine reduced cAMP and IP_3 levels through the cAMP–Epac–Rap2B–PLC–IP_3 pathway. Since the decrease in intracellular inositol was reported to induce autophagy as mentioned above [136], the decrease of IP_3 levels by treatment with clonidine or rilmenidine may induce autophagy and protect against aggregations of HDQ74 and mutant αSYN and against neuronal cell death. On the other hand, IP_3 is widely known to bind to the IP_3 receptor (IP_3R) located on the ER and to thereby facilitate Ca^{2+} release from the organelle. Likewise, L-type Ca^{2+} channel antagonists are known to decrease intracytosolic Ca^{2+}. Minoxidil was reported to decrease whole-cell L-type Ca^{2+} channel currents in a dose-dependent manner [143]. Thus, reductions in intracytosolic Ca^{2+} are common effects exerted by I1R agonists, L-type Ca^{2+} channel antagonists, and ATP-sensitive K^+ channel openers. Rubinsztein and colleagues further revealed the inhibition of calpain activity by reductions of intracytosolic Ca^{2+} induced autophagy [142]. Moreover, calpain-mediated activation of the α subunit of heterotrimeric G protein enhanced adenylate cyclase activity, which then increased cAMP levels [144]. The increased cAMP levels with calpain resulted in the inhibition of autophagy [142]. On the other hand, calpain was reported to cleave ATG5 and inhibit the formation of autophagosomes [145]. Therefore, calpain-mediated cleavage of ATG5 is involved in another possible mechanism by which activation of calpain inhibits autophagy.

Xestospongin B (Fig. 4.8), an IP_3R antagonist originally purified from the marine sponge *Xestospongia exigua* [143, 146], was reported to induce autophagy. As mentioned above, lowering intracellular IP_3 levels was found to induce autophagy by decreasing intracytosolic Ca^{2+}-mediated inhibition of calpain activity. Therefore, xestospongin B was also predicted to induce autophagy through mechanisms similar to those of previously discovered small molecules. However, Kroemer and colleagues reported that addition of *myo*-inositol suppressed autophagy induced by the IMPase inhibitor L-690,330 [136] but not by xestospongin B. On the other hand, xestospongin B-induced autophagy was suppressed in IP_3R-knockout DT40 cells. These observations indicated that IP_3R but not intracellular IP_3 played an important role in xestospongin B-induced autophagy. Moreover, xestospongin B was revealed to disrupt the interaction between Beclin 1 and the IP_3R/BCL-2 complex. Beclin 1 plays an important role in regulation of autophagy. Thus, xestospongin B disrupts IP_3R–BCL-2–Beclin 1 complexes, thereby releasing Beclin 1 from the IP_3R–BCL-2 complex and resulting in induction of autophagy [147, 148].

4.3.3.2 BCL-2 Inhibitors

Beclin 1 was originally identified as a BCL-2-interacting coiled-coil molecule [149], but it was later found to be a mammalian ortholog of yeast ATG6 with a role in mediating autophagy [150]. Beclin 1 contains three identified structural domains: a BH3 domain at the *N*-terminus, a central coiled-coil domain, and an evolutionally conserved domain. Beclin 1 interacts with several cofactors in the regulation of the lipid kinase Vps34, such as ATG14L, ultraviolet-radiation resistance-associated gene, AMBRA1, IP_3R, PTEN-induced putative kinase 1 (PINK1), and survivin. It also promotes formation of Beclin 1-Vps34–Vps15 core complexes, thereby inducing autophagy [151]. In contrast, the BH3 domain of Beclin 1 is bound to BCL-2 or BCL-xL, suppressing the function of Beclin 1 in the modulation of autophagy. However, this interaction can be disrupted by phosphorylations of BCL-2 and Beclin 1 or by ubiquitination of Beclin 1, resulting in stimulation of autophagy. Therefore, small molecules that inhibit the interaction between Beclin 1 and BCL-2/BCL-xL were considered to induce autophagy. Indeed, ABT-737 (Fig. 4.2), a BCL-2/BCL-xL antagonist that mimics the function of endogenous BH3-only proteins, was reported to induce autophagy [152]. Gossypol (Fig. 4.2), a pan-BCL-2 inhibitor, was also reported to induce autophagy [153]. Another pan-BCL-2 inhibitor, obatoclax (GX15-070) (Fig. 4.2), induces ATG7-dependent autophagy, but loss of Beclin 1 expression does not prevent obatoclax-induced autophagy [154]. Further studies are needed to elucidate the involvement of BCL-2/BCL-xL-Beclin 1 complexes in the regulation of autophagy.

4.3.3.3 Lysosome Inhibitors

It is becoming clear that proteases play an important role in regulating various steps in autophagy. Lysosomal acidic proteases known as cathepsins participate in the execution of autophagic processes. Therefore, efficient degradation of autophagic substrates requires cathepsins. Inhibition of the activity of cathepsins using specific protease inhibitors such as pepstatin A and E64d (which are inhibitors of aspartic proteinases and cysteine proteinases, respectively) prevents degradation of autophagic substrates. Conversely, activation of other types of proteases such as calpains and/or caspases inhibits autophagy. As mentioned above, calpain cleaves ATG5, resulting in inhibition of autophagosome formation [145, 155]. Besides calpain, caspase has been found to cleave Beclin 1 during apoptosis and to delay autophagosome formation in the promotion of the apoptotic response [156, 157]. Moreover, cathepsin A inhibits chaperon-mediated autophagy by cleaving the receptor, lysosome-associated membrane protein-2A (LAMP-2A). Thus, cross-talk between activities of different classes of proteases regulates the autophagic machinery (for a detailed review, see ref. [158]). Controlling the activity of proteases involved in autophagy by developing drugs that target this process must be considered as a future strategy.

Similarly, inhibitors of lysosomal compartment acidification, including inhibitors of the vacuolar proton pump or lysosomotropic agents, can regulate the autophagic processes. Bafilomycin A_1 (Fig. 4.8), a macrolide antibiotic isolated from *Streptomyces* sp. [159], is a potent and specific inhibitor of vacuolar-type H^+-ATPase (V-ATPase) [160, 161]. V-ATPases are localized in organelles of the central vacuolar system of coated vesicles, endosomes, chromaffin granules, and Golgi apparatus, and play an important role by maintaining the acidic environment of these compartments [162]. Bafilomycin A_1 was reported to inhibit fusion between autophagosomes and lysosomes in yeast [163] and in mammalian cells [164]. Chloroquine (Fig. 4.8) also prevents fusion of lysosomes with autophagosomes by raising the lysosomal pH [165].

4.3.3.4 Proteasome Inhibitors

Cellular mechanisms for the degradation of misfolded proteins include the ubiquitin-proteasome system (UPS) and autophagy. UPS is the main non-lysosomal degradation pathway for ubiquitinated proteins. The UPS and the autophagy pathways have long been viewed as complementary degradation systems with no point of intersection. However, many investigators have recently reported that proteasome inhibitors such as MG132, bortezomib, and lactacystin (Fig. 4.8) induce autophagy through ER stress. Misfolded proteins in the ER are degraded by a process known as ER-associated degradation (ERAD) [166]. Studies in mammalian cells and in yeast have led to the formulation of a general model for ERAD in which cytoplasmic proteasomes degrade substrates after retrotranslocation across the ER membrane through a process that requires the Sec61 translocon [167]. Therefore, proteasome inhibitors obstruct ERAD and lead to the accumulation of misfolded proteins in the ER and subsequent induction of ER stress. Furthermore, proteasome inhibitor–induced autophagy is impaired in inositol-requiring enzyme 1 (IRE1)–deficient MEF cells, one of the three ER-stress sensors [168]. Likewise, IRE1 is required for autophagy induced by ER-stress-inducing agents such as thapsigargin and tunicamycin [147]. Therefore, autophagy is likely activated in response to ER stress caused by the accumulation of misfolded proteins during proteasome inhibition. On the other hand, proteasome inhibition was reported to induce autophagy through aggresome formation [169]. Proteasome inhibition results in the accumulation of polyubiquitinated proteins, which are prone to aggregation and formation of structures known as "aggresomes." Aggresomes are ultimately transported to the lysosome for degradation through engulfment by autophagosomes. Components of the aggresome machinery are regulated through the coordinated activity of microtubules, motor dyneins, and histone deacetylase 6 (HDAC6) [170]. Clearance of aggresomal proteins depends on the activation of autophagy in which autophagosomes surround aggresomes; those later fuse with lysosomes and result in degradation of the aggresomal proteins. Thus, aggresome degradation and autophagy are directly linked. Alternatively, autophagy acts as a compensatory degradation system when the UPS is impaired.

4.3.3.5 VCP Inhibitors

Valosin-containing protein (VCP), also known as p97, is one of the well-characterized type II AAA ATPase (ATPases associated with diverse cellular activities). VCP plays critical roles in diverse cellular processes, including ERAD via the UPS [171, 172], cell-cycle regulation [173], and DNA repair [174]. Recently, it was reported that VCP is essential for autophagosome–lysosome fusion (autolysosome formation) in human cell lines [175, 176]. VCP consists of a substrate, a cofactor binding N domain, and two AAA ATPase domains, D1 and D2, and forms a hexameric double-ring structure [177, 178]. Small molecules that bind to each domain of VCP have been reported. We recently reported that xanthohumol (Fig. 4.8), a principal prenylated chalcone from the female inflorescence of the hop plant, binds to the N domain of VCP [179]. Eeyarestatin I (Fig. 4.8) was reported to bind to the D1 domain of VCP without affecting ATPase activity [180]. N^2,N^4-Dibenzylquinazoline-2,4-diamine (DBeQ) (Fig. 4.8) was identified as a selective, potent, reversible, and ATP-competitive VCP inhibitor [181]. NMS-859 (Fig. 4.8), a covalent inhibitor that modifies VCP Cys522, induces local disruption of the ATP binding site. This compound activates the unfolded-protein response, inducing accumulation of polyubiquitinated proteins, interfering with autophagy, and inducing cancer cell death. However, NMS-859 does not induce the formation of aggresomes [182]. The precise molecular mechanism by which VCP regulates the formation of autolysosomes is still ambiguous. Therefore, VCP inhibitors may be useful bioprobes that aid elucidation of the regulatory mechanism of VCP in autolysosome formation.

4.3.3.6 Mitophagy Inducers

To maintain the homeostasis of organelles, cells have multiple quality control systems, including (1) organellar chaperones and proteases that respectively promote proper protein folding and proteolytic removal of terminally damaged proteins; (2) retrotranslocation of misfolded proteins from the organelle to the cytosol for proteasomal degradation; and (3) autophagic degradation of dysfunctional organelles. Selective degradation of aged and damaged organelles is one of the important homeostatic functions of autophagy. A well-studied type of selective autophagy is mitophagy, which mediates the selective removal of mitochondria. In particular, the importance of mitophagy in PD has been widely studied. In 1984, Langston and coworkers found that injection of the neurotoxin MPTP in squirrel monkeys damaged the nigrostriatal dopaminergic pathway, as seen in PD [183]. MPTP reproduces most of the biochemical and pathological hallmarks of PD. MPTP itself is not a toxic compound. However, after MPTP crosses the blood–brain barrier, glial cells metabolize it into the toxic MPP^+ (Fig. 4.8) form through the enzyme monoamine oxidase B (MAO-B). It has been suggested that MPP^+ inhibits mitochondrial respiration complex 1 [184]. In addition, many studies have employed the mitochondrial complex I inhibitor rotenone (Fig. 4.8) to generate an experimental animal model of PD that mimics and elicits PD-like symptoms, such as motor and cognitive decline. On the other hand, when the mitochondrial uncoupler CCCP (carbonyl cyanide *m*-chlorophenylhydrazone) damages mitochondria and induces the loss of mitochondrial membrane potential, PINK1 accumulates and recruits the E3 ubiquitin ligase PARKIN specifically to damaged mitochondria [185]. Then, PARKIN ubiquitinates mitochondrial proteins, including mitofusins, which cause engulfment of mitochondria by a double-layered isolation membrane during mitophagy [186]. Together, mitochondria damaging agents induce PINK1/PARKIN-dependent mitophagy and are widely used as cellular model systems for PD. On the other hand, iron chelators, such as deferiprone (DFP) and deferoxamine (DFO) (Fig. 4.8), were found to induce PINK1/PARKIN-independent mitophagy [187]. Although iron chelators are known to generate ROS, iron chelator–induced mitophagy was not suppressed by ROS scavengers. Thus, iron chelator–induced mitophagy was suggested to occur independent of ROS signaling [187]. Moreover, accumulation of iron was observed in the substantia nigra of patients with PD [188]. These observations raise the possibility that iron-regulated mitophagy might be involved in PD. However, the precious molecular mechanism by which iron levels modulate mitophagy is still unknown. Further screening of mitophagy inducers and inhibitors would provide fruitful information about the mechanism of mitophagy.

References

1. White E (1993) Regulation of apoptosis by the transforming genes of the DNA tumor virus adenovirus. Proc Soc Exp Biol Med 204(1):30–39
2. Yonish-Rouach E, Resnitzky D, Lotem J, Sachs L, Kimchi A, Oren M (1991) Wild-type p53 induces apoptosis of myeloid leukaemic cells that is inhibited by interleukin-6. Nature 352 (6333):345–347. doi:10.1038/352345a0
3. Evan GI, Wyllie AH, Gilbert CS, Littlewood TD, Land H, Brooks M, Waters CM, Penn LZ, Hancock DC (1992) Induction of apoptosis in fibroblasts by c-myc protein. Cell 69 (1):119–128
4. Rao L, Debbas M, Sabbatini P, Hockenbery D, Korsmeyer S, White E (1992) The adenovirus E1A proteins induce apoptosis, which is inhibited by the E1B 19-kDa and Bcl-2 proteins. Proc Natl Acad Sci U S A 89(16):7742–7746
5. Wagner AJ, Kokontis JM, Hay N (1994) Myc-mediated apoptosis requires wild-type p53 in a manner independent of cell cycle arrest and the ability of p53 to induce p21waf1/cip1. Genes Dev 8(23):2817–2830
6. Lin Y, Benchimol S (1995) Cytokines inhibit p53-mediated apoptosis but not p53-mediated G1 arrest. Mol Cell Biol 15(11):6045–6054
7. Sakamuro D, Eviner V, Elliott KJ, Showe L, White E, Prendergast GC (1995) c-Myc induces apoptosis in epithelial cells by both p53-dependent and p53-independent mechanisms. Oncogene 11(11):2411–2418
8. Clarke AR, Purdie CA, Harrison DJ, Morris RG, Bird CC, Hooper ML, Wyllie AH (1993) Thymocyte apoptosis induced by p53-dependent and independent pathways. Nature 362 (6423):849–852. doi:10.1038/362849a0
9. Lowe SW, Schmitt EM, Smith SW, Osborne BA, Jacks T (1993) p53 is required for radiation-induced apoptosis in mouse thymocytes. Nature 362(6423):847–849. doi:10.1038/362847a0
10. Han Z, Chatterjee D, He DM, Early J, Pantazis P, Wyche JH, Hendrickson EA (1995) Evidence for a G2 checkpoint in p53-independent apoptosis induction by X-irradiation. Mol Cell Biol 15(11):5849–5857
11. Lowe SW, Ruley HE, Jacks T, Housman DE (1993) p53-dependent apoptosis modulates the cytotoxicity of anticancer agents. Cell 74(6):957–967
12. Shawver LK, Slamon D, Ullrich A (2002) Smart drugs: tyrosine kinase inhibitors in cancer therapy. Cancer Cell 1(2):117–123
13. Capdeville R, Buchdunger E, Zimmermann J, Matter A (2002) Glivec (STI571, imatinib), a rationally developed, targeted anticancer drug. Nat Rev Drug Discov 1(7):493–502. doi:10.1038/nrd839
14. Burris HA 3rd (2004) Dual kinase inhibition in the treatment of breast cancer: initial experience with the EGFR/ErbB-2 inhibitor lapatinib. Oncologist 9(Suppl 3):10–15
15. Weinstein IB (2002) Cancer. Addiction to oncogenes—the Achilles heal of cancer. Science 297(5578):63–64. doi:10.1126/science.1073096
16. Raj L, Ide T, Gurkar AU, Foley M, Schenone M, Li X, Tolliday NJ, Golub TR, Carr SA, Shamji AF, Stern AM, Mandinova A, Schreiber SL, Lee SW (2011) Selective killing of cancer cells by a small molecule targeting the stress response to ROS. Nature 475 (7355):231–234. doi:10.1038/nature10167
17. Sung JY, Kim J, Paik SR, Park JH, Ahn YS, Chung KC (2001) Induction of neuronal cell death by Rab5A-dependent endocytosis of alpha-synuclein. J Biol Chem 276 (29):27441–27448. doi:10.1074/jbc.M101318200

18. Lee HJ, Suk JE, Patrick C, Bae EJ, Cho JH, Rho S, Hwang D, Masliah E, Lee SJ (2010) Direct transfer of alpha-synuclein from neuron to astroglia causes inflammatory responses in synucleinopathies. J Biol Chem 285(12):9262–9272. doi:10.1074/jbc.M109.081125
19. Volpicelli-Daley LA, Luk KC, Patel TP, Tanik SA, Riddle DM, Stieber A, Meaney DF, Trojanowski JQ, Lee VM (2011) Exogenous alpha-synuclein fibrils induce Lewy body pathology leading to synaptic dysfunction and neuron death. Neuron 72(1):57–71. doi:10.1016/j.neuron.2011.08.033
20. Rubinsztein DC, Gestwicki JE, Murphy LO, Klionsky DJ (2007) Potential therapeutic applications of autophagy. Nat Rev Drug Discov 6(4):304–312. doi:10.1038/nrd2272
21. Rubinsztein DC (2002) Lessons from animal models of Huntington's disease. Trends Genet 18(4):202–209
22. Mattson MP (2004) Pathways towards and away from Alzheimer's disease. Nature 430 (7000):631–639. doi:10.1038/nature02621
23. Lee MS, Kwon YT, Li M, Peng J, Friedlander RM, Tsai LH (2000) Neurotoxicity induces cleavage of p35 to p25 by calpain. Nature 405(6784):360–364. doi:10.1038/35012636
24. Spillantini MG, Bird TD, Ghetti B (1998) Frontotemporal dementia and Parkinsonism linked to chromosome 17: a new group of tauopathies. Brain Pathol 8(2):387–402
25. Ohsumi Y (2001) Molecular dissection of autophagy: two ubiquitin-like systems. Nat Rev Mol Cell Biol 2(3):211–216. doi:10.1038/35056522
26. Yoshimori T (2004) Autophagy: a regulated bulk degradation process inside cells. Biochem Biophys Res Commun 313(2):453–458
27. Mizushima N, Komatsu M (2011) Autophagy: renovation of cells and tissues. Cell 147 (4):728–741. doi:10.1016/j.cell.2011.10.026
28. Oltersdorf T, Elmore SW, Shoemaker AR, Armstrong RC, Augeri DJ, Belli BA, Bruncko M, Deckwerth TL, Dinges J, Hajduk PJ, Joseph MK, Kitada S, Korsmeyer SJ, Kunzer AR, Letai A, Li C, Mitten MJ, Nettesheim DG, Ng S, Nimmer PM, O'Connor JM, Oleksijew A, Petros AM, Reed JC, Shen W, Tahir SK, Thompson CB, Tomaselli KJ, Wang B, Wendt MD, Zhang H, Fesik SW, Rosenberg SH (2005) An inhibitor of Bcl-2 family proteins induces regression of solid tumours. Nature 435(7042):677–681. doi:10.1038/nature03579
29. Adams R, Geissman TA, Edwards JD (1960) Gossypol, a pigment of cottonseed. Chem Rev 60:555–574
30. Kitada S, Leone M, Sareth S, Zhai D, Reed JC, Pellecchia M (2003) Discovery, characterization, and structure–activity relationships studies of proapoptotic polyphenols targeting B-cell lymphocyte/leukemia-2 proteins. J Med Chem 46(20):4259–4264. doi:10.1021/jm030190z
31. Nguyen M, Marcellus RC, Roulston A, Watson M, Serfass L, Murthy Madiraju SR, Goulet D, Viallet J, Belec L, Billot X, Acoca S, Purisima E, Wiegmans A, Cluse L, Johnstone RW, Beauparlant P, Shore GC (2007) Small molecule obatoclax (GX15-070) antagonizes MCL-1 and overcomes MCL-1-mediated resistance to apoptosis. Proc Natl Acad Sci U S A 104 (49):19512–19517. doi:10.1073/pnas.0709443104
32. Hughes CC, Prieto-Davo A, Jensen PR, Fenical W (2008) The marinopyrroles, antibiotics of an unprecedented structure class from a marine *Streptomyces* sp. Org Lett 10(4):629–631. doi:10.1021/ol702952n
33. Doi K, Li R, Sung SS, Wu H, Liu Y, Manieri W, Krishnegowda G, Awwad A, Dewey A, Liu X, Amin S, Cheng C, Qin Y, Schonbrunn E, Daughdrill G, Loughran TP Jr, Sebti S, Wang HG (2012) Discovery of marinopyrrole A (maritoclax) as a selective Mcl-1 antagonist that overcomes ABT-737 resistance by binding to and targeting Mcl-1 for proteasomal degradation. J Biol Chem 287(13):10224–10235. doi:10.1074/jbc.M111.334532
34. Cohen NA, Stewart ML, Gavathiotis E, Tepper JL, Bruekner SR, Koss B, Opferman JT, Walensky LD (2012) A competitive stapled peptide screen identifies a selective small molecule that overcomes MCL-1-dependent leukemia cell survival. Chem Biol 19 (9):1175–1186. doi:10.1016/j.chembiol.2012.07.018

35. Johnstone RW, Ruefli AA, Lowe SW (2002) Apoptosis: a link between cancer genetics and chemotherapy. Cell 108(2):153–164
36. Nakahara T, Kita A, Yamanaka K, Mori M, Amino N, Takeuchi M, Tominaga F, Hatakeyama S, Kinoyama I, Matsuhisa A, Kudoh M, Sasamata M (2007) YM155, a novel small-molecule survivin suppressant, induces regression of established human hormone-refractory prostate tumor xenografts. Cancer Res 67(17):8014–8021. doi:10.1158/0008-5472.CAN-07-1343
37. Kawamura T, Matsubara K, Otaka H, Tashiro E, Shindo K, Yanagita RC, Irie K, Imoto M (2011) Generation of 'Unnatural Natural Product' library and identification of a small molecule inhibitor of XIAP. Bioorg Med Chem 19(14):4377–4385. doi:10.1016/j.bmc.2011.05.009
38. Kroemer G, Galluzzi L, Brenner C (2007) Mitochondrial membrane permeabilization in cell death. Physiol Rev 87(1):99–163. doi:10.1152/physrev.00013.2006
39. Machida K, Hayashi Y, Osada H (2002) A novel adenine nucleotide translocase inhibitor, MT-21, induces cytochrome *c* release by a mitochondrial permeability transition-independent mechanism. J Biol Chem 277(34):31243–31248. doi:10.1074/jbc.M204564200
40. Watabe M, Machida K, Osada H (2000) MT-21 is a synthetic apoptosis inducer that directly induces cytochrome *c* release from mitochondria. Cancer Res 60(18):5214–5222
41. Marchetti P, Zamzami N, Joseph B, Schraen-Maschke S, Mereau-Richard C, Costantini P, Metivier D, Susin SA, Kroemer G, Formstecher P (1999) The novel retinoid 6-[3-(1-adamantyl)-4-hydroxyphenyl]-2-naphtalene carboxylic acid can trigger apoptosis through a mitochondrial pathway independent of the nucleus. Cancer Res 59(24):6257–6266
42. Belzacq AS, El Hamel C, Vieira HL, Cohen I, Haouzi D, Metivier D, Marchetti P, Brenner C, Kroemer G (2001) Adenine nucleotide translocator mediates the mitochondrial membrane permeabilization induced by lonidamine, arsenite and CD437. Oncogene 20(52):7579–7587. doi:10.1038/sj.onc.1204953
43. Hartwell LH, Szankasi P, Roberts CJ, Murray AW, Friend SH (1997) Integrating genetic approaches into the discovery of anticancer drugs. Science 278(5340):1064–1068
44. Guarente L (1993) Synthetic enhancement in gene interaction: a genetic tool come of age. Trends Genet 9(10):362–366
45. Hoeijmakers JH (2001) Genome maintenance mechanisms for preventing cancer. Nature 411 (6835):366–374. doi:10.1038/35077232
46. Tutt A, Ashworth A (2002) The relationship between the roles of BRCA genes in DNA repair and cancer predisposition. Trends Mol Med 8(12):571–576
47. Farmer H, McCabe N, Lord CJ, Tutt AN, Johnson DA, Richardson TB, Santarosa M, Dillon KJ, Hickson I, Knights C, Martin NM, Jackson SP, Smith GC, Ashworth A (2005) Targeting the DNA repair defect in BRCA mutant cells as a therapeutic strategy. Nature 434 (7035):917–921. doi:10.1038/nature03445
48. Bryant HE, Schultz N, Thomas HD, Parker KM, Flower D, Lopez E, Kyle S, Meuth M, Curtin NJ, Helleday T (2005) Specific killing of BRCA2-deficient tumours with inhibitors of poly (ADP-ribose) polymerase. Nature 434(7035):913–917. doi:10.1038/nature03443
49. Wray GM, Hinds CJ, Thiemermann C (1998) Effects of inhibitors of poly(ADP-ribose) synthetase activity on hypotension and multiple organ dysfunction caused by endotoxin. Shock 10(1):13–19
50. Bowman KJ, Newell DR, Calvert AH, Curtin NJ (2001) Differential effects of the poly (ADP-ribose) polymerase (PARP) inhibitor NU1025 on topoisomerase I and II inhibitor cytotoxicity in L1210 cells in vitro. Br J Cancer 84(1):106–112. doi:10.1054/bjoc.2000.1555
51. Skalitzky DJ, Marakovits JT, Maegley KA, Ekker A, Yu XH, Hostomsky Z, Webber SE, Eastman BW, Almassy R, Li J, Curtin NJ, Newell DR, Calvert AH, Griffin RJ, Golding BT (2003) Tricyclic benzimidazoles as potent poly(ADP-ribose) polymerase-1 inhibitors. J Med Chem 46(2):210–213. doi:10.1021/jm0255769
52. Wooster R, Weber BL (2003) Breast and ovarian cancer. N Engl J Med 348(23):2339–2347. doi:10.1056/NEJMra012284

53. Menear KA, Adcock C, Boulter R, Cockcroft XL, Copsey L, Cranston A, Dillon KJ, Drzewiecki J, Garman S, Gomez S, Javaid H, Kerrigan F, Knights C, Lau A, Loh VM Jr, Matthews IT, Moore S, O'Connor MJ, Smith GC, Martin NM (2008) 4-[3-(4-Cyclopropanecarbonylpiperazine-1-carbonyl)-4-fluorobenzyl]-2H-phthalazin-1-one: a novel bioavailable inhibitor of poly(ADP-ribose) polymerase-1. J Med Chem 51 (20):6581–6591. doi:10.1021/jm8001263
54. Torrance CJ, Agrawal V, Vogelstein B, Kinzler KW (2001) Use of isogenic human cancer cells for high-throughput screening and drug discovery. Nat Biotechnol 19(10):940–945. doi:10.1038/nbt1001-940
55. Dolma S, Lessnick SL, Hahn WC, Stockwell BR (2003) Identification of genotype-selective antitumor agents using synthetic lethal chemical screening in engineered human tumor cells. Cancer Cell 3(3):285–296
56. Yagoda N, von Rechenberg M, Zaganjor E, Bauer AJ, Yang WS, Fridman DJ, Wolpaw AJ, Smukste I, Peltier JM, Boniface JJ, Smith R, Lessnick SL, Sahasrabudhe S, Stockwell BR (2007) RAS-RAF-MEK-dependent oxidative cell death involving voltage-dependent anion channels. Nature 447(7146):864–868. doi:10.1038/nature05859
57. Guo W, Wu S, Liu J, Fang B (2008) Identification of a small molecule with synthetic lethality for K-ras and protein kinase C iota. Cancer Res 68(18):7403–7408. doi:10.1158/0008-5472.CAN-08-1449
58. Shaw AT, Winslow MM, Magendantz M, Ouyang C, Dowdle J, Subramanian A, Lewis TA, Maglathin RL, Tolliday N, Jacks T (2011) Selective killing of K-ras mutant cancer cells by small molecule inducers of oxidative stress. Proc Natl Acad Sci U S A 108(21):8773–8778. doi:10.1073/pnas.1105941108
59. Furuta Y, Yoshikawa A (1976) Reversible adrenergic alpha-receptor blocking action of 2,4′-dimethyl-3-piperidino-propiophenone (tolperisone). Jpn J Pharmacol 26(5):543–550
60. Sakitama K, Ozawa Y, Aoto N, Tomita H, Ishikawa M (1997) Effects of a new centrally acting muscle relaxant, NK433 (lanperisone hydrochloride) on spinal reflexes. Eur J Pharmacol 337(2-3):175–187
61. Subramanian A, Tamayo P, Mootha VK, Mukherjee S, Ebert BL, Gillette MA, Paulovich A, Pomeroy SL, Golub TR, Lander ES, Mesirov JP (2005) Gene set enrichment analysis: a knowledge-based approach for interpreting genome-wide expression profiles. Proc Natl Acad Sci U S A 102(43):15545–15550. doi:10.1073/pnas.0506580102
62. Lamb J, Crawford ED, Peck D, Modell JW, Blat IC, Wrobel MJ, Lerner J, Brunet JP, Subramanian A, Ross KN, Reich M, Hieronymus H, Wei G, Armstrong SA, Haggarty SJ, Clemons PA, Wei R, Carr SA, Lander ES, Golub TR (2006) The Connectivity Map: using gene-expression signatures to connect small molecules, genes, and disease. Science 313 (5795):1929–1935. doi:10.1126/science.1132939
63. Lane HA, Beuvink I, Motoyama AB, Daly JM, Neve RM, Hynes NE (2000) ErbB2 potentiates breast tumor proliferation through modulation of p27(Kip1)-Cdk2 complex formation: receptor overexpression does not determine growth dependency. Mol Cell Biol 20 (9):3210–3223
64. Neve RM, Lane HA, Hynes NE (2001) The role of overexpressed HER2 in transformation. Ann Oncol 12(Suppl 1):S9–13
65. Fantin VR, Berardi MJ, Scorrano L, Korsmeyer SJ, Leder P (2002) A novel mitochondriotoxic small molecule that selectively inhibits tumor cell growth. Cancer Cell 2(1):29–42
66. Bateman RL, Rauh D, Tavshanjian B, Shokat KM (2008) Human carbonyl reductase 1 is an *S*-nitrosoglutathione reductase. J Biol Chem 283(51):35756–35762. doi:10.1074/jbc.M807125200
67. Ralat LA, Manevich Y, Fisher AB, Colman RF (2006) Direct evidence for the formation of a complex between 1-cysteine peroxiredoxin and glutathione *S*-transferase pi with activity changes in both enzymes. Biochemistry 45(2):360–372. doi:10.1021/bi0520737
68. Li R, Murray AW (1991) Feedback control of mitosis in budding yeast. Cell 66(3):519–531

69. Sotillo R, Schvartzman JM, Socci ND, Benezra R (2010) Mad2-induced chromosome instability leads to lung tumour relapse after oncogene withdrawal. Nature 464 (7287):436–440. doi:10.1038/nature08803
70. Bian Y, Kitagawa R, Bansal PK, Fujii Y, Stepanov A, Kitagawa K (2014) Synthetic genetic array screen identifies PP2A as a therapeutic target in Mad2-overexpressing tumors. Proc Natl Acad Sci U S A. doi:10.1073/pnas.1315588111
71. Tarleton M, Gilbert J, Sakoff JA, McCluskey A (2012) Synthesis and anticancer activity of a series of norcantharidin analogues. Eur J Med Chem 54:573–581. doi:10.1016/j.ejmech.2012.06.010
72. Yap TA, Yan L, Patnaik A, Fearen I, Olmos D, Papadopoulos K, Baird RD, Delgado L, Taylor A, Lupinacci L, Riisnaes R, Pope LL, Heaton SP, Thomas G, Garrett MD, Sullivan DM, de Bono JS, Tolcher AW (2011) First-in-man clinical trial of the oral pan-AKT inhibitor MK-2206 in patients with advanced solid tumors. J Clin Oncol 29(35):4688–4695. doi:10.1200/JCO.2011.35.5263
73. Dai B, Yoo SY, Bartholomeusz G, Graham RA, Majidi M, Yan S, Meng J, Ji L, Coombes K, Minna JD, Fang B, Roth JA (2013) KEAP1-dependent synthetic lethality induced by AKT and TXNRD1 inhibitors in lung cancer. Cancer Res 73(17):5532–5543. doi:10.1158/0008-5472.CAN-13-0712
74. Chan DA, Sutphin PD, Nguyen P, Turcotte S, Lai EW, Banh A, Reynolds GE, Chi JT, Wu J, Solow-Cordero DE, Bonnet M, Flanagan JU, Bouley DM, Graves EE, Denny WA, Hay MP, Giaccia AJ (2011) Targeting GLUT1 and the Warburg effect in renal cell carcinoma by chemical synthetic lethality. Sci Transl Med 3(94):94ra70. doi:10.1126/scitranslmed.3002394
75. DeBoer C, Meulman PA, Wnuk RJ, Peterson DH (1970) Geldanamycin, a new antibiotic. J Antibiot (Tokyo) 23(9):442–447
76. Sasaki K, Rinehart KL Jr, Slomp G, Grostic MF, Olson EC (1970) Geldanamycin I structure assignment. J Am Chem Soc 92(26):7591–7593
77. Whitesell L, Mimnaugh EG, De Costa B, Myers CE, Neckers LM (1994) Inhibition of heat shock protein HSP90-pp60v-src heteroprotein complex formation by benzoquinone ansamycins: essential role for stress proteins in oncogenic transformation. Proc Natl Acad Sci U S A 91(18):8324–8328
78. Gallo KA (2006) Targeting HSP90 to halt neurodegeneration. Chem Biol 13(2):115–116. doi:10.1016/j.chembiol.2006.02.003
79. Auluck PK, Bonini NM (2002) Pharmacological prevention of Parkinson disease in *Drosophila*. Nat Med 8(11):1185–1186. doi:10.1038/nm1102-1185
80. McLean PJ, Klucken J, Shin Y, Hyman BT (2004) Geldanamycin induces Hsp70 and prevents alpha-synuclein aggregation and toxicity in vitro. Biochem Biophys Res Commun 321(3):665–669. doi:10.1016/j.bbrc.2004.07.021
81. Petrucelli L, Dickson D, Kehoe K, Taylor J, Snyder H, Grover A, De Lucia M, McGowan E, Lewis J, Prihar G, Kim J, Dillmann WH, Browne SE, Hall A, Voellmy R, Tsuboi Y, Dawson TM, Wolozin B, Hardy J, Hutton M (2004) CHIP and Hsp70 regulate tau ubiquitination, degradation and aggregation. Hum Mol Genet 13(7):703–714. doi:10.1093/hmg/ddh083
82. Waza M, Adachi H, Katsuno M, Minamiyama M, Tanaka F, Sobue G (2006) Alleviating neurodegeneration by an anticancer agent: an Hsp90 inhibitor (17-AAG). Ann N Y Acad Sci 1086:21–34. doi:10.1196/annals.1377.012
83. Cysyk RL, Parker RJ, Barchi JJ Jr, Steeg PS, Hartman NR, Strong JM (2006) Reaction of geldanamycin and C17-substituted analogues with glutathione: product identifications and pharmacological implications. Chem Res Toxicol 19(3):376–381. doi:10.1021/tx050237e
84. Kitson RR, Chang CH, Xiong R, Williams HE, Davis AL, Lewis W, Dehn DL, Siegel D, Roe SM, Prodromou C, Ross D, Moody CJ (2013) Synthesis of 19-substituted geldanamycins with altered conformations and their binding to heat shock protein Hsp90. Nat Chem 5 (4):307–314. doi:10.1038/nchem.1596

85. Feany MB, Bender WW (2000) A *Drosophila* model of Parkinson's disease. Nature 404 (6776):394–398. doi:10.1038/35006074
86. Auluck PK, Chan HY, Trojanowski JQ, Lee VM, Bonini NM (2002) Chaperone suppression of alpha-synuclein toxicity in a *Drosophila* model for Parkinson's disease. Science 295 (5556):865–868. doi:10.1126/science.1067389
87. Patury S, Miyata Y, Gestwicki JE (2009) Pharmacological targeting of the Hsp70 chaperone. Curr Top Med Chem 9(15):1337–1351
88. Sassa H, Takaishi Y, Terada H (1990) The triterpene celastrol as a very potent inhibitor of lipid peroxidation in mitochondria. Biochem Biophys Res Commun 172(2):890–897
89. Westerheide SD, Bosman JD, Mbadugha BN, Kawahara TL, Matsumoto G, Kim S, Gu W, Devlin JP, Silverman RB, Morimoto RI (2004) Celastrols as inducers of the heat shock response and cytoprotection. J Biol Chem 279(53):56053–56060. doi:10.1074/jbc. M409267200
90. Murakami M, Oketani K, Fujisaki H, Wakabayashi T, Ohgo T (1981) Antiulcer effect of geranylgeranylacetone, a new acyclic polyisoprenoid on experimentally induced gastric and duodenal ulcers in rats. Arzneimittelforschung 31(5):799–804
91. Pinder RM, Brogden RN, Sawyer PR, Speight TM, Spencer R, Avery GS (1976) Carbenoxolone: a review of its pharmacological properties and therapeutic efficacy in peptic ulcer disease. Drugs 11(4):245–307
92. Yamanaka K, Takahashi N, Ooie T, Kaneda K, Yoshimatsu H, Saikawa T (2003) Role of protein kinase C in geranylgeranylacetone-induced expression of heat-shock protein 72 and cardioprotection in the rat heart. J Mol Cell Cardiol 35(7):785–794
93. Nagayama S, Jono H, Suzaki H, Sakai K, Tsuruya E, Yamatsu I, Isohama Y, Miyata T, Kai H (2001) Carbenoxolone, a new inducer of heat shock protein 70. Life Sci 69(24):2867–2873
94. Deng YN, Shi J, Liu J, Qu QM (2013) Celastrol protects human neuroblastoma SH-SY5Y cells from rotenone-induced injury through induction of autophagy. Neurochem Int 63 (1):1–9. doi:10.1016/j.neuint.2013.04.005
95. Kilpatrick K, Novoa JA, Hancock T, Guerriero CJ, Wipf P, Brodsky JL, Segatori L (2013) Chemical induction of Hsp70 reduces alpha-synuclein aggregation in neuroglioma cells. ACS Chem Biol 8(7):1460–1468. doi:10.1021/cb400017h
96. Aronson AL (1980) Pharmacotherapeutics of the newer tetracyclines. J Am Vet Med Assoc 176(10 Spec No):1061-1068
97. Chen M, Ona VO, Li M, Ferrante RJ, Fink KB, Zhu S, Bian J, Guo L, Farrell LA, Hersch SM, Hobbs W, Vonsattel JP, Cha JH, Friedlander RM (2000) Minocycline inhibits caspase-1 and caspase-3 expression and delays mortality in a transgenic mouse model of Huntington disease. Nat Med 6(7):797–801. doi:10.1038/77528
98. Choi Y, Kim HS, Shin KY, Kim EM, Kim M, Park CH, Jeong YH, Yoo J, Lee JP, Chang KA, Kim S, Suh YH (2007) Minocycline attenuates neuronal cell death and improves cognitive impairment in Alzheimer's disease models. Neuropsychopharmacology 32(11):2393–2404. doi:10.1038/sj.npp.1301377
99. Sancho M, Herrera AE, Gortat A, Carbajo RJ, Pineda-Lucena A, Orzaez M, Perez-Paya E (2011) Minocycline inhibits cell death and decreases mutant Huntingtin aggregation by targeting Apaf-1. Hum Mol Genet 20(18):3545–3553. doi:10.1093/hmg/ddr271
100. Moresco EM, Koleske AJ (2003) Regulation of neuronal morphogenesis and synaptic function by Abl family kinases. Curr Opin Neurobiol 13(5):535–544
101. Ko HS, Lee Y, Shin JH, Karuppagounder SS, Gadad BS, Koleske AJ, Pletnikova O, Troncoso JC, Dawson VL, Dawson TM (2010) Phosphorylation by the c-Abl protein tyrosine kinase inhibits parkin's ubiquitination and protective function. Proc Natl Acad Sci U S A 107 (38):16691–16696. doi:10.1073/pnas.1006083107
102. Alvarez AR, Sandoval PC, Leal NR, Castro PU, Kosik KS (2004) Activation of the neuronal c-Abl tyrosine kinase by amyloid-beta-peptide and reactive oxygen species. Neurobiol Dis 17 (2):326–336. doi:10.1016/j.nbd.2004.06.007

103. Alvarez AR, Klein A, Castro J, Cancino GI, Amigo J, Mosqueira M, Vargas LM, Yevenes LF, Bronfman FC, Zanlungo S (2008) Imatinib therapy blocks cerebellar apoptosis and improves neurological symptoms in a mouse model of Niemann–Pick type C disease. FASEB J 22(10):3617–3627. doi:10.1096/fj.07-102715
104. Wolff NC, Richardson JA, Egorin M, Ilaria RL Jr (2003) The CNS is a sanctuary for leukemic cells in mice receiving imatinib mesylate for Bcr/Abl-induced leukemia. Blood 101 (12):5010–5013. doi:10.1182/blood-2002-10-3059
105. Hebron ML, Lonskaya I, Moussa CE (2013) Nilotinib reverses loss of dopamine neurons and improves motor behavior via autophagic degradation of alpha-synuclein in Parkinson's disease models. Hum Mol Genet 22(16):3315–3328. doi:10.1093/hmg/ddt192
106. Otomo M, Takahashi K, Miyoshi H, Osada K, Nakashima H, Yamaguchi N (2008) Some selective serotonin reuptake inhibitors inhibit dynamin I guanosine triphosphatase (GTPase). Biol Pharm Bull 31(8):1489–1495
107. Takahashi K, Miyoshi H, Otomo M, Osada K, Yamaguchi N, Nakashima H (2010) Suppression of dynamin GTPase activity by sertraline leads to inhibition of dynamin-dependent endocytosis. Biochem Biophys Res Commun 391(1):382–387. doi:10.1016/j.bbrc.2009.11.067
108. Lee HJ, Suk JE, Bae EJ, Lee JH, Paik SR, Lee SJ (2008) Assembly-dependent endocytosis and clearance of extracellular alpha-synuclein. Int J Biochem Cell Biol 40(9):1835–1849. doi:10.1016/j.biocel.2008.01.017
109. Lee SJ (2008) Origins and effects of extracellular alpha-synuclein: implications in Parkinson's disease. J Mol Neurosci 34(1):17–22. doi:10.1007/s12031-007-0012-9
110. Zhang J, Ferguson SS, Barak LS, Menard L, Caron MG (1996) Dynamin and beta-arrestin reveal distinct mechanisms for G protein-coupled receptor internalization. J Biol Chem 271 (31):18302–18305
111. Konno M, Hasegawa T, Baba T, Miura E, Sugeno N, Kikuchi A, Fiesel FC, Sasaki T, Aoki M, Itoyama Y, Takeda A (2012) Suppression of dynamin GTPase decreases alpha-synuclein uptake by neuronal and oligodendroglial cells: a potent therapeutic target for synucleinopathy. Mol Neurodegener 7:38. doi:10.1186/1750-1326-7-38
112. Klionsky DJ, Cregg JM, Dunn WA Jr, Emr SD, Sakai Y, Sandoval IV, Sibirny A, Subramani S, Thumm M, Veenhuis M, Ohsumi Y (2003) A unified nomenclature for yeast autophagy-related genes. Dev Cell 5(4):539–545
113. Noda T, Ohsumi Y (1998) Tor, a phosphatidylinositol kinase homologue, controls autophagy in yeast. J Biol Chem 273(7):3963–3966
114. Blommaart EF, Luiken JJ, Blommaart PJ, van Woerkom GM, Meijer AJ (1995) Phosphorylation of ribosomal protein S6 is inhibitory for autophagy in isolated rat hepatocytes. J Biol Chem 270(5):2320–2326
115. Kim DH, Sarbassov DD, Ali SM, King JE, Latek RR, Erdjument-Bromage H, Tempst P, Sabatini DM (2002) mTOR interacts with raptor to form a nutrient-sensitive complex that signals to the cell growth machinery. Cell 110(2):163–175
116. Hamasaki M, Furuta N, Matsuda A, Nezu A, Yamamoto A, Fujita N, Oomori H, Noda T, Haraguchi T, Hiraoka Y, Amano A, Yoshimori T (2013) Autophagosomes form at ER-mitochondria contact sites. Nature 495(7441):389–393. doi:10.1038/nature11910
117. Thoreen CC, Kang SA, Chang JW, Liu Q, Zhang J, Gao Y, Reichling LJ, Sim T, Sabatini DM, Gray NS (2009) An ATP-competitive mammalian target of rapamycin inhibitor reveals rapamycin-resistant functions of mTORC1. J Biol Chem 284(12):8023–8032. doi:10.1074/jbc.M900301200
118. Balgi AD, Fonseca BD, Donohue E, Tsang TC, Lajoie P, Proud CG, Nabi IR, Roberge M (2009) Screen for chemical modulators of autophagy reveals novel therapeutic inhibitors of mTORC1 signaling. PLoS One 4(9):e7124. doi:10.1371/journal.pone.0007124
119. Folkes AJ, Ahmadi K, Alderton WK, Alix S, Baker SJ, Box G, Chuckowree IS, Clarke PA, Depledge P, Eccles SA, Friedman LS, Hayes A, Hancox TC, Kugendradas A, Lensun L, Moore P, Olivero AG, Pang J, Patel S, Pergl-Wilson GH, Raynaud FI, Robson A, Saghir N,

Salphati L, Sohal S, Ultsch MH, Valenti M, Wallweber HJ, Wan NC, Wiesmann C, Workman P, Zhyvoloup A, Zvelebil MJ, Shuttleworth SJ (2008) The identification of 2-(1H-indazol-4-yl)-6-(4-methanesulfonyl-piperazin-1-ylmethyl)-4-morpholin-4-yl-t hieno [3,2-d]pyrimidine (GDC-0941) as a potent, selective, orally bioavailable inhibitor of class I PI3 kinase for the treatment of cancer. J Med Chem 51(18):5522–5532. doi:10.1021/jm800295d
120. Yaguchi S, Fukui Y, Koshimizu I, Yoshimi H, Matsuno T, Gouda H, Hirono S, Yamazaki K, Yamori T (2006) Antitumor activity of ZSTK474, a new phosphatidylinositol 3-kinase inhibitor. J Natl Cancer Inst 98(8):545–556. doi:10.1093/jnci/djj133
121. Kong D, Dan S, Yamazaki K, Yamori T (2010) Inhibition profiles of phosphatidylinositol 3-kinase inhibitors against PI3K superfamily and human cancer cell line panel JFCR39. Eur J Cancer 46(6):1111–1121. doi:10.1016/j.ejca.2010.01.005
122. Wang Y, Liu J, Qiu Y, Jin M, Chen X, Fan G, Wang R, Kong D (2016) ZSTK474, a specific class I phosphatidylinositol 3-kinase inhibitor, induces G1 arrest and autophagy in human breast cancer MCF-7 cells. Oncotarget 7(15):19897–19909. doi:10.18632/oncotarget.7658
123. Blommaart EF, Krause U, Schellens JP, Vreeling-Sindelarova H, Meijer AJ (1997) The phosphatidylinositol 3-kinase inhibitors wortmannin and LY294002 inhibit autophagy in isolated rat hepatocytes. Eur J Biochem 243(1-2):240–246
124. Seglen PO, Gordon PB (1982) 3-Methyladenine: specific inhibitor of autophagic/lysosomal protein degradation in isolated rat hepatocytes. Proc Natl Acad Sci U S A 79(6):1889–1892
125. Liu J, Xia H, Kim M, Xu L, Li Y, Zhang L, Cai Y, Norberg HV, Zhang T, Furuya T, Jin M, Zhu Z, Wang H, Yu J, Hao Y, Choi A, Ke H, Ma D, Yuan J (2011) Beclin1 controls the levels of p53 by regulating the deubiquitination activity of USP10 and USP13. Cell 147 (1):223–234. doi:10.1016/j.cell.2011.08.037
126. MacPherson JD, Gillespie TD, Dunkerley HA, Maurice DH, Bennett BM (2006) Inhibition of phosphodiesterase 5 selectively reverses nitrate tolerance in the venous circulation. J Pharmacol Exp Ther 317(1):188–195. doi:10.1124/jpet.105.094763
127. Hardie DG (2007) AMP-activated/SNF1 protein kinases: conserved guardians of cellular energy. Nat Rev Mol Cell Biol 8(10):774–785. doi:10.1038/nrm2249
128. Kim J, Kundu M, Viollet B, Guan KL (2011) AMPK and mTOR regulate autophagy through direct phosphorylation of Ulk1. Nat Cell Biol 13(2):132–141. doi:10.1038/ncb2152
129. Fryer LG, Parbu-Patel A, Carling D (2002) The Anti-diabetic drugs rosiglitazone and metformin stimulate AMP-activated protein kinase through distinct signaling pathways. J Biol Chem 277(28):25226–25232. doi:10.1074/jbc.M202489200
130. Hawley SA, Gadalla AE, Olsen GS, Hardie DG (2002) The antidiabetic drug metformin activates the AMP-activated protein kinase cascade via an adenine nucleotide-independent mechanism. Diabetes 51(8):2420–2425
131. Zhou G, Myers R, Li Y, Chen Y, Shen X, Fenyk-Melody J, Wu M, Ventre J, Doebber T, Fujii N, Musi N, Hirshman MF, Goodyear LJ, Moller DE (2001) Role of AMP-activated protein kinase in mechanism of metformin action. J Clin Invest 108(8):1167–1174. doi:10.1172/JCI13505
132. Buzzai M, Jones RG, Amaravadi RK, Lum JJ, DeBerardinis RJ, Zhao F, Viollet B, Thompson CB (2007) Systemic treatment with the antidiabetic drug metformin selectively impairs p53-deficient tumor cell growth. Cancer Res 67(14):6745–6752. doi:10.1158/0008-5472.CAN-06-4447
133. Vingtdeux V, Giliberto L, Zhao H, Chandakkar P, Wu Q, Simon JE, Janle EM, Lobo J, Ferruzzi MG, Davies P, Marambaud P (2010) AMP-activated protein kinase signaling activation by resveratrol modulates amyloid-beta peptide metabolism. J Biol Chem 285 (12):9100–9113. doi:10.1074/jbc.M109.060061
134. Luchsinger JA, Mayeux R (2004) Dietary factors and Alzheimer's disease. Lancet Neurol 3 (10):579–587. doi:10.1016/S1474-4422(04)00878-6

135. Carmichael J, Sugars KL, Bao YP, Rubinsztein DC (2002) Glycogen synthase kinase-3beta inhibitors prevent cellular polyglutamine toxicity caused by the Huntington's disease mutation. J Biol Chem 277(37):33791–33798. doi:10.1074/jbc.M204861200
136. Sarkar S, Floto RA, Berger Z, Imarisio S, Cordenier A, Pasco M, Cook LJ, Rubinsztein DC (2005) Lithium induces autophagy by inhibiting inositol monophosphatase. J Cell Biol 170 (7):1101–1111. doi:10.1083/jcb.200504035
137. Coyle JT, Duman RS (2003) Finding the intracellular signaling pathways affected by mood disorder treatments. Neuron 38(2):157–160
138. Gould TD, Chen G, Manji HK (2002) Mood stabilizer psychopharmacology. Clin Neurosci Res 2(3-4):193–212. doi:10.1016/S1566-2772(02)00044-0
139. Atack JR, Cook SM, Watt AP, Fletcher SR, Ragan CI (1993) In vitro and in vivo inhibition of inositol monophosphatase by the bisphosphonate L-690,330. J Neurochem 60(2):652–658
140. Coghlan MP, Culbert AA, Cross DA, Corcoran SL, Yates JW, Pearce NJ, Rausch OL, Murphy GJ, Carter PS, Roxbee Cox L, Mills D, Brown MJ, Haigh D, Ward RW, Smith DG, Murray KJ, Reith AD, Holder JC (2000) Selective small molecule inhibitors of glycogen synthase kinase-3 modulate glycogen metabolism and gene transcription. Chem Biol 7 (10):793–803
141. Williams RS, Cheng L, Mudge AW, Harwood AJ (2002) A common mechanism of action for three mood-stabilizing drugs. Nature 417(6886):292–295. doi:10.1038/417292a
142. Williams A, Sarkar S, Cuddon P, Ttofi EK, Saiki S, Siddiqi FH, Jahreiss L, Fleming A, Pask D, Goldsmith P, O'Kane CJ, Floto RA, Rubinsztein DC (2008) Novel targets for Huntington's disease in an mTOR-independent autophagy pathway. Nat Chem Biol 4 (5):295–305. doi:10.1038/nchembio.79
143. Quirion JC, Sevenet T, Husson HP, Weniger B, Debitus C (1992) Two new alkaloids from *Xestospongia* sp., a New Caledonian sponge. J Nat Prod 55(10):1505–1508
144. Sato-Kusubata K, Yajima Y, Kawashima S (2000) Persistent activation of Gs alpha through limited proteolysis by calpain. Biochem J 347(Pt 3):733–740
145. Yousefi S, Perozzo R, Schmid I, Ziemiecki A, Schaffner T, Scapozza L, Brunner T, Simon HU (2006) Calpain-mediated cleavage of Atg5 switches autophagy to apoptosis. Nat Cell Biol 8(10):1124–1132. doi:10.1038/ncb1482
146. Gafni J, Munsch JA, Lam TH, Catlin MC, Costa LG, Molinski TF, Pessah IN (1997) Xestospongins: potent membrane permeable blockers of the inositol 1,4,5-trisphosphate receptor. Neuron 19(3):723–733
147. Criollo A, Vicencio JM, Tasdemir E, Maiuri MC, Lavandero S, Kroemer G (2007) The inositol trisphosphate receptor in the control of autophagy. Autophagy 3(4):350–353
148. Vicencio JM, Ortiz C, Criollo A, Jones AW, Kepp O, Galluzzi L, Joza N, Vitale I, Morselli E, Tailler M, Castedo M, Maiuri MC, Molgo J, Szabadkai G, Lavandero S, Kroemer G (2009) The inositol 1,4,5-trisphosphate receptor regulates autophagy through its interaction with Beclin 1. Cell Death Differ 16(7):1006–1017. doi:10.1038/cdd.2009.34
149. Liang XH, Kleeman LK, Jiang HH, Gordon G, Goldman JE, Berry G, Herman B, Levine B (1998) Protection against fatal Sindbis virus encephalitis by beclin, a novel Bcl-2-interacting protein. J Virol 72(11):8586–8596
150. Liang XH, Jackson S, Seaman M, Brown K, Kempkes B, Hibshoosh H, Levine B (1999) Induction of autophagy and inhibition of tumorigenesis by beclin 1. Nature 402 (6762):672–676. doi:10.1038/45257
151. He C, Levine B (2010) The Beclin 1 interactome. Curr Opin Cell Biol 22(2):140–149. doi:10.1016/j.ceb.2010.01.001
152. Huang S, Sinicrope FA (2010) Celecoxib-induced apoptosis is enhanced by ABT-737 and by inhibition of autophagy in human colorectal cancer cells. Autophagy 6(2):256–269
153. Voss V, Senft C, Lang V, Ronellenfitsch MW, Steinbach JP, Seifert V, Kogel D (2010) The pan-Bcl-2 inhibitor (−)-gossypol triggers autophagic cell death in malignant glioma. Mol Cancer Res 8(7):1002–1016. doi:10.1158/1541-7786.MCR-09-0562

154. McCoy F, Hurwitz J, McTavish N, Paul I, Barnes C, O'Hagan B, Odrzywol K, Murray J, Longley D, McKerr G, Fennell DA (2010) Obatoclax induces Atg7-dependent autophagy independent of beclin-1 and BAX/BAK. Cell Death Dis 1:e108. doi:10.1038/cddis.2010.86
155. Xia HG, Zhang L, Chen G, Zhang T, Liu J, Jin M, Ma X, Ma D, Yuan J (2010) Control of basal autophagy by calpain1 mediated cleavage of ATG5. Autophagy 6(1):61–66
156. Djavaheri-Mergny M, Maiuri MC, Kroemer G (2010) Cross talk between apoptosis and autophagy by caspase-mediated cleavage of Beclin 1. Oncogene 29(12):1717–1719. doi:10.1038/onc.2009.519
157. Luo S, Rubinsztein DC (2010) Apoptosis blocks Beclin 1-dependent autophagosome synthesis: an effect rescued by Bcl-xL. Cell Death Differ 17(2):268–277. doi:10.1038/cdd.2009.121
158. Kaminskyy V, Zhivotovsky B (2012) Proteases in autophagy. Biochim Biophys Acta 1824(1):44–50. doi:10.1016/j.bbapap.2011.05.013
159. Werner G, Hagenmaier H, Drautz H, Baumgartner A, Zahner H (1984) Metabolic products of microorganisms. 224. Bafilomycins, a new group of macrolide antibiotics. Production, isolation, chemical structure and biological activity. J Antibiot (Tokyo) 37(2):110–117
160. Bowman EJ, Siebers A, Altendorf K (1988) Bafilomycins: a class of inhibitors of membrane ATPases from microorganisms, animal cells, and plant cells. Proc Natl Acad Sci U S A 85(21):7972–7976
161. Hanada H, Moriyama Y, Maeda M, Futai M (1990) Kinetic studies of chromaffin granule H^+-ATPase and effects of bafilomycin A1. Biochem Biophys Res Commun 170(2):873–878
162. Mellman I, Fuchs R, Helenius A (1986) Acidification of the endocytic and exocytic pathways. Annu Rev Biochem 55:663–700. doi:10.1146/annurev.bi.55.070186.003311
163. Takeshige K, Baba M, Tsuboi S, Noda T, Ohsumi Y (1992) Autophagy in yeast demonstrated with proteinase-deficient mutants and conditions for its induction. J Cell Biol 119(2):301–311
164. Yamamoto A, Tagawa Y, Yoshimori T, Moriyama Y, Masaki R, Tashiro Y (1998) Bafilomycin A1 prevents maturation of autophagic vacuoles by inhibiting fusion between autophagosomes and lysosomes in rat hepatoma cell line, H-4-II-E cells. Cell Struct Funct 23(1):33–42
165. Suzuki T, Nakagawa M, Yoshikawa A, Sasagawa N, Yoshimori T, Ohsumi Y, Nishino I, Ishiura S, Nonaka I (2002) The first molecular evidence that autophagy relates rimmed vacuole formation in chloroquine myopathy. J Biochem 131(5):647–651
166. Brodsky JL, McCracken AA (1999) ER protein quality control and proteasome-mediated protein degradation. Semin Cell Dev Biol 10(5):507–513. doi:10.1006/scdb.1999.0321
167. Kopito RR (1997) ER quality control: the cytoplasmic connection. Cell 88(4):427–430
168. Ding WX, Ni HM, Gao W, Yoshimori T, Stolz DB, Ron D, Yin XM (2007) Linking of autophagy to ubiquitin-proteasome system is important for the regulation of endoplasmic reticulum stress and cell viability. Am J Pathol 171(2):513–524. doi:10.2353/ajpath.2007.070188
169. Pandey UB, Nie Z, Batlevi Y, McCray BA, Ritson GP, Nedelsky NB, Schwartz SL, DiProspero NA, Knight MA, Schuldiner O, Padmanabhan R, Hild M, Berry DL, Garza D, Hubbert CC, Yao TP, Baehrecke EH, Taylor JP (2007) HDAC6 rescues neurodegeneration and provides an essential link between autophagy and the UPS. Nature 447(7146):859–863. doi:10.1038/nature05853
170. Lee JY, Koga H, Kawaguchi Y, Tang W, Wong E, Gao YS, Pandey UB, Kaushik S, Tresse E, Lu J, Taylor JP, Cuervo AM, Yao TP (2010) HDAC6 controls autophagosome maturation essential for ubiquitin-selective quality-control autophagy. EMBO J 29(5):969–980. doi:10.1038/emboj.2009.405
171. Ye Y, Meyer HH, Rapoport TA (2001) The AAA ATPase Cdc48/p97 and its partners transport proteins from the ER into the cytosol. Nature 414(6864):652–656. doi:10.1038/414652a
172. Jarosch E, Taxis C, Volkwein C, Bordallo J, Finley D, Wolf DH, Sommer T (2002) Protein dislocation from the ER requires polyubiquitination and the AAA-ATPase Cdc48. Nat Cell Biol 4(2):134–139. doi:10.1038/ncb746

173. Mouysset J, Deichsel A, Moser S, Hoege C, Hyman AA, Gartner A, Hoppe T (2008) Cell cycle progression requires the CDC-48UFD-1/NPL-4 complex for efficient DNA replication. Proc Natl Acad Sci U S A 105(35):12879–12884. doi:10.1073/pnas.0805944105
174. Partridge JJ, Lopreiato JO Jr, Latterich M, Indig FE (2003) DNA damage modulates nucleolar interaction of the Werner protein with the AAA ATPase p97/VCP. Mol Biol Cell 14(10):4221–4229. doi:10.1091/mbc.E03-02-0111
175. Ju JS, Fuentealba RA, Miller SE, Jackson E, Piwnica-Worms D, Baloh RH, Weihl CC (2009) Valosin-containing protein (VCP) is required for autophagy and is disrupted in VCP disease. J Cell Biol 187(6):875–888. doi:10.1083/jcb.200908115
176. Tresse E, Salomons FA, Vesa J, Bott LC, Kimonis V, Yao TP, Dantuma NP, Taylor JP (2010) VCP/p97 is essential for maturation of ubiquitin-containing autophagosomes and this function is impaired by mutations that cause IBMPFD. Autophagy 6(2):217–227
177. Rouiller I, DeLaBarre B, May AP, Weis WI, Brunger AT, Milligan RA, Wilson-Kubalek EM (2002) Conformational changes of the multifunction p97 AAA ATPase during its ATPase cycle. Nat Struct Biol 9(12):950–957. doi:10.1038/nsb872
178. DeLaBarre B, Brunger AT (2003) Complete structure of p97/valosin-containing protein reveals communication between nucleotide domains. Nat Struct Biol 10(10):856–863. doi:10.1038/nsb972
179. Sasazawa Y, Kanagaki S, Tashiro E, Nogawa T, Muroi M, Kondoh Y, Osada H, Imoto M (2012) Xanthohumol impairs autophagosome maturation through direct inhibition of valosin-containing protein. ACS Chem Biol 7(5):892–900. doi:10.1021/cb200492h
180. Wang Q, Shinkre BA, Lee JG, Weniger MA, Liu Y, Chen W, Wiestner A, Trenkle WC, Ye Y (2010) The ERAD inhibitor Eeyarestatin I is a bifunctional compound with a membrane-binding domain and a p97/VCP inhibitory group. PLoS One 5(11):e15479. doi:10.1371/journal.pone.0015479
181. Chou TF, Brown SJ, Minond D, Nordin BE, Li K, Jones AC, Chase P, Porubsky PR, Stoltz BM, Schoenen FJ, Patricelli MP, Hodder P, Rosen H, Deshaies RJ (2011) Reversible inhibitor of p97, DBeQ, impairs both ubiquitin-dependent and autophagic protein clearance pathways. Proc Natl Acad Sci U S A 108(12):4834–4839. doi:10.1073/pnas.1015312108
182. Magnaghi P, D'Alessio R, Valsasina B, Avanzi N, Rizzi S, Asa D, Gasparri F, Cozzi L, Cucchi U, Orrenius C, Polucci P, Ballinari D, Perrera C, Leone A, Cervi G, Casale E, Xiao Y, Wong C, Anderson DJ, Galvani A, Donati D, O'Brien T, Jackson PK, Isacchi A (2013) Covalent and allosteric inhibitors of the ATPase VCP/p97 induce cancer cell death. Nat Chem Biol 9(9):548–556. doi:10.1038/nchembio.1313
183. Langston JW, Forno LS, Rebert CS, Irwin I (1984) Selective nigral toxicity after systemic administration of 1-methyl-4-phenyl-1,2,5,6-tetrahydropyrine (MPTP) in the squirrel monkey. Brain Res 292(2):390–394
184. Mizuno Y, Saitoh T, Sone N (1987) Inhibition of mitochondrial alpha-ketoglutarate dehydrogenase by 1-methyl-4-phenylpyridinium ion. Biochem Biophys Res Commun 143 (3):971–976
185. Matsuda N, Sato S, Shiba K, Okatsu K, Saisho K, Gautier CA, Sou YS, Saiki S, Kawajiri S, Sato F, Kimura M, Komatsu M, Hattori N, Tanaka K (2010) PINK1 stabilized by mitochondrial depolarization recruits Parkin to damaged mitochondria and activates latent Parkin for mitophagy. J Cell Biol 189(2):211–221. doi:10.1083/jcb.200910140
186. Tanaka A, Cleland MM, Xu S, Narendra DP, Suen DF, Karbowski M, Youle RJ (2010) Proteasome and p97 mediate mitophagy and degradation of mitofusins induced by Parkin. J Cell Biol 191(7):1367–1380. doi:10.1083/jcb.201007013
187. Allen GF, Toth R, James J, Ganley IG (2013) Loss of iron triggers PINK1/Parkin-independent mitophagy. EMBO Rep 14(12):1127–1135. doi:10.1038/embor.2013.168
188. Gerlach M, Ben-Shachar D, Riederer P, Youdim MB (1994) Altered brain metabolism of iron as a cause of neurodegenerative diseases? J Neurochem 63(3):793–807

Chapter 5
Adaptive and Innate Immune Systems

Takao Kataoka

Abstract Adaptive and innate immune systems play essential roles in the recognition and elimination of microbial pathogens, such as viruses, bacteria, fungi, and parasites. In adaptive immunity, T cells and B cells induce immune responses through their respective antigen receptors, called T-cell receptors and B-cell receptors, both of which specifically recognize antigens or antigenic peptides derived from microbial pathogens. In innate immunity, germline-encoded pattern recognition receptors (PRRs) serve as sensors that detect the pathogen-associated molecular patterns found in microbial pathogens. PRRs have been classified into different families: the toll-like receptor family, nucleotide binding domain and leucine-rich repeat-containing receptor family, retinoic acid-inducible gene I-like receptor family, and C-type lectin receptor family. Upon the recognition of microbial pathogens, antigen receptors or PRRs initiate various signaling pathways, leading to the activation of different transcriptional programs that provoke immune responses in order to eliminate invaded pathogens. In this chapter, I describe small-molecule modulators that target specific steps in the intracellular signaling pathways induced by antigen receptors, PRRs, and inflammatory cytokine receptors.

Keywords Antigen receptor • C-type lectin receptor • Fas • Granzyme • Inflammasome • Nuclear factor κB • Perforin • RIG-I-like receptor • Toll-like receptor • Tumor necrosis factor receptor

Abbreviations

ASC:	apoptosis-associated speck-like protein containing a CARD
BCL10:	B-cell lymphoma 10
BCR:	B-cell receptors
BTK:	Burton's tyrosine kinase

T. Kataoka (✉)
Department of Applied Biology, Kyoto Institute of Technology, Matsugasaki, Sakyo-ku, Kyoto 606-8585, Japan
e-mail: takao.kataoka@kit.ac.jp

© Springer Japan KK 2017

H. Osada (ed.), *Bioprobes*, DOI 10.1007/978-4-431-56529-1_5

CARD: caspase recruitment domain
CARMA1: caspase recruitment domain-containing membrane-associated guanylate kinase protein 1
CLR: C-type lectin receptor
CRIDs: cytokine release inhibitory drugs
CTLs: cytotoxic T lymphocytes
DAG: diacylglycerol
DISC: death-inducing signaling complex
EGTA: ethylene glycol-bis(β-aminoethylether)-*N*,*N*,*N*′,*N*′-tetraacetic acid
ER: endoplasmic reticulum
FADD: Fas-associated death domain protein
FasL: Fas ligand
IFN: interferon
IκB: inhibitor of NF-κB
IKK: IκB kinase
IL: interleukin
IL-1R: IL-1 receptor
IP_3: inositol-1,4,5-trisphosphate
IPS-1: IFN-β promoter stimulator 1
IRAK: IL-1 receptor-associated kinase
IRFs: IFN regulatory factors
ITAMs: immunoreceptor tyrosine-based activation motifs
ITK: interleukin-2-inducible T-cell kinase
LBP: LPS binding protein
LGP2: laboratory of genetics and physiology 2
LPS: lipopolysaccharide
MACPF: membrane attack complex perforin
MALT1: mucosa-associated lymphoid tissue lymphoma-translocation gene 1
MD-2: myeloid differentiation factor 2
MDA5: melanoma differentiation-associated gene 5
MyD88: myeloid differentiation primary response gene 88
NFAT: nuclear factor of activated T cells
NF-κB: nuclear factor κB
NK: natural killer
NLR: nucleotide binding domain and leucine-rich repeat-containing receptor
PAMPs: pathogen-associated molecular patterns
PDK1: 3-phosphoinositide-dependent kinase 1
PI3K: phosphatidylinositol 3-kinase
PIP_3: phosphatidylinositol-3,4,5-trisphosphate
PKC: protein kinase C
PLC: phospholipase C
PTKs: protein tyrosine kinases
PRRs: pattern recognition receptors

PYD: pyrin domain
RIG-I: retinoic acid-inducible gene I
RIP: receptor-interacting protein
RLR: retinoic acid-inducible gene I-like receptor
ROS: reactive oxygen species
SYK: spleen tyrosine kinase
TAK1: TGF-β-activated kinase 1
TBK1: TANK binding kinase 1
TCR: T-cell receptors
TIR: toll/IL-1R
TIRAP: TIR domain-containing adaptor protein
TLR: toll-like receptor
TNF: tumor necrosis factor
TNF-R1: TNF receptor 1
TRAF: TNF receptor-associated factor
TRAM: TRIF-related adaptor molecule
TRIF: TIR domain-containing adaptor inducing IFN-β
Z-VRPR-fmk: benzyloxycarbonyl-Val-Arg-Pro-Arg-fluoromethylketone

5.1 Introduction

Two lines of the host defense system, referred to as adaptive immunity and innate immunity, play essential roles in the recognition and elimination of microbial pathogens. The adaptive immune system is composed of T cells and B cells, which induce highly specific antigen-dependent responses via antigen receptors called T-cell receptors (TCR) and B-cell receptors (BCR). TCR and BCR trigger multiple signaling pathways that induce the activation of several transcription factors, including nuclear factor κB (NF-κB) and nuclear factor of activated T cells (NFAT) [1–3]. In the innate immune system, germline-encoded pattern recognition receptors (PRRs) serve as microbial sensors by detecting relatively invariant molecular patterns found in microorganisms, referred to as pathogen-associated molecular patterns (PAMPs) [4]. Secreted PRRs bind to microbes and thereby activate the complement system and phagocytosis, whereas innate immune cells harbor transmembrane PRRs, including the toll-like receptor (TLR) family and C-type lectin receptor (CLR) family, as well as cytosolic PRRs, including the retinoic acid-inducible gene I (RIG-I)-like receptor (RLR) family and nucleotide binding domain and leucine-rich repeat-containing receptor (NLR) family [5]. Upon the recognition of PAMPs, PRRs induce the activation of transcription factors, including NF-κB and interferon (IFN) regulatory factors (IRFs) [4].

5.2 Antigen Receptors

5.2.1 Antigen Receptor Signaling

TCR and BCR are composed of structurally distinct proteins, but induce highly analogous signaling processes (Fig. 5.1). When stimulated with peptide-MHC and antigens, TCR and BCR activate a number of signaling proteins, including the Src family of protein tyrosine kinases (PTKs) [1]. The Tec family interleukin (IL)-2-inducible T-cell kinase (ITK) in T cells and Burton's tyrosine kinase (BTK) in B cells are tethered to the plasma membrane through binding to phosphatidylinositol-3,4,5-trisphosphate (PIP_3), which is converted from PIP_2 by phosphatidylinositol 3-kinase (PI3K), leading to the phosphorylation of phospholipase (PLC) γ1 and PLCγ2, respectively [3, 6, 7]. PLCγ1 and PLCγ2 hydrolyze PIP_2 to inositol-1,4,5-trisphosphate (IP_3) and diacylglycerol (DAG) [3]. Upon the binding of IP_3 to the IP_3 receptor in the endoplasmic reticulum (ER), Ca^{2+} is released from ER stores and ER-store depletion stimulates Ca^{2+} entry through plasma membrane Ca^{2+} channels [8]. This results in the activation of Ca^{2+}-dependent signaling proteins, including the protein phosphatase 2B (calcineurin) and its target, NFAT [8]. Protein kinase C (PKC) is translocated from the cytosol to the plasma membrane, and is activated by the binding of DAG [9]. Caspase recruitment domain-containing membrane-associated guanylate kinase protein 1 (CARMA1), B-cell lymphoma 10 (BCL10), and mucosa-associated lymphoid tissue lymphoma-translocation gene 1 (MALT1) are required for the activation of NF-κB downstream of TCR and BCR [1, 10]. The -BCL10-MALT1 complex is recruited to phosphorylated CARMA1, and the resultant complex recruits ubiquitin ligase tumor necrosis factor (TNF) receptor-associated factor (TRAF) 6 [1, 10]. MALT1-bound TRAF6 is considered to mediate the K63-linked polyubiquitination of BCL10 and MALT1, which promotes the recruitment of the inhibitor of the NF-κB (IκB) kinase (IKK) complex [10].

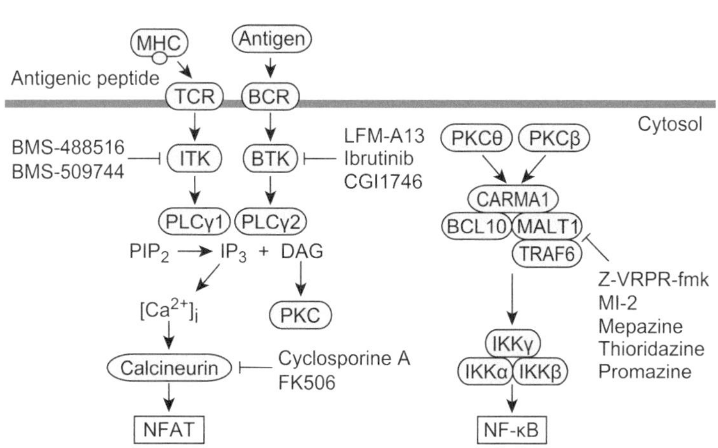

Fig. 5.1 Antigen receptor signaling pathway

5.2.2 Calcineurin Inhibitors

NFAT family transcription factors are regulated by Ca^{2+} signaling and are heavily phosphorylated in the regulatory domain and localized in the cytosol [11]. An increase in $[Ca^{2+}]_i$ promotes the binding of Ca^{2+}-calmodulin to calcineurin and the activation of its phosphatase activity [8]. NFAT is dephosphorylated by calcineurin and then undergoes nuclear translocation and transcriptional activation [8, 11]. Calcineurin is essential for T-cell activation and differentiation; thus, the inhibition of calcineurin blocks adaptive immune responses. The cyclic undecapeptide cyclosporine A and macrolide lactone FK506 (tacrolimus) (Fig. 5.2) are widely used as potent immunosuppressants in clinical settings, including organ transplantation, as well as in basic research [12]. Cyclosporine A and FK506 bind to the immunophilins cyclophilin A and FK506 binding protein 12, respectively, and these cyclosporine A- and FK506-immunophilin complexes inhibit calcineurin activity and subsequent NFAT translocation to the nucleus [12]. NFAT is known to interact with calcineurin through distinct binding sites: the main binding site PxIxIT and secondary binding site LxVP [13]. A previous study reported that immunosuppressant-immunophilin complexes inhibit calcineurin by interfering with substrate recognition at the LxVP site [13]. Small-molecule inhibitors that target different steps in the calcineurin-NFAT pathway, including direct inhibitors of calcineurin, have been identified [12].

5.2.3 ITK Inhibitors

The Tec family of PTKs, including ITK and BTK, consist of an N-terminal pleckstrin homology domain, Tec homology domain, SH3 and SH2 domains, and catalytic kinase domain (SH1 domain) [14]. The pleckstrin homology domain of Tec family kinases mediates PIP_3 binding and is required for their recruitment to the plasma membrane [3, 14]. The Tec homology domain includes the BTK motif, which binds Zn^{2+} and stabilizes the pleckstrin homology domain, and proline-rich repeats, whereas the SH3 domain and SH2 domain interact with proline-rich regions and phosphorylated tyrosine residues, respectively [3, 14]. ITK inhibitors are regarded as attractive therapeutics for autoimmune and allergic diseases. Structurally different types of small-molecule inhibitors have been developed [3]. BMS-488516 and BMS-509744 (Fig. 5.2), 2-aminothiazoles, were identified in a kinase assay using the recombinant ITK kinase domain and have been shown to block T-cell activation [15]. Other types of selective ITK inhibitors, including benzimidazoles, 3-aminopyrid-2-ones, and thienopyrazole-indoles, have been identified by high-throughput screening and structure-based drug design [3].

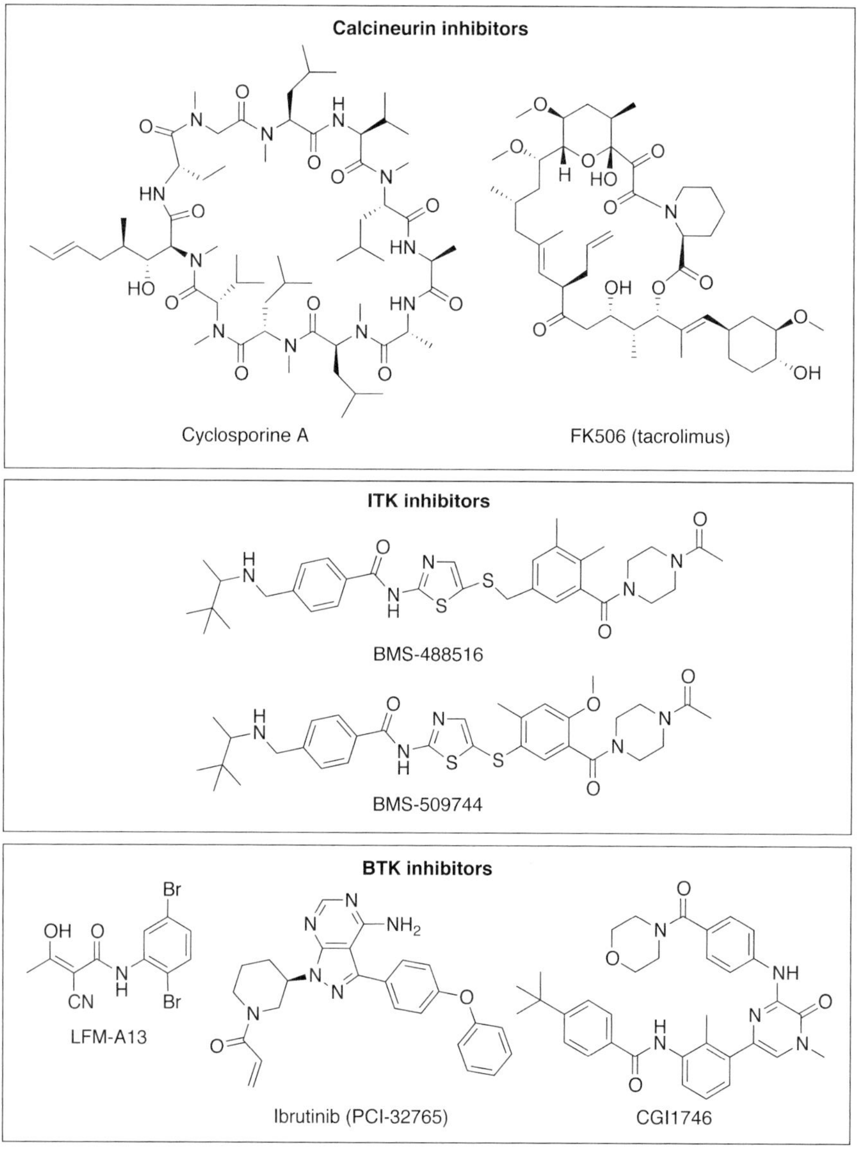

Fig. 5.2 Structures of calcineurin inhibitors, interleukin-2-inducible T-cell kinase (ITK) inhibitors, and Burton's tyrosine kinase (BTK) inhibitors

5.2.4 BTK Inhibitors

Novel BTK inhibitors have been identified, and some are regarded as potential therapeutics for B-cell neoplastic disorders and autoimmune diseases [3, 6]. Three-dimensional homology modeling of the BTK kinase domain and advanced docking procedures have led to the rational design of leflunomide metabolite (LFM) analogs, e.g., LFM-A13 (Fig. 5.2) [16]. However, LFM-A13 was later shown to inhibit polo-like kinase or Janus kinase 2 [3, 6]. Based on the structure of the initial lead compound identified by a FRET-based biochemical enzymology assay, ibrutinib (PCI-32765) (Fig. 5.2) was synthesized and shown to bind BTK covalently, possibly via a cysteine residue (Cys-481) [17]. Ibrutinib potently inhibits BTK kinase activity and has high selectivity for BTK against a wide range of kinases [18]. CGI1746 (Fig. 5.2) has been shown to inhibit the phosphorylation processes essential for BTK activation by stabilizing the inactive nonphosphorylated conformation [19]. CGI1746 suppresses B-cell-mediated and myeloid cell-mediated arthritis [19]. Other types of BTK inhibitors have also been reported [3, 6].

5.2.5 MALT1 Inhibitors

MALT1 contains a caspase-like domain and acts as a scaffold protein that interacts with several binding partners, including BCL10 [20]. MALT1 cleaves multiple substrates, and the tetrapeptide LRSR is the cleavage site in the C-terminus of BCL10 [21]. The peptide-based inhibitor benzyloxycarbonyl-Val-Arg-Pro-Arg-fluoromethylketone (Z-VRPR-fmk) (Fig. 5.3) serves as a MALT1 inhibitor, and inhibits NF-κB reporter activity induced by anti-CD3 and anti-CD28 antibodies or by phorbol ester and ionomycin [21]. Two groups have performed high-throughput screening for MALT1 inhibitors using an in vitro protease cleavage assay. Nineteen

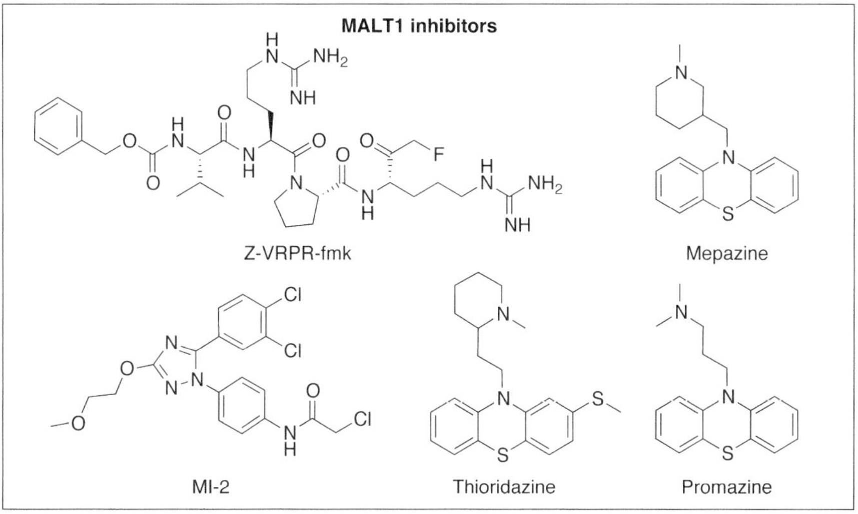

Fig. 5.3 Structures of mucosa-associated lymphoid tissue lymphoma-translocation gene 1 (MALT1) inhibitors

compounds (MI-1 to MI-19) have been shown to inhibit the activity of the recombinant form of LZ-MALT1 (340–789) [22]. MI-2 (Fig. 5.3) directly binds the paracaspase domain of MALT1 (329–728), possibly by covalent modifications to the active site Cys-464, and irreversibly inhibits MALT1 activity [22]. Distinct derivatives of medicinally active phenothiazines, i.e., mepazine, thioridazine, and promazine (Fig. 5.3) have been shown to inhibit the cleavage activity of recombinant full-length MALT1 fused to glutathione *S*-transferase [23]. Mepazine and thioridazine inhibit the MALT1-mediated induction of IL-2 by T cells [23].

5.3 Perforin/Granzymes

5.3.1 *Perforin*

Cytotoxic T lymphocytes (CTLs) and natural killer (NK) cells eliminate virus-infected cells and tumor cells. CTLs and NK cells kill target cells using two main and distinct mechanisms: exocytosis of lytic granules containing perforin and granzymes and engaging with the cell-surface Fas ligand (FasL) (Fig. 5.4) [24]. Perforin is a multidomain protein that contains an N-terminal membrane attack complex perforin (MACPF) domain and C-terminal C2 domain that mediates Ca^{2+}-dependent membrane binding [25]. Perforin monomers polymerize into transmembrane channels by direct ionic attraction between the opposite faces of adjacent perforin monomers in the MACPF domain [26]. Moreover, perforin monomers assemble into pores with sizes in a range that permits granzymes to pass readily through the membrane [25]. Perforin is translated as an inactive form and is processed at the C-terminus to become an active form in acidic compartments [27]. A previous study reported that the extreme C-terminal region of perforin

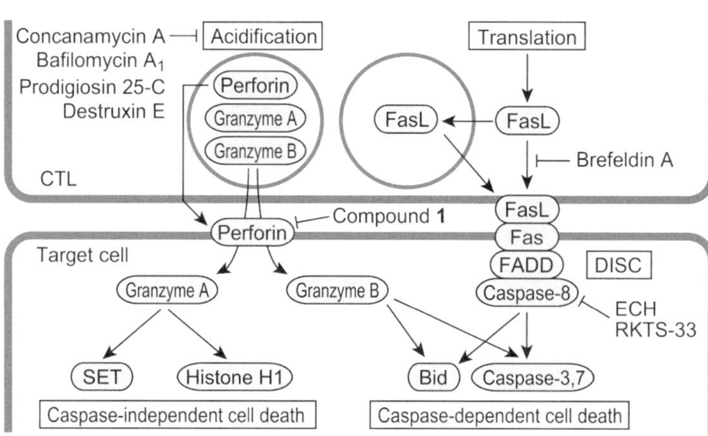

Fig. 5.4 Perforin- and FasL-dependent cytotoxic pathways

regulates rapid trafficking from the ER to the Golgi apparatus, and that N-linked glycosylation is critical for perforin transport from the Golgi apparatus to secretory granules [28].

5.3.2 *Granzymes*

Granzymes belong to the serine protease family with different substrate specificity: granzyme A is a trypsin-like protease that induces caspase-independent cell death, while granzyme B is a unique protease that cleaves after aspartic acid residues in a similar manner to caspases [29, 30]. Granzyme B predominantly promotes apoptosis by Bid cleavage and subsequent mitochondrial permeabilization via its truncated form as well as the direct processing and activation of effector caspases, and also cleaves a number of other proteins [29, 30]. Granzyme A promotes caspase-independent cell death by cleaving components of the SET complex and activating nucleases that make single-stranded DNA lesions as well as by directly cleaving substrates, including the linker histone H1 [29, 30].

5.3.3 *Perforin Inhibitors*

Lytic granules are exocytosed to the immunological synapse between target cells and CTLs or NK cells, leading to target cell death in concert with the actions of granzymes [31]. Due to the requirement of Ca^{2+} for perforin activity, ethylene glycol-bis(β-aminoethylether)-*N*,*N*,*N*′,*N*′-tetraacetic acid (EGTA) (Fig. 5.5) has been used as a calcium chelator to block perforin-dependent cytotoxicity. Small-molecule perforin inhibitors have been identified by high-throughput screening. Dihydrofuro[3,4-*c*]pyridinones, e.g., compound **1** (Fig. 5.5), have been shown to block the perforin-dependent lysis of nucleated cells and perforin-dependent cell lysis by NK cells [32].

5.3.4 *Granzyme Inhibitors*

Synthetic inhibitors, including isocoumarins, peptide chloromethyl ketones, and peptide phosphonates, have been used as valuable tools to elucidate the function of granzymes [33]. Small-molecule granzyme B inhibitors with the 1,2,3 triazole moiety have been identified and shown to block CTL-mediated apoptosis [34]. The biotinylated and substrate specific-based diphenyl phosphonates, bio-x-IGN(AmPhg)^{P}-$(OPh)_2$ and bio-x-IEPD^{P}-$(OPh)_2$ (Fig. 5.5), have been synthesized as irreversible inhibitors of granzyme A and granzyme B, respectively [35]. This

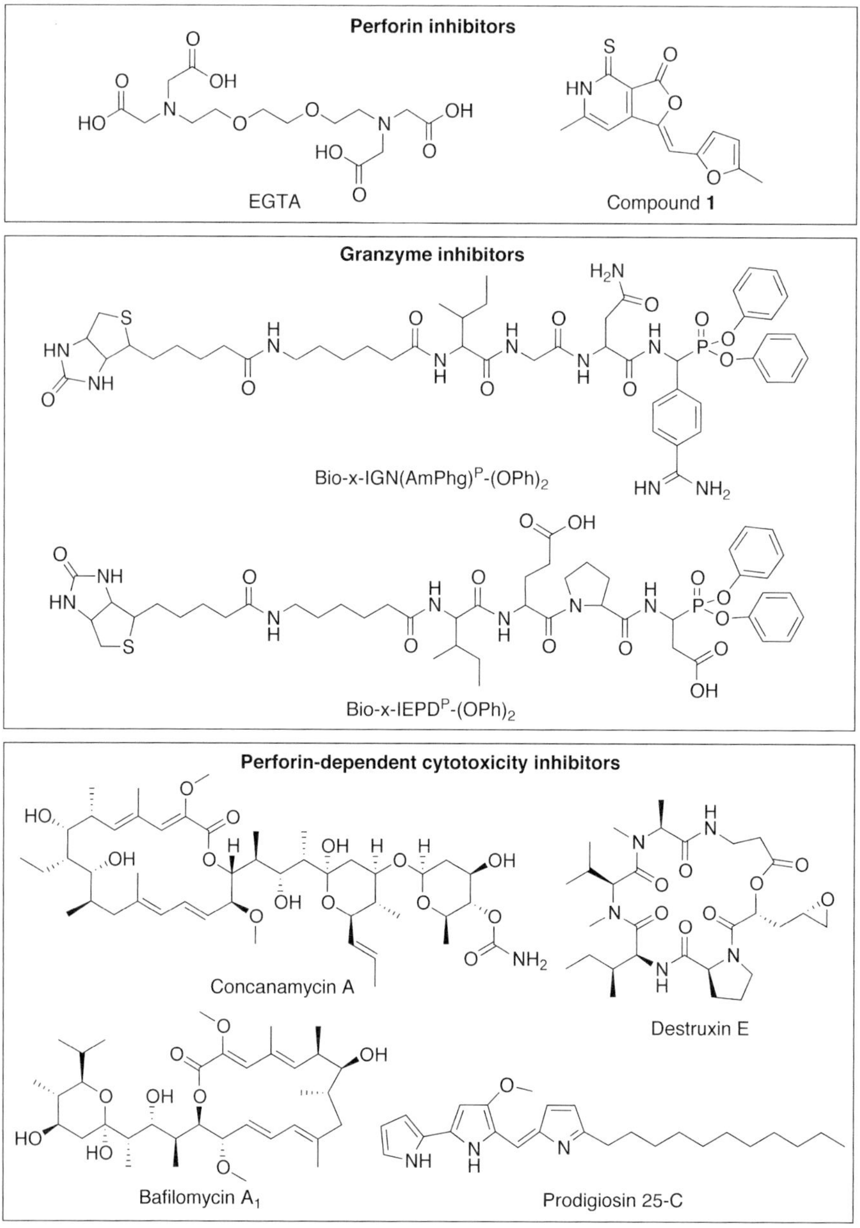

Fig. 5.5 Structures of perforin inhibitors, granzyme inhibitors, and perforin-dependent cytotoxicity inhibitors

study showed that granzyme A and granzyme B play minor and major roles in NK-cell-dependent target cell lysis, respectively [35].

5.3.5 *Perforin-Dependent Cytotoxicity Inhibitors*

Perforin is stored in lytic granules in which the internal pH is maintained at 5.1–5.4 [36]. We previously showed that acidification is essential to maintain the structure and function of lytic granules [37]. The 18-membered macrolide concanamycin A (Fig. 5.5) has been identified as a specific inhibitor of vacuolar-type H^+-ATPase. Concanamycin A increases the internal pH of lytic granules in CTLs to approximately neutral [38]. It has also been shown to induce the degradation of perforin by serine proteases and the Ca^{2+}-dependent inactivation of perforin in lytic granules [37, 39]. The 16-membered macrolide bafilomycin A_1 (Fig. 5.5), an inhibitor of vacuolar-type H^+-ATPase, also induces the proteolytic degradation of perforin, while perforin is predominantly inactivated without proteolysis in treatments with the H^+/Cl^- symporter prodigiosin 25-C (Fig. 5.5) and the cyclic depsipeptide destruxin E (Fig. 5.5), which inhibits vacuolar-type H^+-ATPase [40]. Thus, perforin may undergo Ca^{2+}-dependent inactivation more easily than proteolytic degradation when the internal pH of lytic granules is raised to neutral. Concanamycin A has been shown to inhibit perforin-dependent cytotoxic pathways by a wide range of CTLs without affecting the FasL-dependent cytotoxic pathway [41]. It is also widely used as a specific inhibitor of perforin-dependent cytotoxicity mediated by CTLs and NK cells.

5.4 FasL

5.4.1 *Fas Signaling*

FasL is synthesized and transported to the cell surface via multiple mechanisms, which differ from perforin and granzymes. Membrane-bound FasL oligomerizes the death receptor Fas, thereby inducing apoptosis in target cells (Fig. 5.4) [42]. Fas contains a death domain that is essential for transmitting apoptosis signals in the cytoplasmic region, whereas the adaptor protein Fas-associated death domain protein (FADD) contains a death domain and death effector domain [42]. When stimulated with FasL, FADD was shown to be recruited to Fas via the death domain interaction, and, in turn, recruited procaspase-8 via the death effector domain interaction, leading to the formation of the death-inducing signaling complex (DISC) [43]. Procaspase-8 contains two N-terminal death effector domains and a caspase domain composed of large and small subunits [43]. Procaspase-8 undergoes dimerization and self-processing to generate active heterotetramers consisting of two large and two small subunits [44]. Active caspase-8 directly converts effector caspase-3 and caspase-7 to their active forms by direct cleavage [44]. Active caspase-8 is also able to cleave the Bcl-2 family member Bid to its truncated form, which induces the release of mitochondrial proapoptotic proteins (e.g., cytochrome *c*) [42]. Cytochrome *c* triggers the formation of the apoptosome complex, which recruits and activates procaspase-9 [45]. Active caspase-9 then cleaves effector caspases and induces apoptosis [44].

5.4.2 Caspase Inhibitors

Based on the substrate specificity of caspase-8, peptide-based inhibitors consisting of a tetrapeptide in the cleavage site, e.g., Z-Ile-Glu(OMe)-Thr-Asp(OMe)-fmk (Z-IETD-fmk) (Fig. 5.6), have been used as caspase-8 inhibitors. However, Z-IETD-fmk was previously shown to manifest cross-reactivity toward caspase-3 and caspase-9 [46]. Moreover, caspase-8 and granzyme B cleave similar substrates, such as Bid and effector caspases. Emricasan (IDN-6556) (Fig. 5.6) is an irreversible pan-caspase inhibitor, but is not specific to caspase-8. It has been shown to inhibit Fas-mediated apoptosis in cellular assays and Fas-induced liver injury in mice [47]. We previously identified nonpeptide inhibitors that target the activation of procaspase-8 and selectively block Fas-dependent apoptosis [48–50]. The mycotoxin penicillic acid (Fig. 5.6) did not affect the recruitment of FADD or procaspase-8 in the DISC, but blocked the processing of procaspase-8 [48], indicating that penicillic acid targets the self-processing of procaspase-8 in the DISC. A mass spectrometric analysis revealed that penicillic acid directly binds to the active center cysteine in the large subunit of caspase-8 via the α,β-unsaturated lactone group [48]. An epoxycyclohexenone derivative designated as ECH (Fig. 5.6) has also been shown to inhibit Fas-dependent apoptosis and the activation of caspase-8 [50].

5.4.3 FasL-Dependent Cytotoxicity Inhibitors

A previous study demonstrated that ECH selectively blocked the processing of procaspase-8 in the DISC and bound covalently to procaspase-8 as well as the caspase-8 large subunit [50]. ECH is the most likely to bind to the active center cysteine in the caspase-8 large subunit via the α,β-unsaturated ketone and epoxide moieties. ECH markedly inhibited the FasL-dependent cytotoxic pathway mediated by perforin-negative mouse $CD4^+$ CTL and mouse $CD8^+$ CTL pretreated with concanamycin A, but had a negligible influence on the perforin-dependent cytotoxic pathway mediated by mouse $CD8^+$ CTL [51]. One of the ECH derivatives,

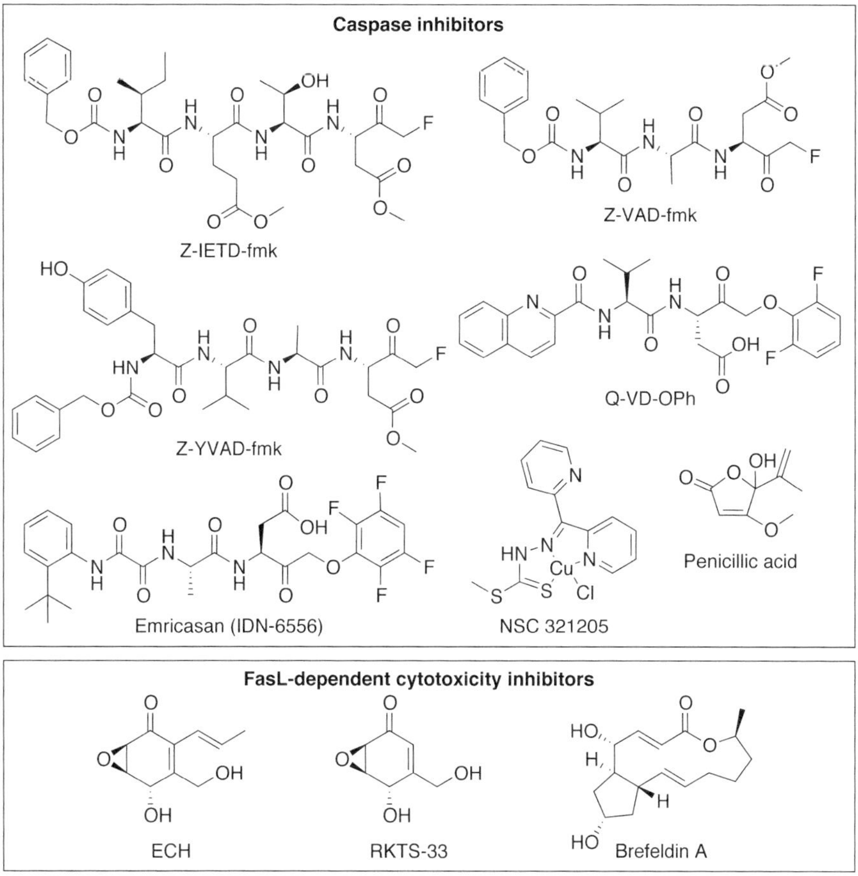

Fig. 5.6 Structures of caspase inhibitors and FasL-dependent cytotoxicity inhibitors

designated as RKTS-33 (Fig. 5.6), was also found to inhibit Fas-dependent apoptosis [49]. RKTS-33 has been shown to inhibit FasL-dependent CTL-mediated cytotoxicity, but not perforin-dependent CTL-mediated cytotoxicity [52]. Thus, ECH and RKTS-33 are considered to be caspase-8 inhibitors that specifically block FasL-dependent CTL-mediated cytotoxicity. TCR-induced FasL expression is sensitive to translation inhibitors. Moreover, FasL is stored in intracellular granules [53]. Thus, upon engaging with target cells, it is likely that FasL is newly synthesized and transported to the cell surface or that a pre-existing pool of FasL is released by degranulation. We previously reported that brefeldin A (Fig. 5.6) efficiently blocked the FasL-dependent killing pathways mediated by CTL [41, 54]. Therefore, brefeldin A may be used to block the FasL-dependent CTL-mediated killing pathway.

5.5 TLR Family

5.5.1 TLR Signaling

TLR family proteins are type I integral transmembrane proteins that contain N-terminal leucine-rich repeats in the extracellular domain, which are required for the recognition of PAMPs [55]. TLRs are localized on the cell surface (e.g., TLR2, TLR4, and TLR6) or within intracellular vesicles (e.g., TLR3, TLR7, and TLR9) [56]. They possess cytoplasmic tails that exhibit high homology with IL-1 receptors (IL-1R) (known as the toll/IL-1R (TIR) domain), and serve as a platform in interactions with adaptor proteins [55]. The adaptor protein myeloid differentiation primary response gene 88 (MyD88) is recruited to all TLRs, except for TLR3, and interacts with IL-1R-associated kinase (IRAK)-4, which, in turn, phosphorylates other IRAK family members, thereby promoting the recruitment and activation of TRAF6 (Fig. 5.7) [4, 56]. TRAF6 catalyzes the synthesis of K63-linked polyubiquitin chains, which are used as a scaffold to recruit TGF-β-activated kinase 1 (TAK1) and IKK, leading to the activation of NF-κB [57]. The adaptor protein TIR domain-containing adaptor inducing IFN-β (TRIF) is recruited to TLR3 and TLR4 in response to their stimulation [4, 56]. In addition to TRAF6, TRIF recruits TRAF3 and two IKK-related kinases, TANK binding kinase 1 (TBK1) and IKKε (also known as IKK-*i*) [4, 56]. Following its phosphorylation by these kinases, IRF3 translocates to the nucleus and initiates the transcriptional activation of type I IFNs [4, 56].

5.5.2 TLR Agonists

TLR agonists are considered to act as vaccines and adjuvants to prevent infection. A large number of TLR agonists have been identified [58]: Pam_3CSK_4 (a TLR1-TLR2

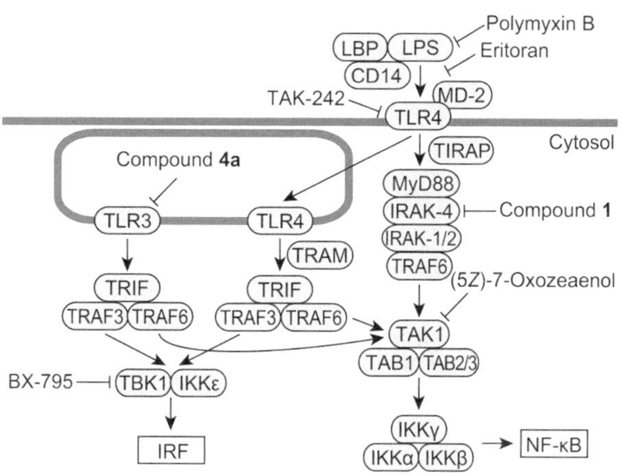

Fig. 5.7 Toll-like receptor (TLR) signaling pathway

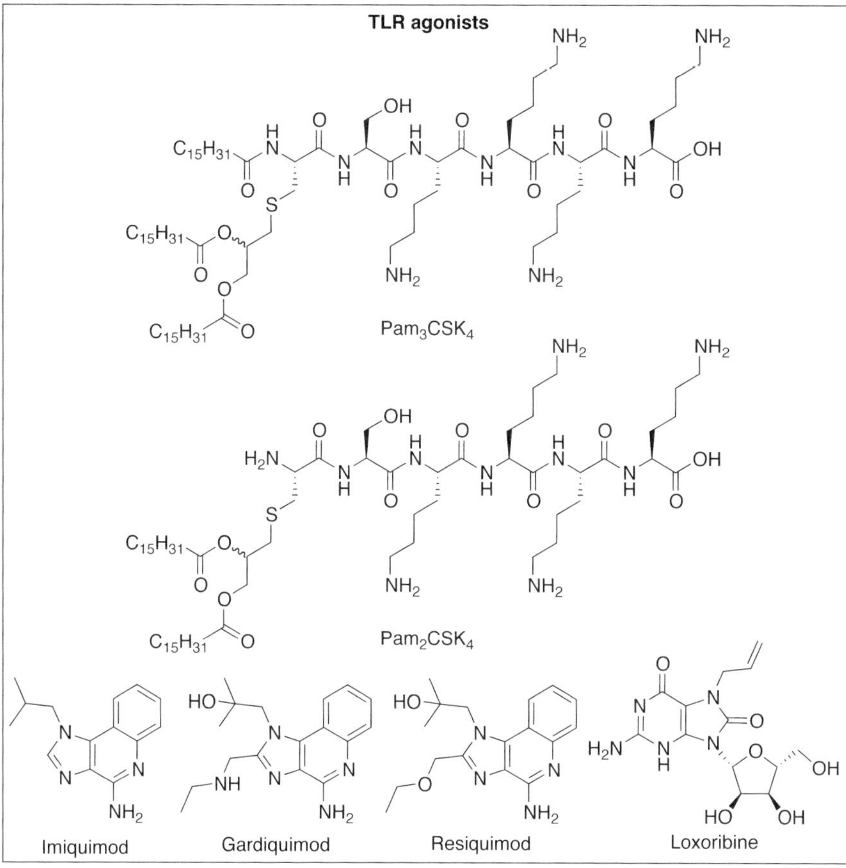

Fig. 5.8 Structures of toll-like receptor (TLR) agonists

agonist) and Pam_2CSK_4 (a TLR2-TLR6 agonist) as synthetic acylated lipopeptides; imiquimod (a TLR7 agonist), gardiquimod (a TLR7 agonist), and resiquimod (a TLR7 and TLR8 agonist) as imidazoquinolines; loxoribine (a TLR7 agonist) as a guanosine analog (Fig. 5.8).

5.5.3 TLR Signaling Inhibitors

Polymyxin B (Fig. 5.9), a cationic cyclic peptide antibiotic, is used as a TLR4 antagonist because it binds and neutralizes the lipid A moiety of lipopolysaccharide (LPS) [59]. E5531 is a LPS antagonist that is synthesized on the basis of the structures of nontoxic lipid A derived from *Rhodobacter capsulatus*, and its derivative eritoran (E5564) (Fig. 5.9) has been developed further. The crystal structure of

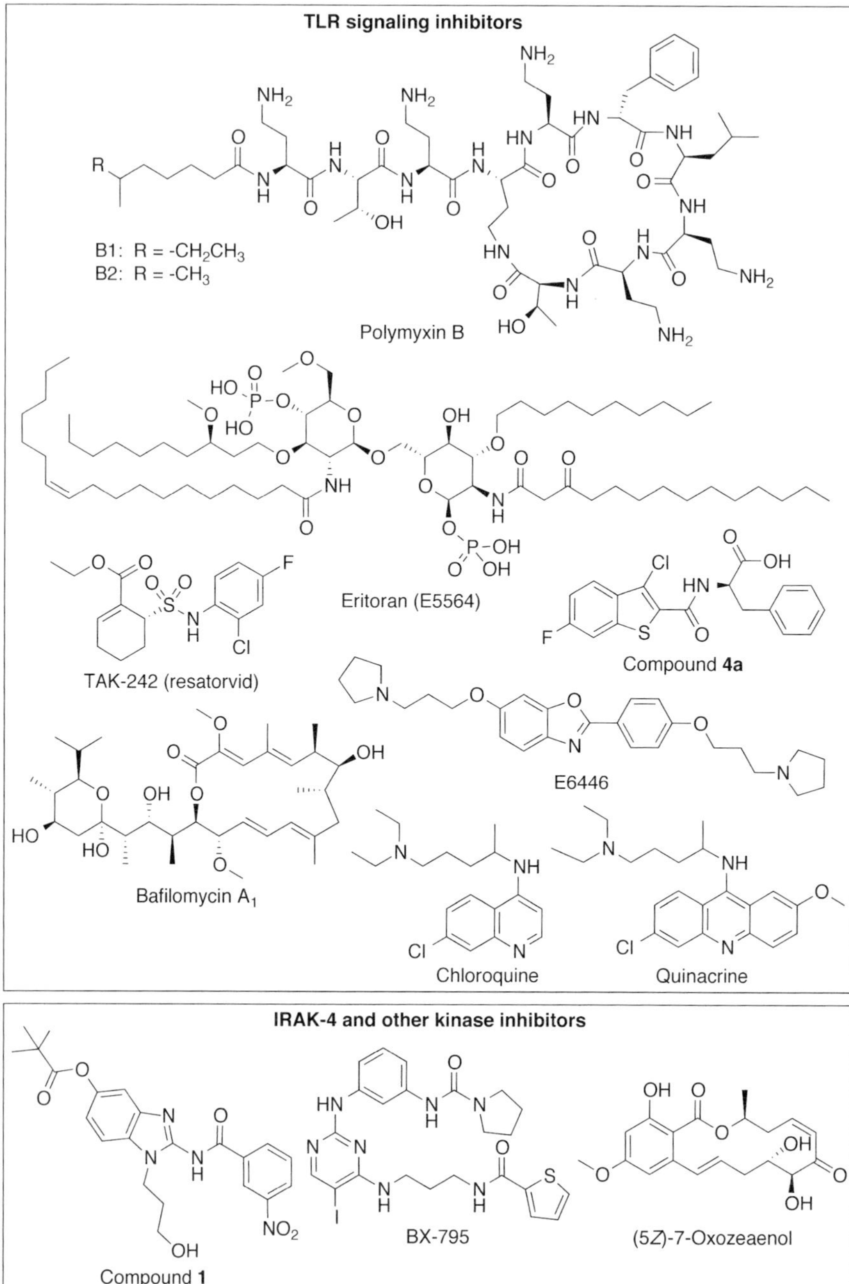

Fig. 5.9 Structures of toll-like receptor (TLR) signaling inhibitors and IL-1 receptor-associated kinase (IRAK)-4 and other kinase inhibitors

the TLR4-myeloid differentiation factor (MD-2) complex with bound eritoran revealed that eritoran binds to the hydrophobic pocket in MD-2 and does not interact directly with TLR4 [60]. Eritoran is currently known as a potent, well-tolerated TLR4 antagonist, and has been shown to protect mice from lethal influenza infection [61]. TAK-242 (resatorvid) (Fig. 5.9), a cyclohexene derivative, inhibits LPS-induced signaling pathways and has been reported to bind selectively to TLR4 from the 10 human TLRs by interfering with the interaction of the TIR domain-containing adaptor protein (TIRAP) and TRIF-related adaptor molecule (TRAM) with TLR4 [62]. Oxidized phospholipids, such as 1-palmitoyl-2-arachidonyl-*sn*-glycero-3-phoshorylcholine, were previously reported to inhibit LPS-induced signaling by blocking the interaction of LPS with the LPS binding protein (LBP) and CD14 [63]. Several types of TLR antagonists other than TLR4 antagonists have also been reported. Small-molecule TLR3 antagonists, including compound **4a** (Fig. 5.9), have been identified by in silico screening and high-throughput cell assays [64]. While imidazoquinolines act as TLR7 agonists, minor modifications on the imidazoquinone scaffold have led to the development of small-molecule TLR7 and TLR8 antagonists [65]. While oligodeoxynucleotide-based compounds have been developed as antagonists for TLR9, which recognizes CpG DNA, a synthetic small molecule called E6446 (Fig. 5.9) has been shown to inhibit the activation of TLR9 [66]. TLR9 is localized within intracellular vesicles [56]. Bafilomycin A_1 (Fig. 5.9), an inhibitor of vacuolar-type H^+-ATPase, inhibits the activation of TLR9 [67], indicating that an acidic pH is a prerequisite for the activation of TLR9 by CpG DNA in endosomes. Unlike bafilomycin A_1, chloroquine and quinacrine (Fig. 5.9) (known to be weak bases that increase the pH of intracellular vesicles) have been shown to directly inhibit the interaction between TLR9 and CpG DNA [68].

5.5.4 IRAK-4 and Other Kinase Inhibitors

IRAK-4 is a common kinase that is indispensable for signaling pathways downstream of TLRs and IL-1R [57]. *N*-acyl 2-aminobenzimidazoles have been identified as potent IRAK-4 inhibitors by high-throughput screening, and the crystal structure of the IRAK-4 kinase domain in complex with one of these IRAK-4 inhibitors designated as compound **1** (Fig. 5.9) revealed that compound **1** binds to the ATP binding site, which does not have a deep pocket in the back, but has a featured front pocket [69]. Other types of IRAK-4 inhibitors have also been reported [70]. BX-795 (Fig. 5.9), containing an aminopyrimidine backbone, was originally identified as a 3-phosphoinositide-dependent kinase 1 (PDK1) inhibitor by screening based on a kinase assay [71]. BX-795 was later shown to inhibit TBK1 and IKKε more potently than PDK1 [72]. Using an in vitro kinase assay, a previous study demonstrated that (5*Z*)-7-oxozeaenol (Fig. 5.9), a resorcylic acid lactone, inhibited TAK1 [73]. (5*Z*)-7-Oxozeaenol has also been identified as an irreversible inhibitor that blocks the kinase and intrinsic ATPase activities of TAK1 [74].

5.6 NLR Family

5.6.1 NLR Signaling

Inflammasomes play a critical role in the maturation of the proinflammatory cytokines IL-1β and IL-18, as well as pyroptosis (Fig. 5.10) [75, 76]. IL-1β and IL-18 are both produced as inactive precursors and are converted into their active forms by caspase-1, which is produced as a proenzyme and undergoes autoactivation in inflammasomes [75, 76]. Caspase-1 induces pyroptosis, which results in cellular lysis and the release of cytosolic contents [75, 76]. NLRs are intracellular PRRs that assemble into inflammasomes, caspase-1-activating molecular platforms activated upon the recognition of PAMPs or host-derived danger signals (danger-associated molecular patterns) [4]. NLR family proteins possess a central nucleotide binding and oligomerization domain (NOD or NACHT domain) flanked by C-terminal leucine-rich repeats, an N-terminal caspase recruitment domain (CARD), or a pyrin domain (PYD) [77]. The adaptor protein apoptosis-associated speck-like protein containing a CARD (ASC) and procaspase-1 are present in NLR inflammasomes in which the PYD of ASC interacts with that of NLRs, and the CARD of ASC binds to that of procaspase-1 [77]. After their activation, NLRs assemble into multiprotein complexes through the NACHT domain and promote the recruitment and autoproteolytic cleavage of procaspase-1 into active heterodimers [77].

5.6.2 Inflammasome Signaling Inhibitors

The NLR family NLRP3 is activated by multiple cellular events in response to various stimuli: potassium efflux, reactive oxygen species (ROS) generation, and phagolysosomal destabilization and the release of endogenous mediators into the

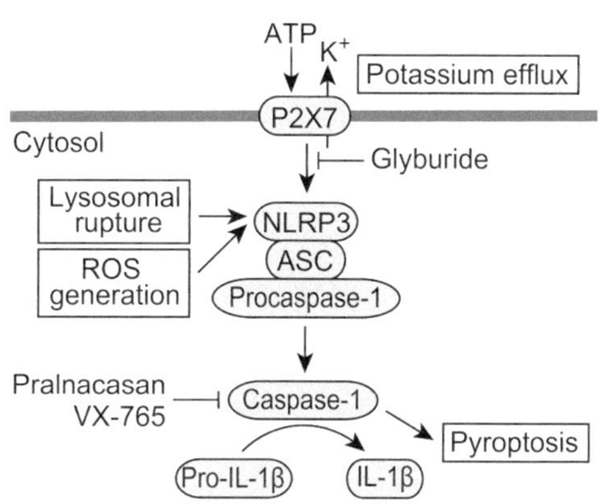

Fig. 5.10 Nucleotide binding domain and leucine-rich repeat-containing receptor (NLR) signaling pathway

cytosol [75]. As inflammasome activators, previous studies reported that the activation of the NLRP3 inflammasome is induced by maitotoxin (a ladder-shaped polyether), nigericin (a polyether ionophore), gramicidin (a channel-forming ionophore), and ATP (Fig. 5.11) [78, 79]. Glyburide (Fig. 5.12), an inhibitor of K_{ATP} channels, is widely used in the treatment of type 2 diabetes, and has been shown to inhibit ATP-dependent IL-1β production and NLRP3 inflammasomes downstream of the P2X7 receptor [80]. 7-Desacetoxy-6,7-dehydrogedunin (Fig. 5.12) was found, using phenotypic screening, to protect macrophages from pyroptotic cell death induced by anthrax lethal toxin [81]. Protein kinase R has been identified as the target of 7-desacetoxy-6,7-dehydrogedunin, revealing that it is involved in caspase-1 activation and pyroptosis [81]. Diarylsulfonylurea-based compounds, designated as cytokine release inhibitory drugs (CRIDs), were previously shown to block IL-1β processing and target glutathione *S*-transferase Omega 1-1 [82]. MCC950 (Fig. 5.12), also known as CRID3, has recently been reported to be a specific inhibitor of the activation of NLRP3 that reduces the production of IL-1β and ameliorates the severity of experimental autoimmune encephalomyelitis in mice [83]. A recent study also demonstrated that β-hydroxybutyrate (Fig. 5.12), a ketone body, specifically inhibited NLRP3 by preventing K^+ efflux, ASC oligomerization, and speck formation [84]. Using small-molecule inhibitor screening, the EGFR tyrosine kinase inhibitors gefitinib and erlotinib (Fig. 5.12) were found to inhibit the tyrosine kinase activity of receptor-interacting protein (RIP) 2 and NOD2-mediated cytokine release in response to muramyl dipeptide [85]. Other types of inflammasome inhibitors have been reported [86].

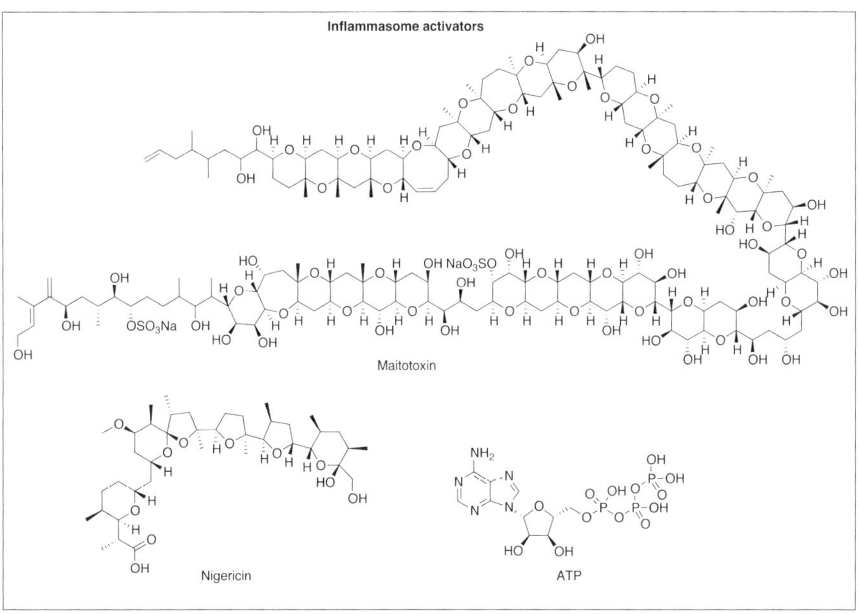

Fig. 5.11 Structures of inflammasome activators

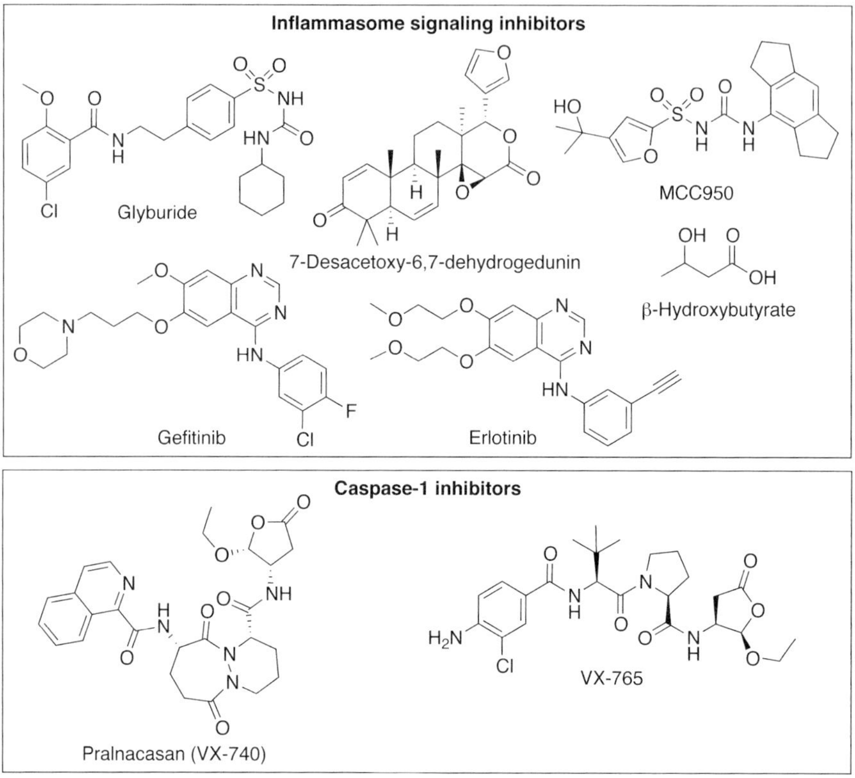

Fig. 5.12 Structures of inflammasome signaling inhibitors and caspase-1 inhibitors

5.6.3 Caspase-1 Inhibitors

Procaspase-1 is recruited into inflammasomes and undergoes autoproteolytic cleavage, leading to the formation of active caspase-1, which cleaves IL-1β and IL-18 [75, 76]. Many peptide-based caspase inhibitors have been developed based on substrate specificity [87]. Z-Tyr-Val-Ala-Asp(OMe)-fmk (Z-YVAD-fmk) (Fig. 5.6) may be more specific to caspase-1 than Z-Val-Ala-Asp(OMe)-fmk (Z-VAD-fmk) (Fig. 5.6), which is widely used as a broad-spectrum inhibitor of caspases, and this has been attributed to the substrate specificity of Ac-YVAD-CHO [87]. Q-VD-OPh (Fig. 5.6) has been reported to inhibit caspase-1, caspase-3, caspase-8, and caspase-9 [88]. Z-VAD-fmk, Z-YVAD-fmk, and Q-VD-OPh were found to inhibit IL-1β cleavage in NLRP3-dependent inflammasomes [89]. In addition to these peptide-based inhibitors, other types of caspase-1 inhibitors, including pralnacasan (VX-740) and VX-765 (Fig. 5.12), have been reported [90, 91]. Pralnacasan is an orally bioavailable prodrug that is rapidly hydrolyzed to an active caspase-1 inhibitor [90]. VX-765 is also an orally available prodrug that is converted to a potent and selective inhibitor of caspase-1, called VRT-043198, which inhibits casapse-1 and caspase-4 with at least 100-fold higher potency than other caspases [91]. Caspase-1, but not caspase-7, contains Cys-331 in the cavity of the dimer interface away from the active site [92]. Using this allosteric site as an anchor for disulfide trapping, a thienopyrazole referred to as compound **34** was shown to conjugate to the Cys-331 of caspase-1 [92]. A group of compounds sharing common chemical scaffolds, e.g., NSC 321205 (Fig. 5.6), have been found to inhibit the activation of caspase-3 by high-throughput screening based on a reconstituted assay using a fluorogenic substrate in the presence of purified recombinant apoptosome components and dATP [93]. These compounds serve as allosteric inhibitors, and have been shown to inhibit the LPS-induced secretion of IL-1β [93].

5.7 RLR Family

5.7.1 RLR Signaling

The RLR family proteins, RIG-I, melanoma differentiation-associated gene 5 (MDA5), and laboratory of genetics and physiology 2 (LGP2), are localized in the cytoplasm and play major roles in the sensing of RNA virus infection (Fig. 5.13) [4]. These members are composed of a central DExD/H box RNA helicase domain, which mediates RNA binding and ATP hydrolysis, and a C-terminal domain containing a repressor domain (in RIG-I and MDA5, but not in LGP2) that is necessary for autoregulation [94]. RIG-I and MDA5, but not LGP2, contain two N-terminal CARDs that are essential to trigger downstream signaling pathways [94]. Upon activation, IFN-β promoter stimulator 1 (IPS-1) is recruited to RIG-I and MDA5 via a homotypic CARD interaction, and, in turn, recruits downstream adaptor proteins, including TRAF2, TRAF3, and TRAF6 [95]. These adaptor proteins induce the activation of the TBK1/IKKε complex, which mediates the

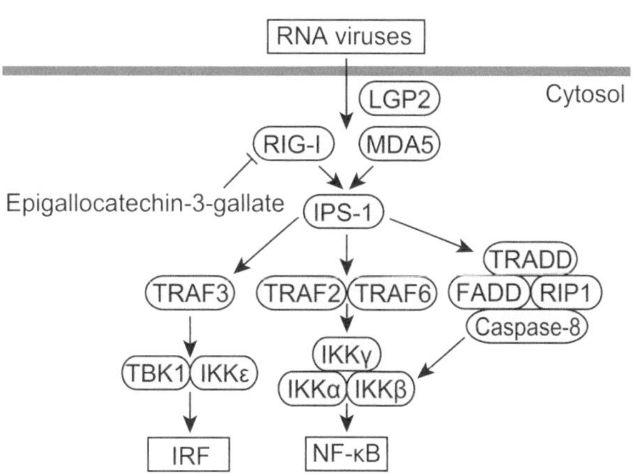

Fig. 5.13 Retinoic acid-inducible gene I-like receptor (RLR) signaling pathway

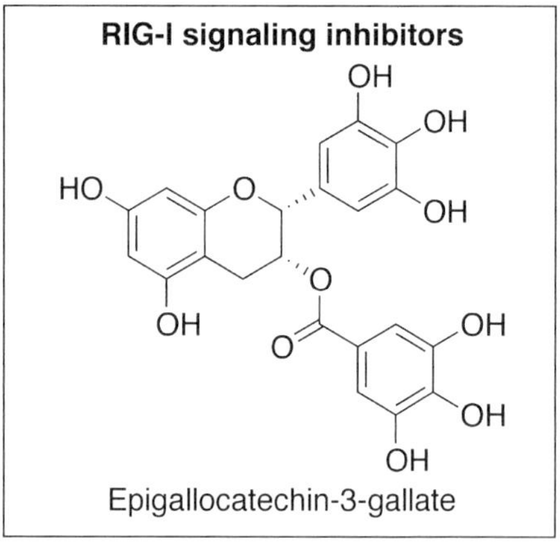

Fig. 5.14 Structures of retinoic acid-inducible gene I (RIG-I) signaling inhibitors

activation of IRF3 and IRF7 as well as the IKKα/IKKβ complex, which, in turn, mediates the activation of NF-κB [95]. RLR signaling shares a number of proteins with TNF signaling, including TNF receptor 1 (TNF-R1)-associated death domain protein (TRADD), FADD, RIP1, and caspase-8 [95].

5.7.2 RLR Signaling Inhibitors

RIG-I primarily recognizes RNA with a triphosphate moiety and blunt-ended base-paired region of ~20 nucleotides at the 5′ end, while MDA5 is considered to act as a dsRNA sensor [96]. Poly(I:C), a synthetic RNA used as a TLR3 agonist, also acts as an RLR agonist by binding to RIG-I and MDA5 [96]. Poly(dA:dT) is transcribed by RNA polymerase III to dsRNA containing a 5′-phosphate moiety and serves as a RIG-I agonist [97]. A previous study reported that phosphorothioated oligonucleotides antagonize RIG-I signaling by binding to RIG-I and blocking its ATPase activity [98]. Epigallocatechin-3-gallate (Fig. 5.14), a major polyphenolic catechin found in green tea, has been shown to bind RIG-I and inhibit its ATPase activity [99].

5.8 CLR Family

5.8.1 CLR Signaling

CLR family proteins, which are characterized by C-type lectin domains, play a critical role in antifungal immunity and recognize carbohydrate structures, including β-glucan and mannan (Fig. 5.15) [100]. Dectin-1 has immunoreceptor tyrosine-

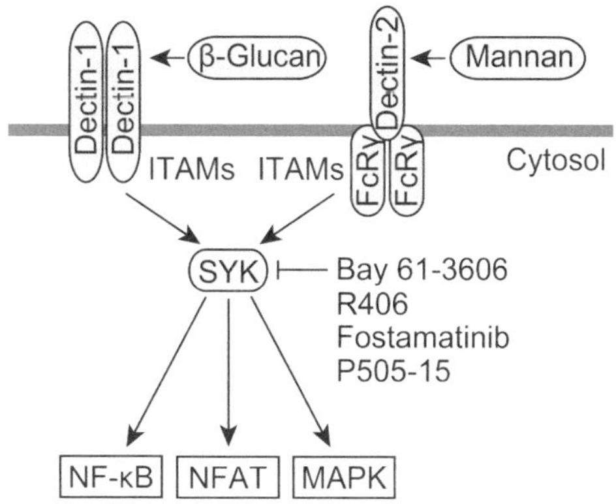

Fig. 5.15 C-type lectin receptor (CLR) signaling pathway

based activation motifs (ITAMs) in its cytoplasmic tail, whereas Dectin-2 and Mincle do not, and it associates with Fc receptor γ (FcRγ), which contains ITAM motifs [101]. Dectin-1, Dectin-2, and Mincle utilize spleen tyrosine kinase (SYK), which mediates the activation of the MAP kinase and NFAT pathways [100–102]. SYK also activates the NF-κB pathway though CARD9, BCL10, and MALT1 [100–102]. Upon ligand binding, ITAMs are rapidly phosphorylated and mediate the recruitment of SYK via an interaction with its SH2 domain, leading to the activation of SYK [102]. SYK mediates downstream signaling by associating with several molecules, including PLCγ isoforms and the regulatory subunits of PI3K [102].

5.8.2 CLR Signaling Inhibitors

SYK inhibitors have therapeutic value for the treatment of autoimmune and allergic diseases as well as neoplastic disorders [103–105]. Piceatannol (Fig. 5.16), a plant-derived natural product, has been used as an SYK inhibitor with less specificity [103, 105]. A previous study reported that piceatannol inhibited PI3K activity and insulin receptor kinase activity [106]. ER-27319 (Fig. 5.16) was shown to block the tyrosine phosphorylation of SYK by inhibiting the interaction between SYK and the ITAM of the FcεRIγ subunit [107]. Other types of small-molecule inhibitors targeting SYK, e.g., Bay 61-3606, R406, fostamatinib (R788), and P505-15 (Fig. 5.16), have also been reported [103–105].

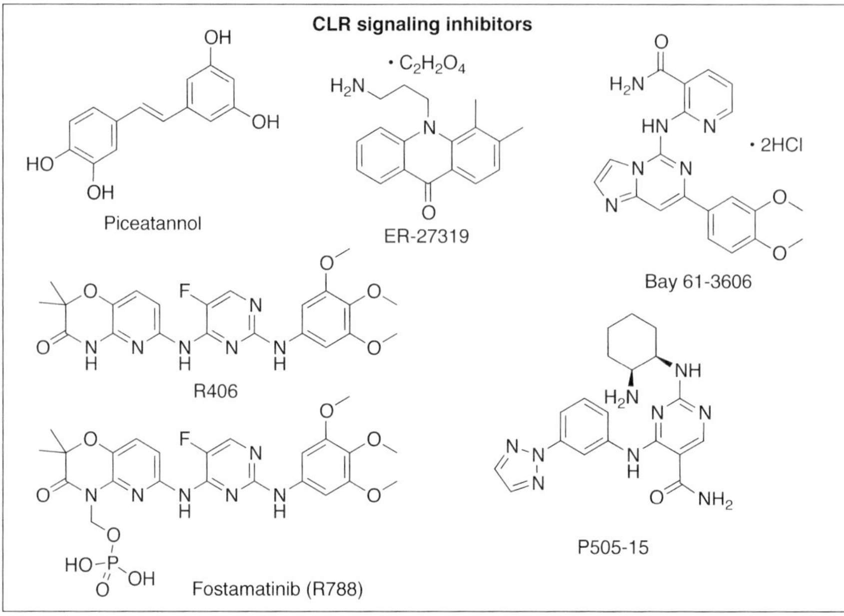

Fig. 5.16 Structures of C-type lectin receptor (CLR) signaling inhibitors

5.9 TNF Receptor Family

5.9.1 NF-κB Signaling

The NF-κB family shares an N-terminal Rel homology domain that is essential for DNA binding and dimerization [108]. NF-κB dimers bind to κB sites and regulate transcriptional activation [108]. In contrast to the TIR domain found in IL-1R and TLRs, TNF-R1 possesses a unique cytoplasmic tail called the death domain, and utilizes different adaptor proteins to induce the activation of NF-κB [108]. When stimulated with TNF-α, TRADD is recruited to TNF-R1, and subsequently interacts with RIP1 and TRAF2, leading to the K63-polyubiquitination of RIP1 and recruitment of the IKK complex and TAK1 complex via ubiquitin chains (Fig. 5.17) [109]. In the cytosol, NF-κB dimers, such as p65/p50, associate with IκBα, which masks the nuclear localization signals of p65, and the activated IKK complex phosphorylates IκBα at N-terminal Ser-32 and Ser-36 [110]. Phosphorylated IκBα immediately undergoes K48-polyubiquitination and degradation by the 26S proteasome [109]. The free NF-κB dimers translocate to the nucleus and induce the transcription of a large number of genes involved in inflammation and tumor promotion [111].

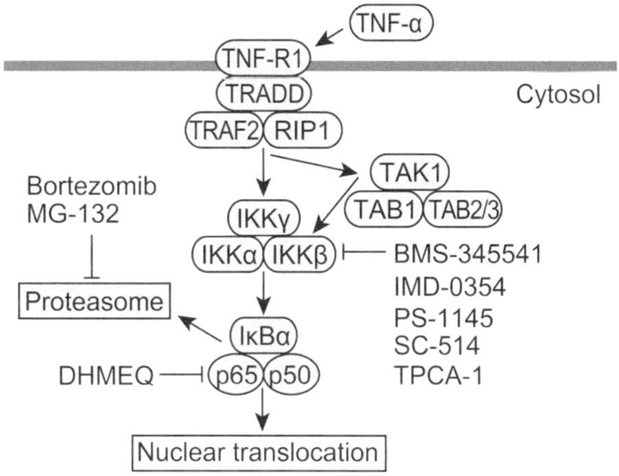

Fig. 5.17 Tumor necrosis factor (TNF) receptor signaling pathway

5.9.2 NF-κB Signaling Inhibitors

Many synthetic IKK inhibitors, e.g., BMS-345541, IMD-0354, PS-1145, SC-514, and TPCA-1 (Fig. 5.18), have been developed, and since some are now commercially available, they have been widely used in biochemical experiments [112]. IKKα and IKKβ are activated via the phosphorylation of two serine residues (Ser-176 and Ser-180 in IKKα and Ser-177 and Ser-181 in IKKβ) in the activation loop of the catalytic domain [110]. Cys-178 and Cys-179 are present in the activation loops of IKKα and IKKβ, respectively [112]. The Cys-179 of IKKβ is known to be highly susceptible to alkylation by thiol-reactive compounds, and their covalent binding inhibits the activation of IKKβ. A large number of compounds, including sesquiterpene lactones, e.g., parthenolide (Fig. 5.18), and epoxyquinoids, e.g., jesterone dimer (Fig. 5.18), have been shown to inhibit IKKβ by covalent modifications to Cys-179 [112, 113]. NF-κB subunits contain a highly conserved cysteine in the N-terminal Rel homology region. It has recently been shown that p65 at Cys-38 undergoes H_2S-dependent sulfhydration, which promotes interaction with the coactivator ribosomal protein S3 [114]. Many compounds, including sesquiterpene lactones, e.g., helenalin (Fig. 5.18), and epoxyquinoids, e.g., DHMEQ (Fig. 5.18), are known to bind covalently to the Cys-38 of p65 and inhibit its nuclear translocation or DNA binding activity [112, 113]. Bortezomib (PS-341) (Fig. 5.18) is a peptide boronic acid that has been reported to potently inhibit proteasome [115]. Bortezomib has been used as a first-in-class proteasome inhibitor in the treatment of multiple myeloma. Proteasome inhibitors, including MG-132 (Fig. 5.18), efficiently block the degradation step of IκB in the NF-κB signaling pathway. Iguratimod (T-614) (Fig. 5.18) has been approved as an antirheumatic drug. It has been shown to inhibit the activation of NF-κB downstream of the degradation of IκB [116]. Bay 11-7082 (Fig. 5.18) was initially identified by screening of cell-based assays and was shown to inhibit TNF-α-induced IκBα phosphorylation [117]. It has been reported to inhibit the ATPase activity of

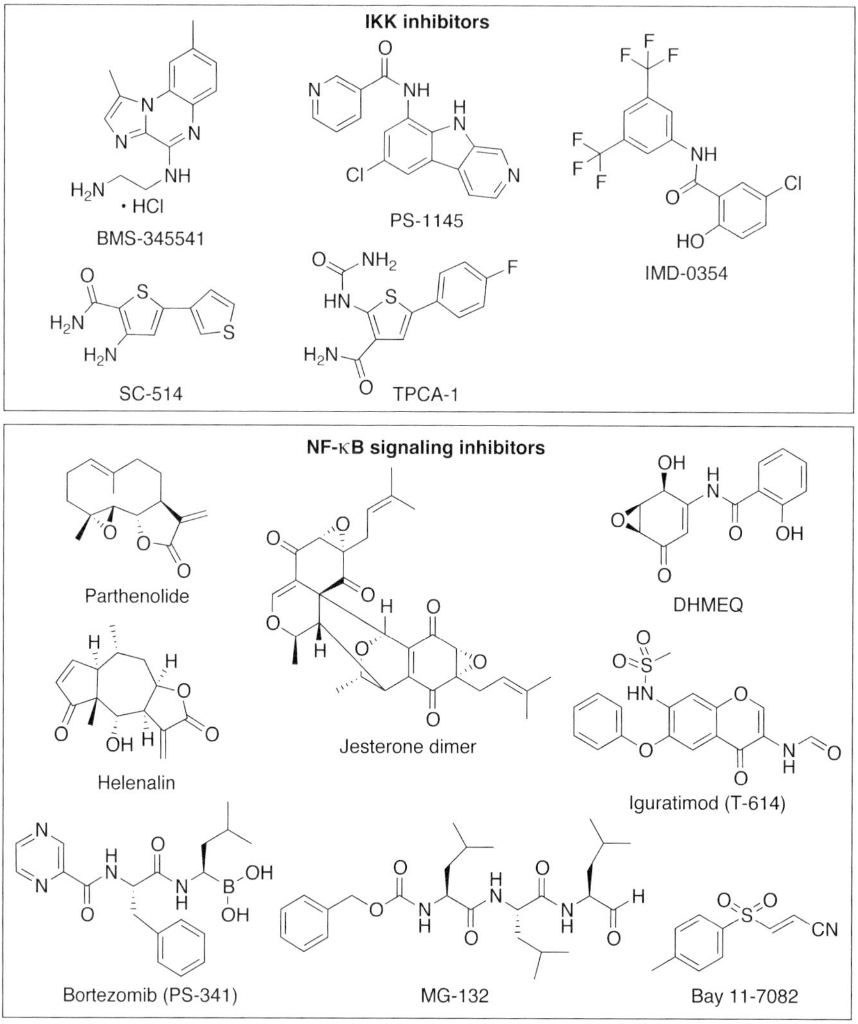

Fig. 5.18 Structures of IκB kinase (IKK) inhibitors and nuclear factor κB (NF-κB) signaling inhibitors

NLRP3 inflammasomes in macrophages in an NF-κB-independent manner [118], and to inactivate protein tyrosine phosphatases by forming a covalent adduct with the active site cysteine [119].

Acknowledgments I sincerely thank Dr. Hiroyuki Koshino for his help in confirming and drawing the structures of the compounds. I am also very grateful to Dr. Kazuo Nagai for his critical reading of this manuscript.

References

1. Schulze-Luehrmann J, Ghosh S (2006) Antigen-receptor signaling to nuclear factor κB. Immunity 25:701–715
2. Brownlie RJ, Zamoyska R (2013) T cell receptor signaling networks: branched, diversified and bounded. Nat Rev Immunol 13:257–269
3. Vargas L, Hamasy A, Nore BF, Smith CIE (2013) Inhibitors of BTK and ITK: state of the new drugs for cancer, autoimmunity and inflammatory diseases. Scand J Immunol 78:130–139
4. Takeuchi O, Akira S (2010) Pattern recognition receptors and inflammation. Cell 140:805–820
5. Iwasaki A, Medzhitov R (2010) Regulation of adaptive immunity by the innate immune system. Science 327:291–295
6. Akinleye A, Chen Y, Mukhi N, Song Y, Liu D (2013) Ibrutinib and novel BTK inhibitors in clinical development. J Hematol Oncol 6:59
7. Conley JM, Gallagher MP, Berg LJ (2016) T cells and gene regulation: the switching on and turning up of genes after T cell receptor stimulation in CD8 T cells. Front Immunol 7:76
8. Oh-hora M, Rao A (2008) Calcium signaling in lymphocytes. Curr Opin Immunol 20:250–258
9. Newton AC (2010) Protein kinase C: poised to signal. Am J Physiol Endocrinol Metab 298: E395–E402
10. Thome M, Charton JE, Pelzer C, Hailfinger S (2010) Antigen receptor signaling to NF-κB via CARMA1, BCL10, and MALT1. Cold Spring Harb Perspect Biol 2:a003004
11. Hogan PG, Chen L, Nardone J, Rao A (2003) Transcriptional regulation by calcium, calcineruin, and NFAT. Genes Dev 17:2205–2232
12. Sieber M, Baumgrass R (2009) Novel inhibitors of the calcineurin/NFATc hub–alternatives to CsA and FK506? Cell Commun Signal 7:25
13. Rodríguez A, Roy J, Martínez-Martínez S, López-Maderuelo MD, Niño-Moreno P, Ortí L, Pantoja-Uceda D, Pineda-Lucena A, Cyert MS, Redondo JM (2009) A conserved docking surface on calcineurin mediates interaction with substrates and immunosuppressants. Mol Cell 33:616–626
14. Yang WC, Collette Y, Nunès JA, Olive D (2000) Tec kinases: a family with multiple roles in immunity. Immunity 12:373–382
15. Lin TA, McIntyre KW, Das J, Liu C, O'Day KD, Penhallow B, Hung CY, Whitney GS, Shuster DJ, Yang X, Townsend R, Postelnek J, Spergel SH, Lin J, Moquin RV, Furch JA, Kamath AV, Zhang H, Marathe PH, Perez-Villar JJ, Doweyko A, Killar L, Dodd JH, Barrish JC, Wityak J, Kanner SB (2004) Selective Itk inhibitors block T-cell activation and murine lung inflammation. Biochemistry 43:11056–11062
16. Mahajan S, Ghosh S, Sudbeck EA, Zheng Y, Downs S, Hupke M, Uckun FM (1999) Rational design and synthesis of a novel anti-leukemic agent targeting Bruton's tyrosine kinase (BTK), LMF-A13 [α-cyano-β-hydroxy-β-methyl-N-(2,5-dibromophenyl)propenamide]. J Biol Chem 274:9587–9599
17. Pan Z, Scheerens H, Li SJ, Schultz BE, Sprengeler PA, Burrill LC, Mendonca RV, Sweeney MD, Scott KCK, Grothaus PG, Jeffery DA, Spoerke JM, Honigberg LA, Young PR, Dalrymple SA, Palmer JT (2007) Discovery of selective irreversible inhibitors for Burton's tyrosine kinase. ChemMedChem 2:58–61
18. Honigberg LA, Smith AM, Sirisawad M, Verner E, Loury D, Chang B, Li S, Pan Z, Thamm DH, Miller RA, Buggy JJ (2010) The Bruton tyrosine kinase inhibitor PCI-32765 blocks B-cell activation and is efficacious in models of autoimmune disease and B-cell malignancy. Proc Natl Acad Sci USA 107:13075–13080
19. Di Paolo JA, Huang T, Balazs M, Barbosa J, Barck KH, Bravo BJ, Carano RAD, Darrow J, Davies DR, DeForge LE, Diehl L, Ferrando R, Gallion SL, Giannetti AM, Gribling P, Hurez V, Hymowitz SG, Jones R, Kropf JE, Lee WP, Maciejewski PM, Mitchell SA,

Rong H, Staker BL, Whitney JA, Yeh S, Young WB, Yu C, Zhang J, Reif K, Currie KS (2011) Specific Btk inhibition suppresses B cell- and myeloid cell-mediated arthritis. Nat Chem Biol 7:41–50
20. Thome M (2008) Multifunctional roles for MALT1 in T-cell activation. Nat Rev Immunol 8:495–500
21. Rebeaud F, Hailfinger S, Posevitz-Fejfar A, Tapernoux M, Moser R, Rueda D, Gaide O, Guzzardi M, Iancu EM, Rufer N, Fasel N, Thome M (2008) The proteolytic activity of the paracaspase MALT1 is key in T cell activation. Nat Immunol 9:272–281
22. Fontan L, Yang C, Kabaleeswaran V, Volpon L, Osborne MJ, Beltran E, Garcia M, Cerchietti L, Shaknovich R, Yang SN, Fang F, Gascoyne RD, Martinez-Climent JA, Glickman JF, Borden K, Wu H, Melnick A (2012) MALT1 small molecule inhibitors specifically suppress ABC-DLBCL in vitro and in vivo. Cancer Cell 22:812–824
23. Nagel D, Spranger S, Vincendeau M, Grau M, Raffegerst S, Kloo B, Hlahla D, Neuenschwander M, Peter von Kries J, Hadian K, Dörken B, Lenz P, Lenz G, Schendel DJ, Krappmann D (2012) Pharmacologic inhibition of MALT1 protease by phenothiazines as a therapeutic approach for the treatment of aggressive ABC-DLBCL. Cancer Cell 22:825–837
24. Barry M, Bleackley RC (2002) Cytotoxic T lymphocytes: all roads lead to death. Nat Rev Immunol 2:401–409
25. Law RHP, Lukoyanova N, Voskoboinik I, Caradoc-Davies TT, Baran K, Dunstone MA, D'Angelo ME, Orlova EV, Coulibaly F, Verschoor S, Browne KA, Ciccone A, Kuiper MJ, Bird PI, Trapani JA, Saibil HR, Whisstock JC (2010) The structural basis for membrane binding and pore formation by lymphocyte perforin. Nature 468:447–451
26. Baran K, Dunstone M, Chia J, Ciccone A, Browne KA, Clarke CJP, Lukoyanova N, Saibil H, Whisstock JC, Voskoboinik I, Trapani JA (2009) The molecular basis for perforin oligomerization and transmembrane pore assembly. Immunity 30:684–695
27. Uellner R, Zvelebil MJ, Hopkins J, Jones J, MacDougall LK, Morgan BP, Podack E, Waterfield MD, Griffiths GM (1997) Perforin is activated by a proteolytic cleavage during biosynthesis which reveals a phospholipid-binding C2 domain. EMBO J 16:7287–7296
28. Brennan AJ, Chia J, Browne KA, Ciccone A, Ellis S, Lopez JA, Susanto O, Verschoor S, Yagita H, Whisstock JC, Trapani JA, Voskoboinik I (2011) Protection from endogenous perforin: glycans and the C terminus regulate exocytic trafficking in cytotoxic lymphocytes. Immunity 34:879–892
29. Cullen SP, Brunet M, Martin SJ (2010) Granzymes in cancer and immunity. Cell Death Differ 17:616–623
30. Chowdhury D, Lieberman J (2008) Death by a thousand cuts: granzyme pathways of programmed cell death. Annu Rev Immunol 26:389–420
31. Krzewski K, Coligan JE (2012) Human NK cell lytic granules and regulation of their exocytosis. Front Immunol 3:335
32. Lena G, Trapani JA, Sutton VR, Ciccone A, Browne KA, Smyth MJ, Denny WA, Spicer JA (2008) Dihydrofuro[3,4-*c*]pyridinones as inhibitors of the cytolytic effects of the pore-forming glycoprotein perforin. J Med Chem 51:7614–7624
33. Kam CM, Hudig D, Powers JC (2000) Granzymes (lymphocyte serine proteases): characterization with natural and synthetic substrates and inhibitors. Biochim Biophys Acta 1477:307–323
34. Willoughby CA, Bull HG, Garcia-Calvo M, Jiang J, Chapman KT, Thornberry NA (2002) Discovery of potent, selective human granzyme B inhibitors that inhibit CTL mediated apoptosis. Bioorg Med Chem Lett 12:2197–2200
35. Mahrus S, Craik CS (2005) Selective chemical functional probes of granzymes A and B reveal granzyme B is a major effector of natural killer cell-mediated lysis of target cells. Chem Biol 12:567–577
36. Dell'Angelica EC, Mullins C, Caplan S, Bonifacino JS (2000) Lysosome-related organelles. FASEB J 14:1265–1278

37. Kataoka T, Takaku K, Magae J, Shinohara N, Takayama H, Kondo S, Nagai K (1994) Acidification is essential for maintaining the structure and function of lytic granules of CTL. J Immunol 153:3938–3947
38. Kataoka T, Sato M, Kondo S, Nagai K (1996) Estimation of pH and the number of lytic granules in a $CD8^+$ CTL clone treated with an inhibitor of vacuolar type H^+-ATPase, concanamycin A. Biosci Biotechnol Biochem 60:1729–1731
39. Kataoka T, Togashi K, Takayama H, Takaku K, Nagai K (1997) Inactivation and proteolytic degradation of perforin within lytic granules upon neutralization of acidic pH. Immunology 91:493–500
40. Togashi K, Kataoka T, Nagai K (1997) Characterization of a series of vacuolar type H^+-ATPase inhibitors on CTL-mediated cytotoxicity. Immunol Lett 55:139–144
41. Kataoka T, Shinohara N, Takayama H, Takaku K, Kondo S, Yonehara S, Nagai K (1996) Concanamycin A, a powerful tool for characterization and estimation of contribution of perforin- and Fas-based lytic pathways in cell-mediated cytotoxicity. J Immunol 156:3678–3686
42. Strasser A, Jost PJ, Nagata S (2009) The many roles of FAS receptor signaling in the immune system. Immunity 20:180–192
43. Lavrik IN, Krammer PH (2012) Regulation of CD95/Fas signaling at the DISC. Cell Death Differ 19:36–41
44. Boatright KM, Salvesen GS (2003) Mechanisms of caspase activation. Curr Opin Immunol 15:725–731
45. Schafer ZT, Kornbluth S (2006) The apoptosome: physiological, developmental, and pathological modes of regulation. Dev Cell 10:549–561
46. Berger AB, Sexton KB, Bogyo M (2006) Commonly used caspase inhibitors designed based on substrate specificity profiles lack selectivity. Cell Res 16:961–963
47. Linton SD, Aja T, Armstrong RA, Bai X, Chen LS, Chen N, Ching B, Contreras P, Diaz JL, Fisher CD, Fritz LC, Gladstone P, Groessl T, Gu X, Herrmann J, Hirakawa BP, Hoglen NC, Jahangiri KG, Kalish VJ, Karanewsky DS, Kodandapani L, Krebs J, McQuiston J, Meduna SP, Nalley K, Robinson ED, Sayers RO, Sebring K, Spada AP, Ternansky RJ, Tomaselli KJ, Ullman BR, Valentino KL, Weeks S, Winn D, Wu JC, Yeo P, Zhang CZ (2005) First-in-class pan caspase inhibitor developed for the treatment of liver disease. J Med Chem 48:6779–6782
48. Bando M, Hasegawa M, Tsuboi Y, Miyake Y, Shiina M, Ito M, Handa H, Nagai K, Kataoka T (2003) The mycotoxin penicillic acid inhibits Fas ligand-induced apoptosis by blocking self-processing of caspase-8 in death-inducing signaling complex. J Biol Chem 278:5786–5793
49. Kakeya H, Miyake Y, Shoji M, Kishida S, Hayashi Y, Kataoka T, Osada H (2003) Novel non-peptide inhibitors targeting death receptor-mediated apoptosis. Bioorg Med Chem Lett 13:3743–3746
50. Miyake Y, Kakeya H, Kataoka T, Osada H (2003) Epoxycyclohexenone inhibits Fas-mediated apoptosis by blocking activation of pro-caspase-8 in the death-inducing signaling complex. J Biol Chem 278:11213–11220
51. Mitsui T, Miyake Y, Kakeya H, Osada H, Kataoka T (2004) ECH, an epoxycyclohexenone derivative that specifically inhibits Fas ligand-dependent apoptosis in CTL-mediated cytotoxicity. J Immunol 172:3428–3436
52. Mitsui T, Miyake Y, Kakeya H, Hayashi Y, Osada H, Kataoka T (2005) RKTS-33, an epoxycyclohexenone derivative that specifically inhibits Fas ligand-dependent apoptosis in CTL-mediated cytotoxicity. Biosci Biotechnol Biochem 69:1923–1928
53. He JS, Ostergaard HL (2007) CTLs contain and use intracellular stores of FasL distinct from cytolytic granules. J Immunol 179:2339–2348
54. Kataoka T, Taniguchi M, Yamada A, Suzuki H, Hamada S, Magae J, Nagai K (1996) Identification of low molecular weight probes on perforin- and Fas-based killing mediated by cytotoxic T lymphocytes. Biosci Biotechnol Biochem 60:1726–1728
55. Akira S, Takeda K (2004) Toll-like receptor signaling. Nat Rev Immunol 4:499–511

56. Kawai T, Akira S (2011) Toll-like receptors and their cross talk with other innate receptors in infection and immunity. Immunity 34:637–650
57. Bhoj VG, Chen ZJ (2009) Ubiquitylation in innate and adaptive immunity. Nature 458:430–437
58. Lee CC, Avalos AM, Ploegh HL (2012) Accessory molecules for toll-like receptors and their function. Nat Rev Immunol 12:168–179
59. Muta T, Takeshige K (2001) Essential roles of CD14 and lipopolysaccharide-binding protein for activation of toll-like receptor (TLR)2 as well as TLR4. Eur J Biochem 268:4580–4589
60. Kim HM, Park BS, Kim JI, Kim SE, Lee J, Oh SC, Enkhbayar P, Matsushima N, Lee H, Yoo OJ, Lee JO (2007) Crystal structure of the TLR4-MD-2 complex with bound endotoxin antagonist eritoran. Cell 130:906–917
61. Shirey KA, Lai W, Scott AJ, Lipsky M, Mistry P, Pletneva LM, Karp CL, McAlees J, Gioannini TL, Weiss J, Chen WH, Ernst RK, Rossignol DP, Gusovsky F, Blanco JC, Vogel SN (2013) The TLR4 antagonist eritoran protects mice from lethal influenza infection. Nature 497:498–502
62. Matsunaga N, Tsuchimori N, Matsumoto T, Ii M (2011) TAK-242 (resatorvid), a small-molecule inhibitor of toll-like receptor (TLR) 4 signaling, binds selectively to TLR4 and interferes with interactions between TLR4 and its adaptor molecules. Mol Pharmacol 79:34–41
63. Bochkov VN, Kadl A, Huber J, Gruber F, Binder BR, Leitinger N (2002) Protective role of phospholipid oxidation products in endotoxin-induced tissue damage. Nature 419:77–81
64. Cheng K, Wang X, Yin H (2011) Small-molecule inhibitors of the TLR3/dsRNA complex. J Am Chem Soc 133:3764–3767
65. Shukla NM, Mutz CA, Malladi SS, Warshakoon HJ, Balakrishna R, David SA (2012) Toll-like receptor (TLR)-7 and -8 modulatory activities of dimeric imdazoquinolines. J Med Chem 55:1106–1116
66. Franklin BS, Ishizaka ST, Lamphier M, Gusovsky F, Hansen H, Rose J, Zheng W, Ataíde MA, Oliveira RB, Golenbock DT, Gazzinelli RT (2011) Therapeutical targeting of nucleic acid-sensing toll-like receptors prevents experimental cerebral malaria. Proc Natl Acad Sci USA 108:3689–3694
67. Lund J, Sato A, Akira S, Medzhitov R, Iwasaki A (2003) Toll-like receptor 9-mediated recognition of herpes simplex virus by plasmacytoid dendritic cells. J Exp Med 198:513–520
68. Kužnik A, Benčina M, Švajger U, Jeras M, Rozman B, Jerala R (2011) Mechanism of endosomal TLR inhibition by antimalarial drugs and imidazoquinolines. J Immunol 186:4794–4804
69. Wang Z, Liu J, Sudom A, Ayres M, Li S, Wesche H, Powers JP, Walker NPC (2006) Crystal structures of IRAK-4 kinase in complex with inhibitors: a serine/threonine kinase with tyrosine as a gatekeeper. Structure 14:1835–1844
70. Wang Z, Weshe H, Stevens T, Walker N, Yeh WC (2009) IRAK-4 inhibitors for inflammation. Curr Top Med Chem 9:724–737
71. Feldman RI, Wu JM, Polokoff MA, Kochanny MJ, Dinter H, Zhu D, Biroc SL, Alicke B, Bryant J, Yuan S, Buckman BO, Lentz D, Ferrer M, Whitlow M, Adler M, Finster S, Chang Z, Arnaiz DO (2005) Novel small molecule inhibitors of 3-phosphoinositide-dependent kinase-1. J Biol Chem 280:19867–19874
72. Clark K, Plater L, Peggie M, Cohen P (2009) Use of the pharmacological inhibitor BX-795 to study the regulation and physiological roles of TBK1 and IκB kinase ε. J Biol Chem 284:14136–14146
73. Ninomiya-Tsuji J, Kajino T, Ono K, Ohtomo T, Matsumoto M, Shiina M, Mihara M, Tsuchiya M, Matsumoto K (2003) A resorcylic acid lactone, 5Z-7-oxozeaenol, prevents inflammation by inhibiting the catalytic activity of TAK1 MAPK kinase kinase. J Biol Chem 278:18485–18490

74. Wu J, Powell F, Larsen NA, Lai Z, Byth KF, Read J, Gu RF, Roth M, Toader D, Saeh JC, Chen H (2013) Mechanism and in vitro pharmacology of TAK1 inhibition by (5*Z*)-7-oxozeaenol. ACS Chem Biol 8:643–650
75. Rathinam VAK, Vanaja SK, Fitzgerald KA (2012) Regulation of inflammasome signaling. Nat Immunol 13:333–342
76. Man SM, Kanneganti TD (2016) Converging roles of caspases in inflammasome activation, cell death and innate immunity. Nat Rev Immunol 16:7–21
77. Schroder K, Tschopp J (2010) The inflammasomes. Cell 140:821–832
78. Mariathasan S, Weiss DS, Newton K, McBride J, O'Rourke K, Roose-Girma M, Lee WP, Weinrauch Y, Monack DM, Dixit VM (2006) Cryopyrin activates the inflammasome in response to toxin and ATP. Nature 440:228–232
79. Muñoz-Planillo R, Kuffa P, Martínez-Colón G, Smith BL, Rajendiran TM, Nuñez G (2013) K^+ efflux is the common trigger of NLRP3 inflammasome activation by bacterial toxins and particulate matter. Immunity 38:1142–1153
80. Lamkanfi M, Mueller JL, Vitari AC, Misaghi S, Fedorova A, Deshayes K, Lee WP, Hoffman HM, Dixit VM (2009) Glyburide inhibits the cryopyrin/Nalp3 inflammasome. J Cell Biol 187:61–70
81. Hett EC, Slater LH, Mark KG, Kawate T, Monks BG, Stutz A, Latz E, Hung DT (2013) Chemical genetics reveals a kinase-independent role for protein kinase R in pyroptosis. Nat Chem Biol 9:398–405
82. Laliberte RE, Perregaux DG, Hoth LR, Rosner PJ, Jordan CK, Peese KM, Eggler JF, Dombroski MA, Geoghegan KF, Gabel CA (2003) Glutathione *S*-transferase omega 1-1 is a target of cytokine release inhibitory drugs and may be responsible for their effect on interleukin-1β posttranslational processing. J Biol Chem 278:16567–16578
83. Coll RC, Robertson AAB, Chae JJ, Higgins SC, Muñoz-Planillo R, Inserra MC, Vetter I, Dungan LS, Monks BG, Stutz A, Croker DE, Butler MS, Haneklaus M, Sutton CE, Nuñez G, Latz E, Kastner DL, Mills KHG, Masters SL, Schroder K, Cooper MA, O'Neill LAJ (2015) A small molecule inhibitor of the NLRP3 inflammasome is a potential therapeutic for inflammatory diseases. Nat Med 21:248–255
84. Youm YH, Nguyen KY, Grant RW, Goldberg EL, Bodogai M, Kim D, D'Agostino D, Planavsky N, Lupfer C, Kanneganti TD, Kang S, Horvath TL, Fahmy TM, Crawford PA, Biragyn A, Alnemri E, Dixit VD (2015) Ketone body β-hydroxybutyrate blocks NLRP3 inflammasome-mediated inflammatory disease. Nat Med 21:263–269
85. Tingo-Aranjuez JT, Asara JM, Abbott DW (2010) Inhibition of RIP2's tyrosine kinase activity limits NOD2-driven cytokine responses. Genes Dev 24:2666–2677
86. Baldwin AG, Brough D, Freeman S (2016) Inhibiting the inflammasome: a chemical perspective. J Med Chem 59:1691–1710
87. Ekert PG, Silke J, Vaux DL (1999) Caspase inhibitors. Cell Death Differ 6:1081–1086
88. Callus BA, Vaux DL (2007) Caspase inhibitors: viral, cellular and chemical. Cell Death Differ 14:73–78
89. Groß O, Yazdi AS, Thomas CJ, Masin M, Heinz LX, Guarda G, Quadroni M, Drexler SK, Tschopp J (2012) Inflammasome activators induce interleukin-1α secretion via distinct pathways with differential requirement for the protease function of caspase-1. Immunity 36:388–400
90. Rudolphi K, Gerwin N, Verzijl N, van der Kraan P, van den Berg W (2003) Pralnacasan, an inhibitor of interleukin-1β converting enzyme, reduces joint damage in two murine models of osteoarthritis. Osteoarthr Cartil 11:738–746
91. Wannamaker W, Davies R, Namchuk M, Pollard J, Ford P, Ku G, Decker C, Charifson P, Weber P, Germann UA, Kuida K, Randle JC (2007) (*S*)-1-((*S*)-2-{[1-(4-amino-3-chlorophenyl)-methanoyl]-amino}-3,3-dimethyl-butanoyl)-pyrrolidine-2-caboxylic acid ((2*R*,3*S*)-2-ethoxy-5-oxo-tetrahydro-furan-3-yl)-amide (VX-765), an orally available selective interleukin (IL)-converting enzyme/caspase-1 inhibitor, exhibits potent anti-inflammatory activities by inhibiting the release of IL-1β and IL-18. J Pharmacol Exp Ther 321:509–516

92. Scheer JM, Romanowski MJ, Wells JA (2006) A common allosteric site and mechanism in caspases. Proc Natl Acad Sci USA 103:7595–7600
93. Feldman T, Kabaleeswaran V, Jang SB, Antczak C, Djaballah H, Wu H, Jiang X (2012) A class of allosteric caspase inhibitors identified by high-throughput screening. Mol Cell 47:585–595
94. Loo YM, Gare M Jr (2011) Immune signaling by RIG-I-like receptors. Immunity 34:680–692
95. Yoneyama M, Fujita T (2009) RNA recognition and signal transduction by RIG-I-like receptors. Immunol Rev 227:54–65
96. Goubau D, Deddouche S, Reis e Sousa C (2013) Cytosolic sensing of viruses. Immunity 38:855–869
97. Ablasser A, Bauernfeind F, Hartmann G, Latz E, Fitzgerald KA, Hornung V (2009) RIG-I-dependent sensing of poly(dA:dT) through the induction of an RNA polymerase III-transcribed RNA intermediate. Nat Immunol 10:1065–1072
98. Ranjith-Kumar CT, Murali A, Dong W, Srisathiyanarayanan D, Vaughan R, Ortiz-Alacantara J, Bhardwaj K, Li X, Li P, Kao CC (2009) Agonist and antagonist recognition by RIG-I, a cytoplasmic innate immunity receptor. J Biol Chem 284:1155–1165
99. Ranjith-Kumar CT, Lai Y, Sarisky RT, Kao CC (2010) Green tea catechin, epigallocatechin gallate, suppresses signaling by the dsRNA innate immune receptor RIG-I. PLoS One 9: e12878
100. Hardison SE, Brown GD (2012) C-type lectin receptors orchestrate anti-fungal immunity. Nat Immunol 13:817–822
101. Kerrigan AM, Brown GD (2011) Syk-coupled C-type lectins in immunity. Trends Immunol 32:151–156
102. Mócsai A, Ruland J, Tybulewicz VLJ (2010) The SYK tyrosine kinase: a crucial player in diverse biological functions. Nat Rev Immunol 10:387–402
103. Pamuk ON, Tsokos GC (2010) Spleen tyrosine kinase inhibition in the treatment of autoimmune, allergic and autoinflammatory diseases. Arthritis Res Ther 12:222
104. D'Cruz OJ, Uckun FM (2012) Targeting spleen tyrosine kinase (SYK) for treatment of human disease. J Pharm Drug Deliv Res 1:2
105. Geahlen RL (2014) Getting Syk: spleen tyrosine kinase as a therapeutic target. Trends Pharmacol Sci 35:414–422
106. Kwon JY, Seo SG, Heo YS, Yue S, Cheng JX, Lee KW, Kim KH (2012) Piceatannol, natural polyphenolic stilbene, inhibits adipogenesis via modulation of mitotic clonal expansion and insulin receptor-dependent insulin signaling in early phase of differentiation. J Biol Chem 287:11566–11578
107. Moriya K, Rivera J, Odom S, Sakuma Y, Muramoto K, Yoshiuchi T, Miyamoto M, Yamada K (1997) ER-27319, an acridone-related compound, inhibits release of antigen-induced allergic mediators from mast cells by selective inhibition of Fcε receptor I-mediated activation of Syk. Proc Natl Acad Sci USA 94:12539–12544
108. Hayden MS, Ghosh S (2014) Regulation of NF-κB by TNF family cytokines. Semin Immunol 26:253–266
109. Sasaki K, Iwai K (2015) Roles of linear ubiquitinylation, a crucial regulator of NF-κB and cell death, in the immune system. Immunol Rev 266:175–189
110. Perkins ND (2006) Post-translational modifications regulating the activity and function of the nuclear factor kappa B pathway. Oncogene 25:6717–6730
111. Karin M, Greten FR (2005) NF-κB: linking inflammation and immunity to cancer development and progression. Nat Rev Immunol 5:749–759
112. Kataoka T (2009) Chemical biology of inflammatory cytokine signaling. J Antibiot 62:655–667
113. Gilmore TD, Herscovitch M (2006) Inhibitors of NF-κB signaling: 785 and counting. Oncogene 25:6887–6899

114. Sen N, Paul BD, Gadalla MM, Mustafa AK, Sen T, Xu R, Kim S, Snyder SH (2012) Hydrogen sulfide-linked sulfhydration of NF-κB mediates its antiapoptotic actions. Mol Cell 45:13–24
115. Palombella VJ, Conner EM, Fuseler JW, Destree A, Davis JM, Laroux FS, Wolf RE, Huang J, Brand S, Elliott PJ, Lazarus D, McCormack T, Parent L, Stein R, Adams J, Grisham MB (1998) Role of the proteasome and NF-κB in streptococcal cell wall-induced polyarthritis. Proc Natl Acad Sci USA 95:15671–15676
116. Aikawa Y, Yamamoto M, Yamamoto T, Morimoto K, Takana K (2002) An anti-rheumatic agent T-614 inhibits NF-κB activation in LPS- and TNF-α-stimulated THP-1 cells without interfering with IκBα degradation. Inflamm Res 51:188–194
117. Pierce JW, Schoenleber R, Jesmok G, Best J, Moore SA, Collins T, Gerritsen ME (1997) Novel inhibitors of cytokine-induced IκBα phosphorylation and endothelial cell adhesion molecule expression show anti-inflammatory effects in vivo. J Biol Chem 272:21096–21103
118. Juliana C, Fernandes-Alnemri T, Wu J, Datta P, Solorzano L, Yu JW, Meng R, Quong AA, Latz E, Scott CP, Alnemri ES (2010) Anti-inflammatory compounds parthenolide and Bay 11-7082 are direct inhibitors of the inflammasome. J Biol Chem 285:9792–9802
119. Krishnan N, Bencze G, Cohen P, Tonks NK (2013) The anti-inflammatory compound Bay-11-7082 is a potent inhibitor of protein tyrosine phosphatases. FEBS J 280:2830–2841

Chapter 6
Bioprobes at a Glance

Hiroyuki Koshino and Takeo Usui

Abstract This chapter provides the fact database of bioprobes that have been reported.

1. The "Bioprobes at a Glance" section is ordered alphabetically by entry name.
2. A compound with an asterisk (*) at the entry name is the synthetic compound.
3. Molecular weight is rounded off to two decimal places.
4. Solubility is represented by symbols: +++, easily soluble; ++, soluble; +, sparingly soluble; ±, barely soluble; −, insoluble.

Keywords Discovery • Isolation • Structure • Chemical study • Target identification • Biological activity

H. Koshino
Molecular Structure Characterization Unit, RIKEN CSRS, Hirosawa 2-1, Wako, Saitama 351-0198, Japan

T. Usui (✉)
Faculty of Life and Environmental Sciences, University of Tsukuba, Tennodai 1-1-1, Tsukuba, Ibaraki 305-8572, Japan
e-mail: usui.takeo.kb@u.tsukuba.ac.jp

© Springer Japan KK 2017

H. Osada (ed.), *Bioprobes*, DOI 10.1007/978-4-431-56529-1_6

Compound List

(continued)

(continued)

16F16*

Keywords: [Protein disulfide isomerase (PDI)] [Misfolded proteins]

Structure

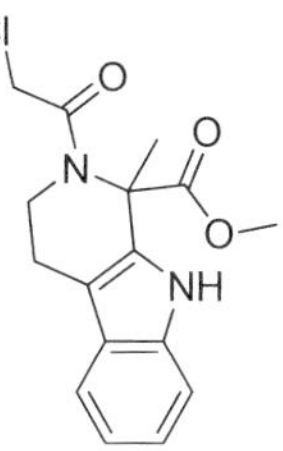

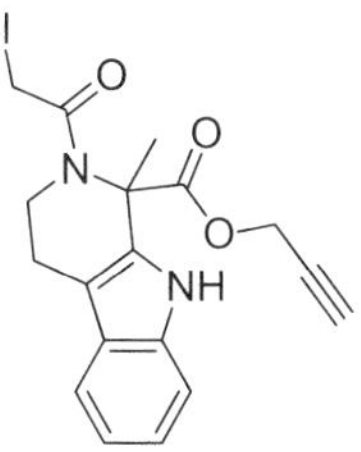

	16F16	16F16A
Molecular formula	$C_{16}H_{17}ClN_2O_3$	$C_{18}H_{17}ClN_2O_3$
Molecular weight	320.77	344.80
Solubility	DMSO, ++; H_2O, ±; MeOH, +	DMSO, ++; H_2O, ±; MeOH, +

Discovery/Isolation/Chemical Studies

16F16 was identified as a small-molecule suppressor of polyglutamine-induced apoptosis in a PC12 cell model of Huntington's disease from 68,887 compounds (small-molecule libraries containing natural products, natural product analogs, synthetic drug–like compounds, and annotated biologically active compounds) [1]. An analog of 16F16 lacking the chloro substituent was inactive, and the alkyne analog of 16F16 (16F16A) suppressed Q103-induced apoptosis, suggesting that 16F16 covalently binds its target proteins. By coupling azidofluorescein to PC12-Q103 lysates treated with 16F16A, two isoforms of protein disulfide isomerase (PDIA1 and PDIA3) were identified as specific targets of 16F16A.

Biological Studies

16F16 restored "normal" cell morphology to induced Q103 (the first exon of the human huntingtin gene (HTT) containing mutant polyglutamine repeats fused to

EGFP)–expressing cells without affecting the accumulation of the HTT-Q103 inclusion body. 16F16 inhibited the catalytic activity of PDI, and suppressed the PDI-induced MOMP (mitochondrial outermembrane permeabilization) and mitochondrial release of cytochrome *c*. In a rat corticostriatal brain-slice model, PDI inhibitors including 16F16 suppressed neurotoxicity induced by mutant huntingtin exon 1 and amyloid β. These results indicate the linking of protein misfolding and apoptotic cell death, and the proapoptotic function of PDI [1].

Biological Activity of 16F16

IC_{50} values for 16F16 and 16F16A in insulin aggregation assay: 74 μM and 108 μM, respectively [2]

GI_{50} for 16F16 and 16F16A against several cancer cell lines: ~ 4 μM [2]

References

1. Hoffstrom BG, Kaplan A, Letso R, Schmid RS, Turmel GJ, Lo DC, Stockwell BR (2010) Inhibitors of protein disulfide isomerase suppress apoptosis induced by misfolded proteins. Nat Chem Biol 6: 900–906
2. Ge J, Zhang CJ, Li L, Chong LM, Wu X, Hao P, Sze SK, Yao SQ (2013) Small molecule probe suitable for in situ profiling and inhibition of protein disulfide isomerase. ACS Chem Biol 8:2577–2585

Aplyronine A and C

Keywords: [Antitumor] [Actin polymerization inhibitor] [Actin-tubulin complex inducer]

Structure

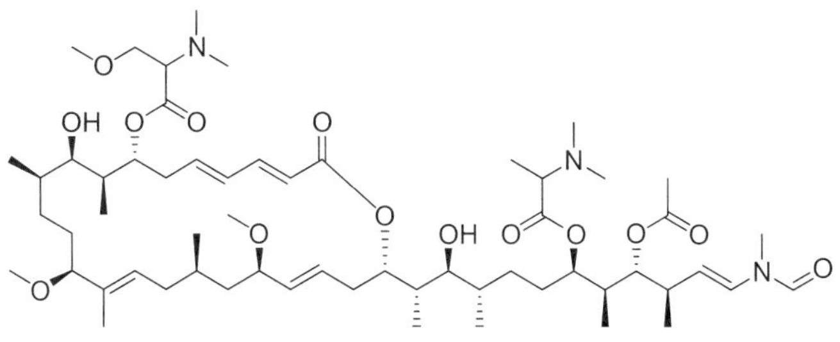

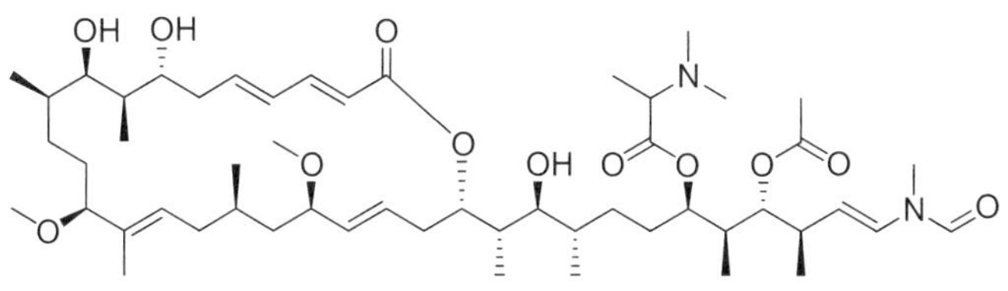

	Aplyronine A	Aplyronine C
Molecular formula	$C_{59}H_{101}N_3O_{14}$	$C_{53}H_{90}N_2O_{12}$
Molecular weight	1076.46	947.31
Solubility	DMSO, +++; H_2O, –; MeOH, +++	DMSO, +++; H_2O, –; MeOH, +++

Discovery/Isolation

Aplyronine A was isolated as a potent antitumor compound from the sea hare *Aplysia kurodai* from the Pacific coast of Mie Prefecture, Japan [1].

Biological Studies

Aplyronine A inhibited actin polymerization and quickly depolymerized F-actin by forming a 1:1 complex with G-actin [2]. X-ray crystallography revealed that aplyronine A binds to a hydrophobic cleft composed of subdomains 1 and 3 of actin by intercalating its aliphatic tail part into the actin molecule [3]. Interestingly, the trimethylserine moiety of aplyronine A extends to the outside of the actin protein. Because aplyronine A shows potent cytotoxicity compared with aplyronine C which lacks trimethylserine moiety, this moiety is thought to be important for cytotoxicity. Kita et al. noticed that both aplyronine A and C disrupt actin network in cells at the same concentration, but only aplyronine A induced apoptosis at low concentrations [4]. Further chemical biological analyses revealed that aplyronine A forms a 1:1:1 heterotrimeric complex with actin and tubulin, in association with actin synergistically binding to tubulin, and inhibits tubulin polymerization. This complex formation results in the inhibition of the spindle formation.

Biological Activity of Aplyronine A and C

IC_{50} values for aplyronine A and C on cell growth of HeLa-S3 cells: 0.039 and 159 ng/ml, respectively [1]

Antitumor activity in vivo of aplyronine A against P388 murine leukemia: 0.08 mg/kg, T/C = 545% [1]

K_d for aplyronine A to G-actin: 0.10 μM [2]

Inhibition of actin network in HeLa cells by aplyronines A and C: 100 nM [4]

Induction of abnormal spindle formation by aplyronine A: 0.1 nM [4]

References

1. Yamada K, Ojika M, Ishigaki T, Yoshida Y, Ekimoto H, Arakawa M (1993) Aplyronine-A, a Potent Antitumor Substance, and the Congeners Aplyronine-B and Aplyronine-C Isolated from the Sea Hare *Aplysia kurodai*. J Am Chem Soc 115:11020–11021
2. Saito S, Watabe S, Ozaki H, Kigoshi H, Yamada K, Fusetani N, Karaki H (1996) Novel actin depolymerizing macrolide aplyronine A. J Biochem 120:552–555
3. Hirata K, Muraoka S, Suenaga K, Kuroda T, Kato K, Tanaka H, Yamamoto M, Takata M, Yamada K, Kigoshi H (2006) Structure basis for antitumor effect of aplyronine A. J Mol Biol 356:945–954
4. Kita M, Hirayama Y, Yoneda K, Yamagishi K, Chinen T, Usui T, Sumiya E, Uesugi M, Kigoshi H (2013) Inhibition of microtubule assembly by a complex of actin and antitumor macrolide aplyronine A. J Am Chem Soc 135:18089–18095

Apoptolidin

Keywords: [Apoptosis inducer] [F_0/F_1 ATPase inhibitor]

Structure

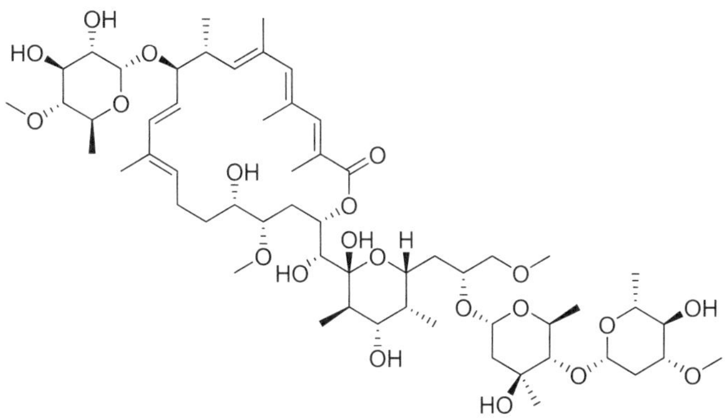

	Apoptolidin
Molecular formula	$C_{58}H_{96}O_{21}$
Molecular weight	1129.39
Solubility	DMSO, +++; H_2O, ±; MeOH, +++

Discovery/Isolation/Chemical Studies

Apoptolidin was isolated as an apoptosis inducer in adenovirus E1A–transformed rat glia cells from *Nocardiopsis* sp. [1]. The structure of apoptolidin was determined by detailed NMR studies including its stereochemistry [2]. The first total synthesis was done by the Nicolau group [3–5].

Biological Studies

Apoptosis induced by 1 μM apoptolidin was independent of p53 status, inhibited by BCL-2, and dependent on the action of caspase-9 in a mouse lymphoma cell line. From a combination of molecular and cell-based pharmacological assays as well as structural comparisons between apoptolidin and other macrocyclic polyketides with known mechanism of action, the target molecule of apoptolidin was revealed as mitochondrial F_0F_1-ATP synthase [6]. Correlation analysis of F_0F_1-ATP synthase inhibitors and gene expression suggested that pyruvate kinase and aspartate aminotransferase are drug sensitivity determinants, suggesting that the mitochondrial apoptotic pathway is activated in response to a shift in balance between aerobic and anaerobic ATP biosynthesis. Indeed, inhibitors of both lactate formation and carbon flux through the Embden–Meyerhof pathway significantly sensitized apoptolidin-resistant tumors to apoptolidin [7].

Biological Activity of Apoptolidin

IC_{50} for glia cells: >100 μg/ml [1]

IC_{50} for RG-E1A-7, RG-E1A19K-2, RG-E1A54K-9, or RG-E1-4 cells: 10–13 ng/ml [1]

K_i for yeast mitochondria F_0F_1-ATPase: 4–5 μM [6]

References

1. Kim JW, Adachi H, Shin-ya K, Hayakawa Y, Seto H (1997) Apoptolidin, a new apoptosis inducer in transformed cells from *Nocardiopsis* sp. J Antibiot (Tokyo) 50:628–630
2. Hayakawa Y, Kim JW, Adachi H, Shin-ya K, Fujita K, Seto H (1998) Structure of apoptolidin, a specific apoptosis inducer in transformed cells. J Am Chem Soc 120:3524–3525
3. Nicolaou KC, Li Y, Fylaktakidou KC, Mitchell HJ, Wei HX, Weyershausen B (2001) Total Synthesis of Apoptolidin: part 1. Retrosynthetic Analysis and Construction of Building Blocks. Angew Chem Int Ed Eng. 40:3849–3854
4. Nicolaou KC, Li Y, Fylaktakidou KC, Mitchell HJ, Sugita K (2001) Total Synthesis of Apoptolidin: part 2. Coupling of Key Building Blocks and Completion of the Synthesis. Angew Chem Int Ed Eng. 40:3854–3857
5. Nicolaou KC, Li Y, Sugita K, Monenschein H, Guntupalli P, Mitchell HJ, Fylaktakidou KC, Vourloumis D, Giannakakou P, O″Brate A (2003) Total synthesis of apoptolidin: completion of the synthesis and analogue synthesis and evaluation. J Am Chem Soc 125:15443–15454
6. Salomon AR, Voehringer DW, Herzenberg LA, Khosla C (2001) Apoptolidin, a selective cytotoxic agent, is an inhibitor of F_0F_1-ATPase. Chem Biol 8:71–80
7. Salomon AR, Voehringer DW, Herzenberg LA, Khosla C (2000) Understanding and exploiting the mechanistic basis for selectivity of polyketide inhibitors of F_0F_1-ATPase. Proc Natl Acad Sci U S A 97:14766–14771

Apratoxin A

Keywords: [Hsc70/Hsp70] [JAK/STAT signaling] [N-glycosylation] [Secretory pathway]

Structure

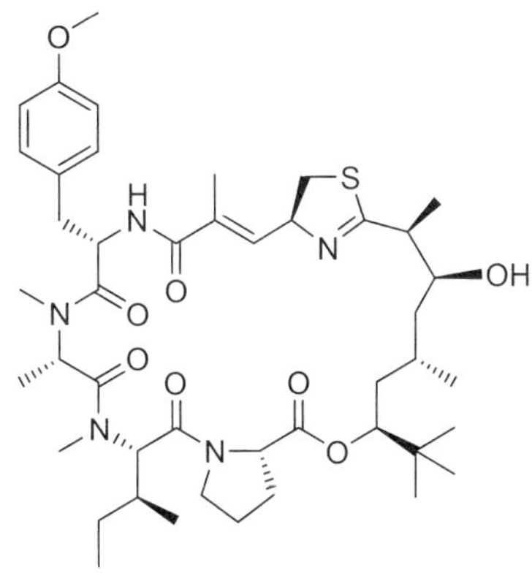

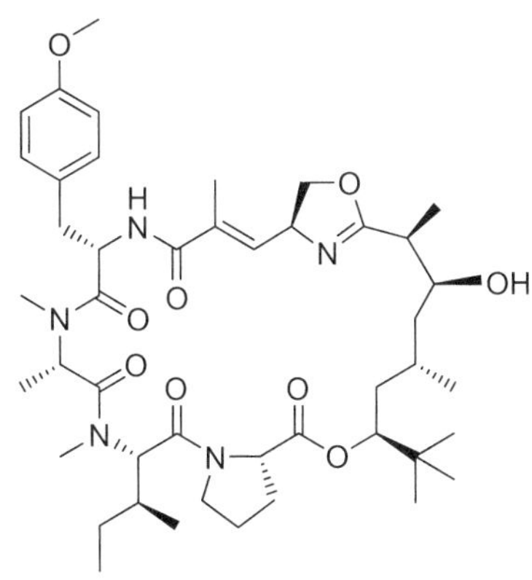

	Apratoxin A	Oz-apratoxin A
Molecular formula	$C_{45}H_{69}N_5O_8S$	$C_{45}H_{69}N_5O_9$
Molecular weight	840.13	824.07
Solubility	DMSO, ? ; H_2O, ? ; MeOH, ++	DMSO, ? ; H_2O, ? ; MeOH, ++

Discovery/Isolation/Chemical Studies

Apratoxin A was isolated as a potent cytotoxin from the marine cyanobacterium *Lyngbya majuscule* Harvey ex Gomont [1].

Using an oxazoline analog of apratoxin A (oz-apraA) which has similar potency with respect to tumor toxicity as aprotoxin A, Hsc70, Hsp70, RARS (arginyl-tRNA synthease protein), and SRP68 were identified as the binding proteins of oz-apraA [2].

Biological Studies

Apratoxin A mediates its antiproliferative activity through the induction of G1-phase cell cycle arrest and apoptosis, which is at least partially initiated through antagonism of FGF signaling via STAT3 [3]. Down-regulation of signal transduction was caused by the degradation of several receptor kinases, which lack *N*-glycosylation. Using in vitro cell-free systems revealed that apratoxin A prevents cotranslational translocation of proteins destined for the secretory pathway [4].

Oz-apraA induces the accumulation of polyubiquitin-positive proteins, and promotes the degradation of Hsp90 clients through chaperone-mediated autophagy. Because apratoxin A binds to Hsc70/Hsp70, it is speculated that apratoxin A inhibits Hsp90 function by stabilizing the interaction of Hsp90 client proteins with Hsc70/Hsp70 and thus prevents their interactions with Hsp90 [2].

Biological Activity of Apratoxin A

IC_{50} for cytotoxicity in vitro against human tumor cell lines: from 0.36 to 0.52 nM [1]
G1 cell cycle arrest for 24- and 48-hour treatment in U2OS cells: 5 nM [3]
Caspase-3 activation and inhibition of STAT3 phosphorylation in U2OS cells: 50 nM [3]
Inhibition of in vitro angiogenesis of HUVECs: 10 nM [3]
Inhibition of *N*-linked glycosylation of several proteins in U2OS cells: 50 nM [4]
Inhibition of cotranslational translocation in vitro: 10 nM [4]
Accumulation of ubiquitinated proteins and formation of aggresome in HEK293 cells by oz-apra A: 100 nM [2]

References

1. Luesch H, Yoshida WY, Moore RE, Paul VJ, Corbett TH (2001) Total structure determination of apratoxin A, a potent novel cytotoxin from the marine cyanobacterium *Lyngbya majuscula*. J Am Chem Soc 123:5418–5423
2. Shen S, Zhang P, Lovchik MA, Li Y, Tang L, Chen Z, Zeng R, Ma D, Yuan J, Yu Q (2009) Cyclodepsipeptide toxin promotes the degradation of Hsp90 client proteins through chaperone-mediated autophagy. J Cell Biol 185:629–639
3. Luesch H, Chanda SK, Raya RM, DeJesus PD, Orth AP, Walker JR, Izpisua Belmonte JC, Schultz PG (2006) A functional genomics approach to the mode of action of apratoxin A. Nat Chem Biol 2:158–167
4. Liu Y, Law BK, Luesch H (2009) Apratoxin a reversibly inhibits the secretory pathway by preventing cotranslational translocation. Mol Pharmacol 76:91–104

Artemisinin

Keywords: [Antimalarial] [PfATP6 inhibitor]

Structure

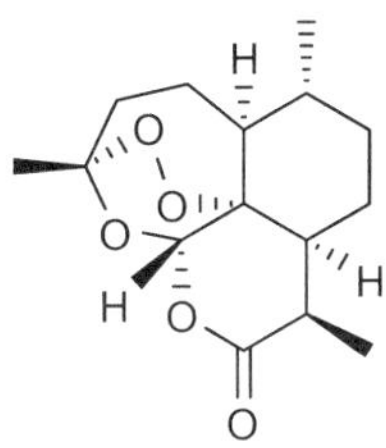

	Artemisinin
Molecular formula	$C_{15}H_{22}O_5$
Molecular weight	282.34
Solubility	DMSO, +++; H_2O, –; MeOH, ++

Discovery/Isolation

Artemisinin was extracted by Youyou Tu, a principal investigator at the Institute of Chinese Materia Medica, China Academy of Chinese Medical Sciences, as the most potent antimalarial compound from sweet wormwood (*Artemisia annua*), which has been used by Chinese herbalists for more than 2000 years [1].

Biological Studies

Artemisinin rapidly kills all asexual stages of *Plasmodium falciparum*. Activated artemisinin form adducts with a variety of biological macromolecules, including haem [2], translationally controlled tumor protein (TCTP) [3], and other higher molecular weight proteins. The crystal structure of the TCTP–arteminisin complex revealed that there are two binding sites on TCTP which are close to amino acid residues important for TCTP function, suggesting that TCTP may be one of the target proteins in *P. falciparum* [4].

Eckstein-Ludwig et al. reported that artemisinin, but not quinine or chloroquine, inhibits the SERCA ortholog (PfATP6) of *P. falciparum* in *Xenopus* oocytes with similar potency to thapsigargin [5]. Together with several results, they concluded that PfATP6 is a target protein of arteminisin; however, there are also controversial reports. Therefore, the actual mechanism of action of arteminisin remains ambiguous.

Biological Activity of Arteminisin

K_i for endogenous Ca^{2+}-ATPase activity in PfATP6-expressing oocytes: 162 ± 31 nM [5]

K_d for PfTCTP measured by SPR: 77 ± 58 μM [4]

References

1. Miller LH, Su X (2011) Artemisinin: discovery from the Chinese herbal garden. Cell 146:855–858
2. Hong YL, Yang YZ, Meshnick SR (1994) The interaction of artemisinin with malarial hemozoin. Mol Biochem Parasitol 63:121–128
3. Bhisutthibhan J, Pan XQ, Hossler PA, Walker DJ, Yowell CA, Carlton J, Dame JB, Meshnick SR (1998) The Plasmodium falciparum translationally controlled tumor protein homolog and its reaction with the antimalarial drug artemisinin. J Biol Chem 273:16192–16198
4. Eichhorn T, Winter D, Buchele B, Dirdjaja N, Frank M, Lehmann WD, Mertens R, Krauth-Siegel RL, Simmet T, Granzin J, Efferth T (2013) Molecular interaction of artemisinin with translationally controlled tumor protein (TCTP) of *Plasmodium falciparum*. Biochem Pharmacol 85:38–45
5. Eckstein-Ludwig U, Webb RJ, Van Goethem ID, East JM, Lee AG, Kimura M, O″Neill PM, Bray PG, Ward SA, Krishna S (2003) Artemisinins target the SERCA of *Plasmodium falciparum*. Nature 424:957–961

AGK2*

Keywords: [SIRT2 inhibitor]

Structure

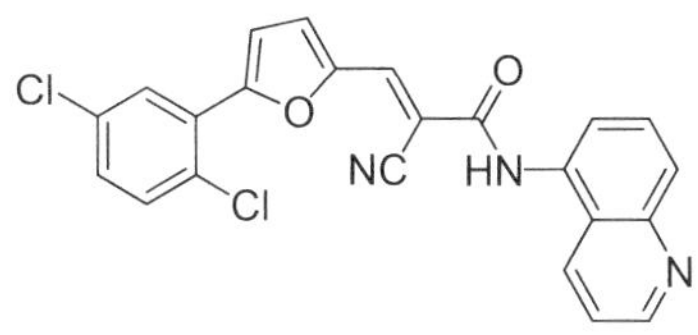

	AGK2
Molecular formula	$C_{23}H_{13}Cl_2N_3O_2$
Molecular weight	434.28
Solubility	DMSO, +; H_2O, ? ; MeOH, ?

Discovery/Isolation

AGK2 is a selective synthetic SIRT2 inhibitor, which was identified among compounds that increase intracellular α-synuclein inclusions from numerous small aggregates to large inclusions [1].

Biological Studies

AGK2 induces hyperacetylation of α-tubulin, a substrate of SIRT2, and rescues α-synuclein toxicity both in vitro and in a *Drosophila* model of Parkinson's disease [1]. Pharmacological inhibition of SIRT2 by AGK2 reduces viability of merlin-mutant mouse Schwann cells [2]. AGK2 induces cell death and decreases the intracellular ATP levels in microglial cells. Both the AGK2-induced cell death and the AGK2-induced ATP decline are mediated by activation of poly (ADP-ribose) polymerase (PARP) [3]. Protective effects of AGK2 on LPD-induced neuroinflammation and brain injury have also been reported [4].

Biological Activity of AGK2

In vitro IC_{50} for SIRT2: 3.5 μM [1]
In vitro IC_{50} for SIRT1 and SIRT3: >50 μM [1]
Increase of α-tubulin acetylation: 10–50 μM [1]

References

1. Outeiro TF, Kontopoulos E, Altmann SM, Kufareva I, Strathearn KE, Amore AM, Volk CB, Maxwell MM, Rochet JC, McLean PJ, Young AB, Abagyan R, Feany MB, Hyman BT, Kazantsev AG (2007) Sirtuin 2 inhibitors rescue α-synuclein-mediated toxicity in models of Parkinson's disease. Science 317:516–519

2. Petrilli A, Bott M, Fernández-Valle C (2013) Inhibition of SIRT2 in merlin/NF2-mutant Schwann cells triggers necrosis. Oncotarget 4:2354–2365
3. Li Y, Nie H, Wu D, Zhang J, Wei X, Ying W (2013) Poly(ADP-ribose) polymerase mediates both cell death and ATP decreases in SIRT2 inhibitor AGK2-treated microglial BV2 cells. Neurosci Lett 544:36–40
4. Wang B, Zhang Y, Cao W, Wei X, Chen J, Ying W (2016) SIRT2 Plays Significant Roles in Lipopolysaccharides-Induced Neuroinflammation and Brain Injury in Mice. Neurochem Res 41:2490–2500

Aurilide

Keywords: [Cytotoxic] [OPA1-mediated apoptosis] [Prohibitin]

Structure

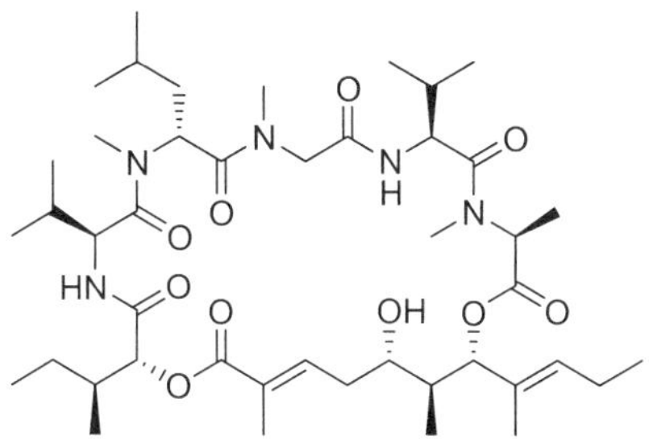

	Aurilide
Molecular formula	$C_{44}H_{75}N_5O_{10}$
Molecular weight	834.11
Solubility	DMSO, +++; H_2O, –; MeOH, +++

Discovery/Isolation/Chemical Studies

Aurilide is a 26-member cyclodepsipeptide isolated from the Japanese sea hare *Dolabella auricularia* [1, 2]. Enantioselective synthesis and SAR studies showed that modifications at the C35 hydroxyl group had a limited impact on the biological activity [2, 3].

Biological Studies

Aurilide induces apoptosis in cultured human cells at the picomolar to nanomolar range. Using an aurilide probe whose C35 hydroxyl group was conjugated with biotin molecule through a protease-cleavable polyproline linker, prohibitin 1 (PHB1) was identified as one of the binding proteins [4]. PHB1 is a mitochondria protein negatively regulating cell proliferation, and aurilide–fluorescein conjugate accumulated in mitochondria in HeLa cells. Aurilide accumulation in mitochondria induces the cleavage of OPA1 (one of the proteins essential for mitochondria fusion) and mitochondria-induced apoptosis.

Biological Activity of Aurilide

IC_{50} for cytotoxicity against HeLa S_3: 0.011 μg/ml [2]
Mitochondria fragmentation and apoptosis induction in HeLa cells: 100 nM [4]

References

1. Suenaga K, Mutou T, Shibata T, Itoh T, Kigoshi H, Yamada K (1996) Isolation and stereostructure of aurilide, a novel cyclodepsipeptide from the Japanese sea hare *Dolabella auricularia*. Tetrahedron Lett 37:6771–6774
2. Suenaga K, Mutou T, Shibata T, Itoh T, Fujita T, Takada N, Hayamizu K, Takagi M, Irifune T, Kigoshi H, Yamada K (2004) Aurilide, a cytotoxic depsipeptide from the sea hare *Dolabella auricularia*: isolation, structure determination, synthesis, and biological activity. Tetrahedron 60:8509–8527
3. Suenaga K, Kajiwara S, Kuribayashi S, Handa T, Kigoshi H (2008) Synthesis and cytotoxicity of aurilide analogs. Bioorg Med Chem Lett 18:3902–3905
4. Sato S, Murata A, Orihara T, Shirakawa T, Suenaga K, Kigoshi H, Uesugi M (2011) Marine natural product aurilide activates the opa1-mediated apoptosis by binding to prohibitin. Chem Biol 18:131–139

Avermectin, Ivermectin

Keywords: [Biocidal] [Glutamate-gated chloride ion channel opener]

Structure

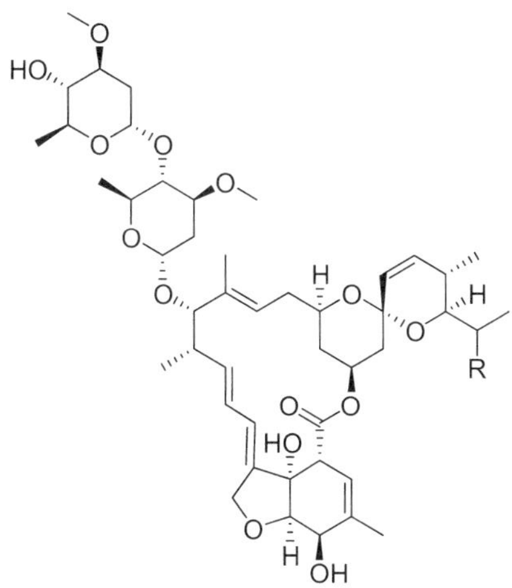

	Avermectin B_{1a} (R = Et)	Avermectin B_{1b} (R = Me)
Molecular formula	$C_{48}H_{72}O_{14}$	$C_{48}H_{70}O_{14}$
Molecular weight	873.09	859.06
Solubility	DMSO, +++; H_2O, –; MeOH, +++	DMSO, +++; H_2O, –; MeOH, +++

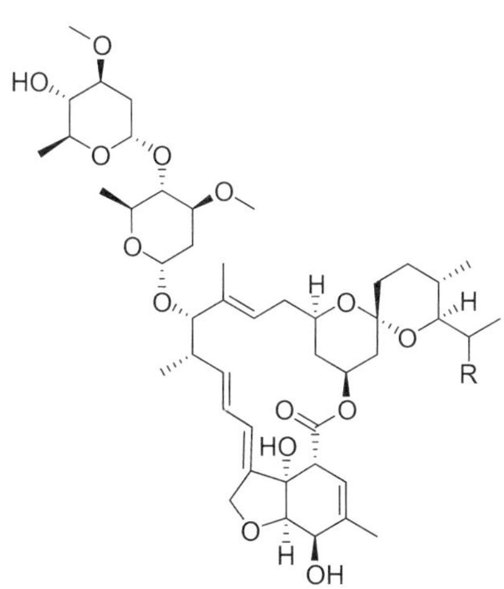

	Ivermectin (8:2 mixture of 22,23-dihydro avermectin B_{1a} and B_{1b})	
	22,23-dihydro avermectin B_{1a} (R = Et)	22,23-dihydro avermectin B_{1b} (R = Me)
Molecular formula	$C_{48}H_{74}O_{14}$	$C_{48}H_{72}O_{14}$
Molecular weight	875.11	861.08
Solubility	DMSO, +++; H_2O, –; MeOH, +++	DMSO, +++; H_2O, –; MeOH, +++

Discovery/Isolation/Chemical Studies

By using a helminth-infected mouse model, the avermectins were isolated from *Streptomyces avermitilis* as a complex of chemically related compounds (A_{1a}, A_{1b}, A_{2a}, A_{2b}, B_{1a}, B_{1b}, B_{2a}, and B_{2b}) which exhibit extraordinarily potent anthelmintic activity [1]. Different from most macrolide or polyene antibiotics, avermectins lack significant antibacterial or antifungal activity [2].

Ivermectin, which is a reduction product of the C22–C23 double bond of avermectin B_{1a} and B_{1b}, improved both the spectrum of activity and safety [3]. Methyl group at terminal next to R group is up stereochemistry for avermectins and ivermectin.

Biological Studies

Because avermectin quickly immobilized parasitic nematodes at low doses, it was thought that its anthelmintic activity may be due to effects on the nervous system. Following this suggestion, Fritz et al. investigated the effects on the lobster stretcher neuromuscular junction, and speculated that avermectins induced the opening of membrane Cl^- channels including those regulated by γ-aminobutyric acid at the inhibitory synapse [4]. In addition to the effects on the GABA receptor–Cl^- channel, Scott and Duce found that ivermectin also affected the glutamate H receptor–Cl^- channel complex [5]. Further investigations identified the avermectin-sensitive glutamate-gated chloride channels by expression cloning using *Xenopus* oocytes [6]. These results strongly suggest that avermectin/ivermectin interacts with glutamate-gated chloride channels and prevents their closure, leading to hyperpolarization of the neuronal membrane.

Biological Activity of Avermectin B_{1a}

>95% effective in reducing the numbers of *Haemonchus placei*, *Ostertagia ostertagi*, *Trichostrongylus axei*, *T. colubriformis*, *Cooperia oncophora*, *C. punctata*, *Oesophagostomum radiatum*, and *Dictyocaulus viviparous* in infected cattle with a single oral dose: 0.1 mg/kg [2]

References

1. Burg RW, Miller BM, Baker EE, Birnbaum J, Currie SA, Hartman R, Kong YL, Monaghan RL, Olson G, Putter I, Tunac JB, Wallick H, Stapley EO, Oiwa R,

Omura S. (1979) Avermectins, new family of potent anthelmintic agents: producing organism and fermentation. Antimicrob Agents Chemother 15:361–367
2. Egerton JR, Ostlind DA, Blair LS, Eary CH, Suhayda D, Cifelli S, Riek RF, Campbell WC. (1979) Avermectins, new family of potent anthelmintic agents: efficacy of the B1a component. Antimicrob Agents Chemother 15:372–378
3. Chabala JC, Mrozik H, Tolman RL, Eskola P, Lusi A, Peterson LH, Woods MF, Fisher MH, Campbell WC, Egerton JR, Ostlind DA. (1980) Ivermectin, a new broad-spectrum antiparasitic agent. J Med Chem 23:1134–1136
4. Fritz LC, Wang CC, Gorio A. (1979) Avermectin B1a irreversibly blocks postsynaptic potentials at the lobster neuromuscular junction by reducing muscle membrane resistance. Proc Natl Acad Sci U S A 76:2062–2066
5. Scott RH, Duce IR. (1985) Effects of 22,23-dihydroavermectin B1a on locust (*Schistocerca gregaria*) muscles may involve several sites of action. Pestic Sci 16:599–604
6. Cully DF, Vassilatis DK, Liu KK, Paress PS, Van der Ploeg LH, Schaeffer JM, Arena JP. (1994) Cloning of an avermectin-sensitive glutamate-gated chloride channel from *Caenorhabditis elegans*. Nature 371:707–711

Bafilomycin (Setamycin)

Keywords: [Antifungal] [V-ATPase inhibitor]

Structure

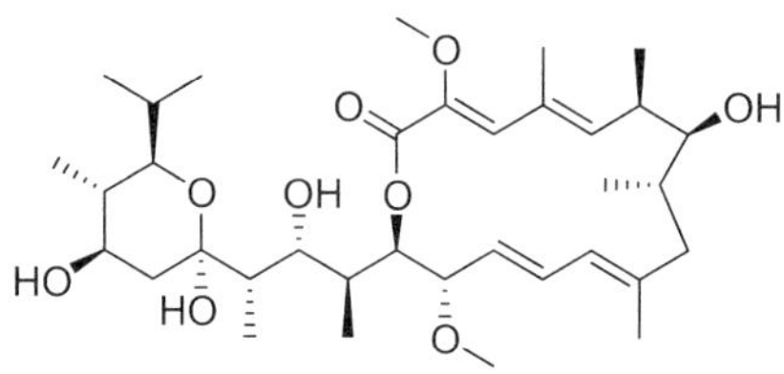

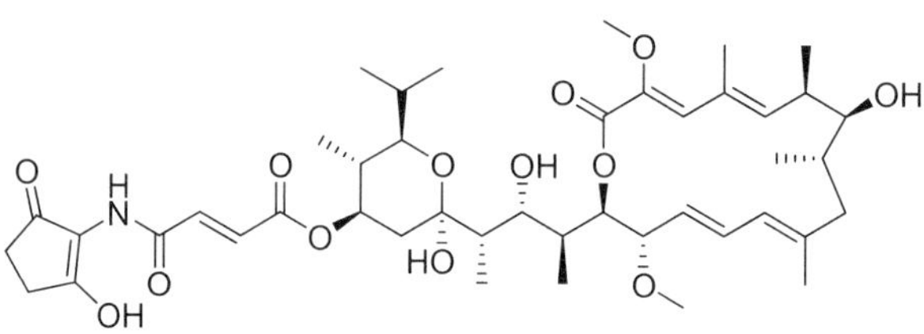

	Bafilomycin A_1	Bafilomycin B_1 (Setamycin)
Molecular formula	$C_{35}H_{58}O_9$	$C_{44}H_{65}NO_{13}$
Molecular weight	622.84	816.00
Solubility	DMSO, +++; H_2O, +; MeOH, +++	DMSO, +++; H_2O, +; MeOH, +++

Methanol inactivates bafilomyins by methylation.

Discovery/Isolation

The first bafilomycin to be isolated was setamycin [1], which was shown to have antifungal and antinematodal activities in 1981 [2]. In 1983 and 1984, L-681,110, another bafilomycin, was isolated as an inhibitor of Na^+/K^+-ATPases from *Streptomyces* sp. MA-5038 [3, 4]. Bafilomycin A_1, A_2, B_1, B_2, C_1, and C_2 were isolated as antibacterial and antifungal compounds from *Streptomyces griseus* strains TÜ1922, TÜ2473, and TÜ2599 [5]. Setamycin and L-681,110 are the same compounds with bafilomycin B_1 and C_1, respectively [2].

Biological Studies

Because bafilomycin C_1 is a moderate Na^+/K^+-ATPase inhibitor, the effects of bafilomycin A_1 on various membrane ATPases were investigated and it was found that this compound is a specific potent inhibitor of vacuolar ATPases (V-ATPases) [6]. Bafilomycin A_1 blocks vacuole acidification and causes protein sorting/secretion deficiencies [7]. V-ATPase inhibition by bafilomycins resulted in several biological phenotypes including inhibitions of bone resorption by osteoclasts [8], of protein degradation in lysosomes [9], and of the accumulation of lipid droplets in macrophages [10].

A biochemical approach using a purified V_0 domain of V-ATPase revealed that the binding site of bafilomycins is located in the V_0 domain [11], and further genetic analysis suggests that bafilomycins locate on the pocket formed by helices 1, 2, and 4 of subunit *c* [12].

Biological Activity of Bafilomycins

MIC for *Sacchromyces sake*: 0.78 μg/ml [1]

I_{50} values for membrane-bound ATPases from *Neurospora* vacuoles, chromaffin granules, and corn vacuoles: 0.4, 0.05, and 0.004 nmol/mg, respectively [6]

IC_{50} for osteoclast bone resorption: 1 nM [8]

Inhibition of oxidized low-density lipoprotein (LDL)–induced accumulation of lipid droplets in macrophage J774: 3 nM [10]

References

1. Omura S, Otoguro K, Nishikiori T, Oiwa R, Iwai Y (1981) Setamycin, a new antibiotic. J Antibiot (Tokyo) 34: 1253–1256
2. Otoguro K, Nakagawa A, Omura S (1988) Setamycin, a 16-membered macrolide antibiotic. Identification and nematocidal activity. J Antibiot (Tokyo) 41: 250–252
3. Hensens OD, Monaghan RL, Huang LY, Albersschonberg G (1983) Structure of the sodium and potassium-ion activated adenosine-triphosphatase inhibitor-L-681,110. J Am Chem Soc 105: 3672–3679
4. Huang L, Albers-Schonberg G, Monaghan RL, Jakubas K, Pong SS, Hensens OD, Burg RW, Ostlind DA, Conroy J, Stapley EO (1984) Discovery, production and purification of the Na^+, K^+ activated ATPase inhibitor, L-681,110 from the fermentation broth of *Streptomyces* sp. MA-5038. J Antibiot (Tokyo) 37: 970–975
5. Werner G, Hagenmaier H, Drautz H, Baumgartner A, Zahner H (1984) Metabolic products of microorganisms. 224. Bafilomycins, a new group of macrolide antibiotics. Production, isolation, chemical structure and biological activity. J Antibiot (Tokyo) 37: 110–117
6. Bowman EJ, Siebers A, Altendorf K (1988) Bafilomycins: a class of inhibitors of membrane ATPases from microorganisms, animal cells, and plant cells. Proc Natl Acad Sci U S A 85: 7972–7976
7. Banta LM, Robinson JS, Klionsky DJ, Emr SD (1988) Organelle assembly in yeast: characterization of yeast mutants defective in vacuolar biogenesis and protein sorting. J Cell Biol 107: 1369–1383
8. Sundquist K, Lakkakorpi P, Wallmark B, Vaananen K (1990) Inhibition of osteoclast proton transport by bafilomycin A_1 abolishes bone resorption. Biochem Biophys Res Commun 168: 309–313

9. Yoshimori T, Yamamoto A, Moriyama Y, Futai M, Tashiro Y (1991) Bafilomycin A_1, a specific inhibitor of vacuolar-type H^+-ATPase, inhibits acidification and protein degradation in lysosomes of cultured cells. J Biol Chem 266: 17707–17712
10. Naganuma S, Kuzuya N, Sakai K, Hasumi K, Endo A (1992) Inhibition of the accumulation of lipid droplets in macrophage J774 by bafilomycin B_1 and destruxin E. Biochim Biophys Acta 1126: 41–48
11. Zhang J, Feng Y, Forgac M (1994) Proton conduction and bafilomycin binding by the V_0 domain of the coated vesicle V-ATPase. J Biol Chem 269: 23518–23523
12. Bowman EJ, Graham LA, Stevens TH, Bowman BJ (2004) The bafilomycin/concanamycin binding site in subunit *c* of the V-ATPases from *Neurospora crassa* and *Saccharomyces cerevisiae*. J Biol Chem 279: 33131–33138

BI2536*

Keywords: [Plk1 inhibitor]

Structure

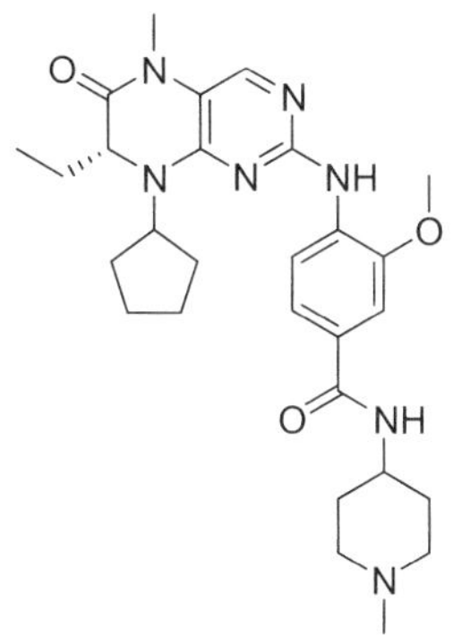

	BI 2536
Molecular formula	$C_{28}H_{39}N_7O_3$
Molecular weight	521.65
Solubility	DMSO, +++; MeOH, +++; H_2O, ?

Discovery/Isolation

BI2536 was screened out as a potent inhibitor of polo-like kinase 1 (Plk1) from a diverse library of organic compounds [1]. BI2536 is a compound inhibiting Plk1 at a concentration less than a nanomole in vitro (0.83 nM) and is active against human, mouse, and rat Plk1 equipotently [1]. It is over 1000-fold more selective to Plk1 when compared to other protein kinases, and some inhibitory activity against closely related kinases of polo-like kinase family members (Plk2 and Plk3) has been reported [2].

Biological Studies

BI2536 inhibits Plk1 at subnanomolar concentration, while it inhibits the activities of Plk2 and Plk3 to a slightly lesser extent with IC_{50} values of 3.5 nM and 9.0 nM, respectively. BI2536 treatment (10–100 nM) leads to inhibition of mitotic properties such as the phosphorylation of Apc6 at mitotic centrosomes and recruitment of γ-tubulin. It also induces the inhibition of cohesin release from chromosome arms, and induction of monopolar spindles in HeLa cells. Treatment of BI2536 on HeLa cells leads to cell cycle arrest at the G2/M border, and subsequently increases cells with a sub-G1 DNA peak indicating the induction of apoptosis in these cells. Actually, accumulation of cleaved PARP p85 fragments was observed in the treated cells. It inhibits the proliferation of 32 human cancer cell lines, with EC_{50} values of 2–25 nM. Because similar inhibitory effects have been observed in nontransformed, immortalized cells, BI2536 might affect the growth of all proliferating cells. Cells that are treated with BI 2536 arrest in M phase with monopolar mitotic spindles, a phenotype that is similar to that induced by an siRNA against Plk1, suggesting that Plk1 is the main cellular target. Plk1 inhibition by BI2536 reduces the growth and viability of anaplastic thyroid carcinoma cells such as OCUT-1, CAL62, SW1736, 8505C, and ACT-1 with EC_{50} values of 1.4–5.6 nM [1, 3].

Biological Activity of BI2536
IC_{50} values for Plk1, Plk2 and Plk3: 0.83 nM, 3.5 nM, and 9.0 nM, respectively [1]

References

1. Steegmaier M, Hoffmann M, Baum A, Lenart P, Petronczki M, Krssak M, Gurtler U, Garin-Chesa P, Lieb S, Quant J, Grauert M, Adolf GR, Kraut N, Peters JM, Rettig WJ (2007) BI2536, a potent and selective inhibitor of polo-like kinase 1, inhibits tumor growth in vivo. Curr Biol: 17:316–322
2. Lenart P, Petronczki M, Steegmaier M, Di Fiore B, Lipp JJ, Hoffmann M, Rettig WJ, Kraut N, Peters JM (2007) The small-molecule inhibitor BI2536 reveals novel insights into mitotic roles of polo-like kinase 1. Curr Biol: 17:304–315
3. Nappi TC, Salerno P, Zitzelsberger H, Carlomagno F, Salvatore G, Santoro M (2009) Identification of Polo-like kinase 1 as a potential therapeutic target in anaplastic thyroid carcinoma. Cancer Res 69:1916–1923

Bisebromoamide

Keywords: [Actin stabilizer]

Structure

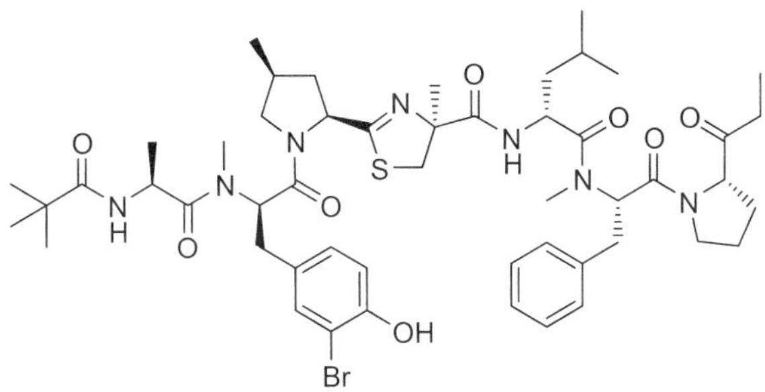

	Bisebromoamide
Molecular formula	$C_{51}H_{72}BrN_7O_8S$
Molecular weight	1023.13
Solubility	DMSO, ++; H_2O, ++; MeOH, ++

Discovery/Isolation
Bisebromoamide is a cytotoxic peptide isolated from the marine cyanobacterium *Lyngbya* sp. [1].

Biological Studies
Automated high-content image analysis revealed that bisebromoamide induced morphological changes in HeLa cells similar to those induced by known actin-stabilizing compounds [2]. Bisebromoamide stimulated actin polymerization and stabilized F-actin in vitro. Furthermore, a bisebromoamide–fluorescein conjugate localized at the actin filament in cells.

Bisebromoamide not only induced cell apoptosis, but inhibited the Raf/MEK/ERK and PI3K/Akt/mTOR pathways in advanced renal cell carcinomas [3].

Biological Activity of Bisebromoamide
IC_{50} for cytotoxicity against HeLa: 0.04 μg/ml [1]
Inhibition of the phosphorylation of ERK in NRK cells by PDGF-stimulation: 0.1–10 μM [1]
F-actin stabilization in vitro: 10 nM [2]
IC_{50} for two cell lines of advanced renal cell carcinoma: 1.5 and 2.1 μM [3]

References

1. Teruya T, Sasaki H, Fukazawa H, Suenaga K (2009) Bisebromoamide, a potent cytotoxic peptide from the marine cyanobacterium *Lyngbya* sp.: isolation, stereostructure, and biological activity. Org Lett 11:5062–5065
2. Sumiya E, Shimogawa H, Sasaki H, Tsutsumi M, Yoshita K, Ojika M, Suenaga K, Uesugi M (2011) Cell-morphology profiling of a natural product library identifies bisebromoamide and miuraenamide A as actin filament stabilizers. ACS Chem Biol 6:425–431
3. Suzuki K, Mizuno R, Suenaga K, Teruya T, Tanaka N, Kosaka T, Oya M (2013) Bisebromoamide, an extract from *Lyngbya* species, induces apoptosis through ERK and mTOR inhibitions in renal cancer cells. Cancer Med 2:32–39

BMS-488516*, BMS-509744*

Keywords: [Itk kinase inhibitor] [immunosuppressive and inflammatory diseases]

Structure

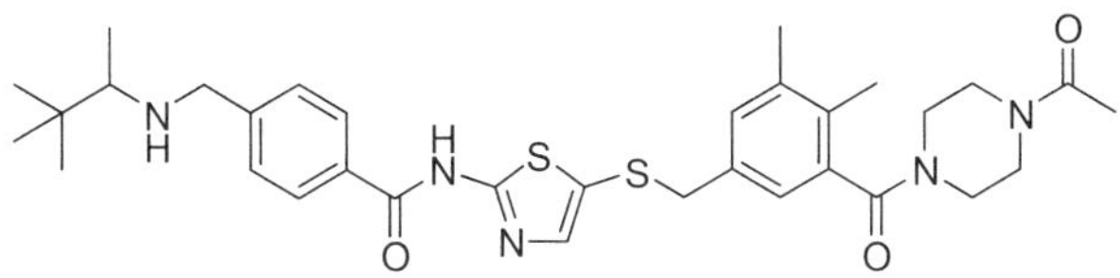

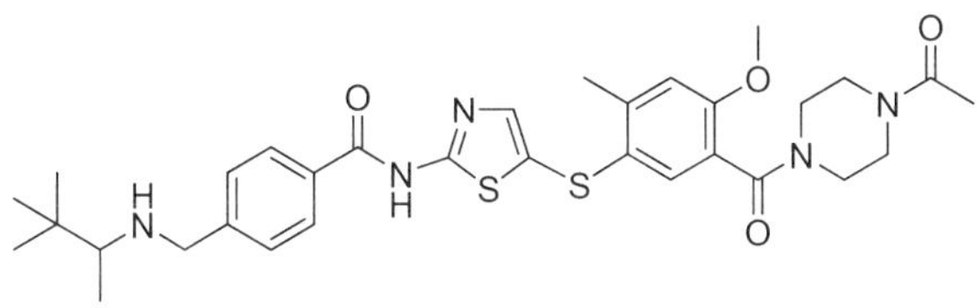

	BMS-488516	BMS-509744
Molecular formula	$C_{33}H_{43}N_5O_3S_2$	$C_{32}H_{41}N_5O_4S_2$
Molecular weight	621.86	623.83
Solubility	DMSO, ++; H_2O, ? ; MeOH, ?	DMSO, ++; H_2O, ? ; MeOH, ?

Discovery/Isolation

BMS-488516 and BMS-509744 were developed from an aminothiazole compound, BMS-285047, which was isolated as an Itk kinase inhibitor by a homogeneous time-resolved fluorescence assay using GST-Itk kinase domain and GST-SLP-76 as a substrate [1].

Biological Studies

Both BMS-488516 and BMS-509744 potently and selectively inhibited Itk kinase activity in an ATP-competitive manner, and reduced TCR-induced functions in vitro and in cells [1]. MS-509744 also significantly diminished lung inflammation in an allergy/asthma model mouse. These results strongly suggest that Itk is a therapeutic target for immunosuppressive and inflammatory diseases. X-ray crystal structures of IL-2-inducible T-cell kinase in complex with BMS-509744 stabilized the activation loop in a substrate-blocking inactive conformation, indicating that structural change is involved in Itk kinase activity [2].

The HIV-1 Nef virulence factor interacts with the Src homology 3 (SH3) domain of Itk kinase, and induces constitutive Itk activation. BMS-509744 potently blocked HIV-1 infectivity and replication, but not that of a Nef-defective mutant, suggesting that Nef provides a mechanistic link between HIV-1 and Itk signaling in the viral life cycle [3].

Biological Activity of BMS-488516 and BMS-509744
IC_{50} values for BMS-488516 and BMS-509744 on in vitro Itk kinase: 96 and 19 nM, respectively [1]
Inhibition of PLCγ phosphorylation by anti-CD3 antibody in Jurkat cells by BMS-509744: 10 μM [1]
IC_{50} values for BMS-488516 and BMS-509744 on LPS-induced TNFR production from PBMCs: 11 and >25 μM, respectively [1]
IC_{50} values for BMS-488516 and BMS-509744 on anti-CD3 antibody–induced IL-2 secretion in murine EL4 cells: 250 and 72 nM, respectively [1]
K_d for BMS-509744 on C477S/E614A/E617A Itk kinase variant measured by SPR: 0.8 nM [2]
IC_{50} for BMS-509744 on HIV infectivity: ~3 μM [3]

References

1. Lin TA, McIntyre KW, Das J, Liu C, O″Day KD, Penhallow B, Hung CY, Whitney GS, Shuster DJ, Yang X, Townsend R, Postelnek J, Spergel SH, Lin J, Moquin RV, Furch JA, Kamath AV, Zhang H, Marathe PH, Perez-Villar JJ, Doweyko A, Killar L, Dodd JH, Barrish JC, Wityak J, Kanner SB (2004) Selective Itk inhibitors block T-cell activation and murine lung inflammation. Biochemistry 43:11056–11062
2. Kutach AK, Villasenor AG, Lam D, Belunis C, Janson C, Lok S, Hong LN, Liu CM, Deval J, Novak TJ, Barnett JW, Chu W, Shaw D, Kuglstatter A (2010) Crystal structures of IL-2-inducible T cell kinase complexed with inhibitors: Insights into rational drug design and activity regulation. Chem Biol Drug Des 76:154–163
3. Tarafdar S, Poe JA, Smithgall TE (2014) The accessory factor Nef links HIV-1 to Tec/Btk kinases in an Src homology 3 domain-dependent manner. J Biol Chem 289:15718–15728

BNS-22*

Keywords: [Antitumor] [Topoisomerase II]

Structure

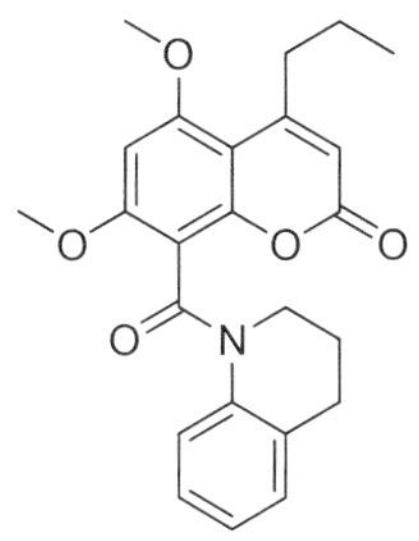

	BNS-22
Molecular formula	$C_{24}H_{25}NO_5$
Molecular weight	407.47
Solubility	DMSO, ++; H_2O, –; MeOH, ++

Discovery/Isolation/Chemical Studies

BNS-22 is a synthesized derivative of the natural product GUT-70, which was isolated from the stem bark of *Calophyllum brasiliense* in Brazil as an antileukemic compound [1]. This compound has antiproliferative activity against human cancer cells. The target molecule of BNS-22 was suggested as DNA topoisomerase II (TOP2) by proteomic profiling as below. An affinity precipitation assay using purified human TOP2α showed that BNS-22 binds directly to TOP2α in a different mode from ICRF-193, a TOP2 catalytic inhibitor [2].

Biological Studies

By proteomic profiling analysis, it was suggested that the target molecule of BNS-22 is the same with ICRF-193 [2]. Furthermore, BNS-22 inhibits kinetoplast DNA decatenation that is mediated by human TOP2α and TOP2β. BNS-22 does not affect DNA damage and antagonizes TOP2 poison–mediated DNA damage, and induces mitotic abnormalities, characterized by impairments in chromosome alignment and segregation, thereby causing polyploidy in HeLa cells.

Biological Activity of BNS-22

IC_{50} values for human TOP2α and TOP2β in vitro: 2.8 and 0.42 μM, respectively [2]

Perturbation of mitotic spindle formation and polyploidy induction in HeLa cells: 3 μM [2]

References

1. Kimura S, Ito C, Jyoko N, Segawa H, Kuroda J, Okada M, Adachi S, Nakahata T, Yuasa T, Filho VC, Furukawa H, Maekawa T (2005) Inhibition of leukemic cell growth by a novel anti-cancer drug (GUT-70) from *Calophyllum brasiliense* that acts by induction of apoptosis. Int J Cancer 113:158–165
2. Kawatani M, Takayama H, Muroi M, Kimura S, Maekawa T, Osada H (2011) Identification of a small-molecule inhibitor of DNA topoisomerase II by proteomic profiling. Chem Biol 18:743–751

Brefeldin A

Keywords: [Antifungal] [Antiviral] [ARF GEF inhibitor] [Transport inhibitor]

Structure

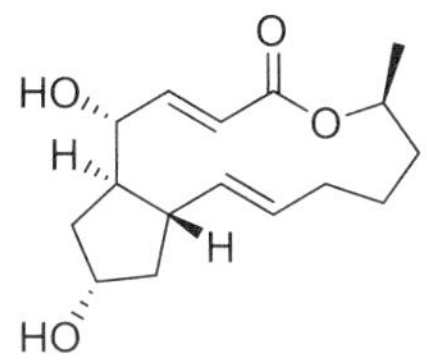

	Brefeldin A
Molecular formula	$C_{16}H_{24}O_4$
Molecular weight	280.36
Solubility	DMSO, +++; H_2O, +; MeOH, ++

Discovery/Isolation
Brefeldin A (primarily named decumbin) was isolated as the chief component of the crystalline bodies produced from *Penicillium decumbens* [1]. Brefeldin A was reisolated as an antifungal antibiotic by Hayashi et al. [2].

Biological Studies
Brefeldin A is a unique agent, which impedes protein transport from the ER to the Golgi complex [3] by inducing the rapid and reversible dissociation of coat protein, β-COP, and small GTPase, ADP-ribosylation factor (ARF) [4, 5].

The inhibition step of brefeldin A was identified biochemically as guanine nucleotide exchange factor (GEF) activity on ARF in Golgi membranes [6, 7]. The first ARF GEF, Gea1, was identified using a genetic selection in yeast, and its GEF activity was inhibited by brefeldin A [8]. Gea1 contains a Sec7 domain that is sufficient for ARF GEF activity, and brefeldin A stabilizes the ARF-GDP-Sec7 domain complex as an inactive, abortive complex [9, 10]. This dead-end complex structure showed that brefeldin A binds at the protein–protein (ARF-GDP/Sec7 domain) interface and freezes the complex in an abortive conformation that cannot proceed to nucleotide dissociation [11, 12].

Brefeldin A is a good bioprobe for investigating the transport system, especially from the endoplasmic reticulum to the Golgi network. Nuchtern et al. showed that the influenza A matrix protein can be effectively presented to class II–restricted T cells by two pathways, one of which is chloroquine sensitive and BFA insensitive, the other being chloroquine insensitive and BFA sensitive [13]. Rubartelli et al. suggested a novel secretory pathway for interleukin-1β, a protein lacking a signal sequence [14]. Brefeldin A also induces an unfolded protein response, protein synthesis inhibition and apoptosis [15–17].

Biological Activity of Brefeldin A

Blockage of secretion in primary culture of rat hepatocyte: 1 μg/ml [3]
70% inhibition of protein synthesis in rat glioma C6 cells: 0.1–1 mg/ml [16]
50–60% inhibition of the membrane-dependent nucleotide-exchange reaction: 10 μM [6, 7]
Redistribution of S2P into the ER in HepG2 cells: 10 μg/ml [15]

References

1. Singleton VL, Bohonos N, Ullstrup AJ (1958) Decumbin, a new compound from a species of *Penicillium*. Nature 181:1072–1073
2. Hayashi T, Takatsuki A, Tamura G (1974) The action mechanism of brefeldin A. I. Growth recovery of *Candida albicans* by lipids from the action of brefeldin A. J Antibiot (Tokyo) 27:65–72
3. Misumi Y, Miki K, Takatsuki A, Tamura G, Ikehara Y (1986) Novel blockade by brefeldin A of intracellular transport of secretory proteins in cultured rat hepatocytes. J Biol Chem 261:11398–11403
4. Donaldson JG, Lippincott-Schwartz J, Bloom GS, Kreis TE, Klausner RD (1990) Dissociation of a 110-kD peripheral membrane protein from the Golgi apparatus is an early event in brefeldin A action. J Cell Biol 111:2295–2306
5. Serafini T, Stenbeck G, Brecht A, Lottspeich F, Orci L, Rothman JE, Wieland FT (1991) A coat subunit of Golgi-derived non-clathrin-coated vesicles with homology to the clathrin-coated vesicle coat protein β-adaptin. Nature 349:215–220
6. Donaldson JG, Finazzi D, Klausner RD (1992) Brefeldin A inhibits Golgi membrane-catalysed exchange of guanine nucleotide onto ARF protein. Nature 360:350–352
7. Helms JB, Rothman JE (1992) Inhibition by brefeldin A of a Golgi membrane enzyme that catalyses exchange of guanine nucleotide bound to ARF. Nature 360:352–354
8. Peyroche A, Paris S, Jackson CL (1996) Nucleotide exchange on ARF mediated by yeast Gea1 protein. Nature 384:479–481
9. Peyroche A, Antonny B, Robineau S, Acker J, Cherfils J, Jackson CL (1999) Brefeldin A acts to stabilize an abortive ARF-GDP-Sec7 domain protein complex: involvement of specific residues of the Sec7 domain. Mol Cell 3:275–285
10. Jackson CL, Casanova JE (2000) Turning on ARF: the Sec7 family of guanine-nucleotide-exchange factors. Trends Cell Biol 10:60–67
11. Mossessova E, Corpina RA, Goldberg J (2003) Crystal structure of ARF1*Sec7 complexed with Brefeldin A and its implications for the guanine nucleotide exchange mechanism. Mol Cell 12:1403–1411
12. Renault L, Guibert B, Cherfils J (2003) Structural snapshots of the mechanism and inhibition of a guanine nucleotide exchange factor. Nature 426:525–530

13. Nuchtern JG, Biddison WE, Klausner RD (1990) Class II MHC molecules can use the endogenous pathway of antigen presentation. Nature 343:74–76
14. Rubartelli A, Cozzolino F, Talio M, Sitia R (1990) A novel secretory pathway for interleukin-1 beta, a protein lacking a signal sequence. EMBO J 9:1503–1510
15. Citterio C, Vichi A, Pacheco-Rodriguez G, Aponte AM, Moss J, Vaughan M (2008) Unfolded protein response and cell death after depletion of brefeldin A-inhibited guanine nucleotide-exchange protein GBF1. Proc Natl Acad Sci U S A 105:2877–2882
16. Fishman PH, Curran PK (1992) Brefeldin A inhibits protein synthesis in cultured cells. FEBS Lett 314:371–374
17. Shinjo S, Mizotani Y, Tashiro E, Imoto M (2013) Comparative analysis of the expression patterns of UPR-target genes caused by UPR-inducing compounds. Biosci Biotechnol Biochem 77:729–735

C646*

Keywords: [p300 inhibitor]

Structure

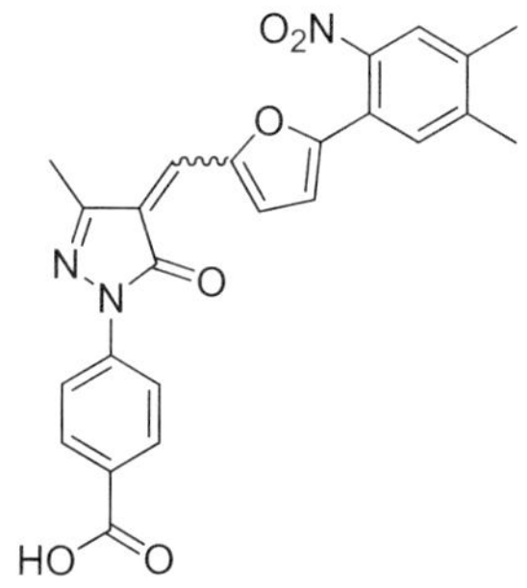

	C646
Molecular formula	$C_{24}H_{19}N_3O_6$
Molecular weight	445.43
Solubility	DMSO, +++; H_2O, ? ; MeOH, ?

Discovery/Isolation

C646 is a synthetic, pyrazolone-containing selective and competitive p300 inhibitor, which was identified by structure-based in silico screening [1].

Biological Studies

C646, but not its inactive analog, reduces histone acetylation levels in cells and inhibits proliferation of melanoma cells [1]. C646 exhibits an antiproliferative effect on acute myeloid leukemia cells with distinct alterations of transcription and histone acetylation associated with genomic integrity [2, 3] and increases radiosensitization of lung cancer cells by enhancing mitotic catastrophe [4]. In addition to the antitumor activity, inhibitory activity against influenza virus infection and antitauopathy activity were also reported [5, 6].

Biological Activity of C646

K_i for p300: 400 nM [1]
Inhibition of p300 enzymatic activity at 10 μM: 86% inhibition [1]
Inhibition of histone acetylation in mouse fibroblasts: 25 μM [1]
IC_{50} for proliferation of melanoma cells: 10–20 μM [1]

References

1. Bowers EM, Yan G, Mukherjee C, Orry A, Wang L, Holbert MA, Crump NT, Hazzalin CA, Liszczak G, Yuan H, Larocca C, Saldanha SA, Abagyan R, Sun Y, Meyers DJ, Marmorstein R, Mahadevan LC, Alani RM, Cole PA (2010) Virtual ligand screening of the p300/CBP histone acetyltransferase: identification of a selective small molecule inhibitor. Chem Biol 17:471–482
2. Gao XN, Lin J, Ning QY, Gao L, Yao YS, Zhou JH, Li YH, Wang LL, Yu L (2013) A histone acetyltransferase p300 inhibitor C646 induces cell cycle arrest and apoptosis selectively in AML1-ETO-positive AML cells. PLoS One 8: e55481
3. Giotopoulos G, Chan WI, Horton SJ, Ruau D, Gallipoli P, Fowler A, Crawley C, Papaemmanuil E, Campbell PJ, Göttgens B, Van Deursen JM, Cole PA, Huntly BJ (2016) The epigenetic regulators CBP and p300 facilitate leukemogenesis and represent therapeutic targets in acute myeloid leukemia. Oncogene 35:279–289
4. Oike T, Komachi M, Ogiwara H, Amornwichet N, Saitoh Y, Torikai K, Kubo N, Nakano T, Kohno T (2014) C646, a selective small molecule inhibitor of histone acetyltransferase p300, radiosensitizes lung cancer cells by enhancing mitotic catastrophe. Radiother Oncol 111:222–227
5. Zhao D, Fukuyama S, Sakai-Tagawa Y, Takashita E, Shoemaker JE, Kawaoka Y (2015) C646, a novel p300/CREB-binding protein-specific inhibitor of histone acetyltransferase, attenuates influenza A virus infection. Antimicrob Agents Chemother 60:1902–1906
6. Min SW, Cho SH, Zhou Y, Schroeder S, Haroutunian V, Seeley WW, Huang EJ, Shen Y, Masliah E, Mukherjee C, Meyers D, Cole PA, Ott M, Gan L. (2010) Acetylation of tau inhibits its degradation and contributes to tauopathy. Neuron 67:953–966

Celastrol

Keywords: [Anti-inflammatory] [Cysteine reactivity]

Structure

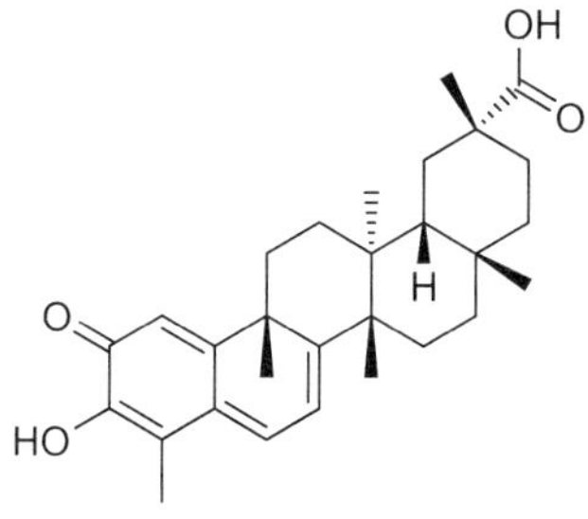

	Celastrol
Molecular formula	$C_{29}H_{38}O_4$
Molecular weight	450.62
Solubility	DMSO, +++; H_2O, –; MeOH, +

Discovery/Isolation/Chemical Studies

Celastrol is a quinone methide purified from traditional Chinese medicine *Celastrus scandens* (Thunder of God Vine) which is generally used for the treatment of inflammatory and autoimmune diseases [1, 2]. The structure was determined by Nakanishi et al. [3]. Celastrol is an electrophilic quinine methide, and sterospecifically reacts with biological sulfur nucleophiles by Michael additions [4].

Biological Studies

Celastrol has numerous pharmacological properties, inhibits the proliferation of a variety of tumor cells, and suppresses tumor initiation, promotion, and metastasis in various cancer models in vivo. Recently celastrol was identified as an inhibitor of androgen receptor–mediated signaling, and a gene expression–based analysis revealed that celastrol is classified as a HSP90 inhibitor [5]. Lin et al. found that a cancerous inhibitor of PP2A (CIP2A) is a target protein of celastrol using biotinylated derivative [6].

There are many reports that celastrol modulates several biological activities of NF-κB, proteasome, topoisomerase II, heat shock response, and so on [7]. Some of them are directly inhibited in vitro; therefore, celastrol probably has several molecular targets in cells.

Biological Activity of Celastrol

Inhibition of androgen receptor signaling in LNCaP cells stimulated with the synthetic androgen R1881: 1.25 μM [5]

CIP2A degradation in various cancer cell lines: 5 μM [6]

References

1. Chou T, Mei P (1936) Study on Chinese herb Lei Gong Teng, *Tripterygium wilfordii* Hook f. I The coloring substance and the sugars. Chin J Phys 10:529–534
2. Gisvold O (1939) The pigments contained in the bark of the root of *Celastrus scandens*. J Am Pharm Soc 28:440–443
3. Nakanishi K, Kakisawa H, Hirata Y (1956) Structure of pristimerin and celastrol. Bull Chem Soc Jpn 29:7–15
4. Klaic L, Trippier PC, Mishra RK, Morimoto RI, Silverman RB (2011) Remarkable stereospecific conjugate additions to the Hsp90 inhibitor celastrol. J Am Chem Soc 133:19634–19637
5. Hieronymus H, Lamb J, Ross KN, Peng XP, Clement C, Rodina A, Nieto M, Du J, Stegmaier K, Raj SM, Maloney KN, Clardy J, Hahn WC, Chiosis G, Golub TR (2006) Gene expression signature-based chemical genomic prediction identifies a novel class of HSP90 pathway modulators. Cancer Cell 10:321–330
6. Liu Z, Ma L, Wen ZS, Hu Z, Wu FQ, Li W, Liu J, Zhou GB (2014) Cancerous inhibitor of PP2A is targeted by natural compound celastrol for degradation in non-small-cell lung cancer. Carcinogenesis 35:905–914
7. Kannaiyan R, Shanmugam MK, Sethi G (2011) Molecular targets of celastrol derived from Thunder of God Vine: potential role in the treatment of inflammatory disorders and cancer. Cancer Lett 303:9–20

Cerulenin

Keywords: [Antifungal] [Lipid biosynthesis inhibitor]

Structure

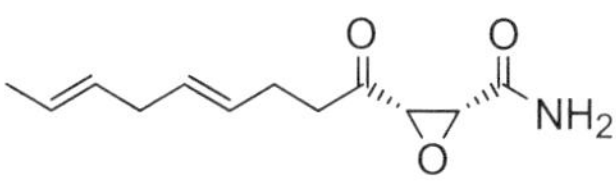

	Cerulenin
Molecular formula	$C_{12}H_{17}NO_3$
Molecular weight	223.27
Solubility	DMSO, ++; H_2O, +; MeOH, ++

Discovery/Isolation

Cerulenin was isolated as an antifungal antibiotic from the culture broth of *Cephalosporium caerulens* [1].

Biological Studies

Cerulenin is a potent and noncompetitive inhibitor of fatty acid biosynthesis through irreversible inhibition of β-ketoacyl acylcarrier protein synthetase [2, 3]. The binding site of cerulenin in fatty acid synthetase has been determined as the highly reactive substrate binding Cys-204 using a combination of radiolabeling studies with [^{3}H]cerulenin, proteolytic and chemical modification [4, 5]. The inhibition mechanism of cerulenin is suggested to be that cerulenin forms a hydroxylactam ring and reacts at its epoxide carbon with the SH group of the cysteine residue in the condensing reaction domain of fatty acid synthetase. The X-ray structure of the complex of cerulenin and β-ketoacyl carrier protein synthetase revealed that cerulenin is bound in a hydrophobic pocket formed at the dimer interface of β-ketoacyl carrier protein synthetases [6].

Cerulenin also shows several inhibitory activities on antigen processing in antigen-presenting cells [7], HIV-1 proteinase in vitro [8], protein palmitoylation, and insulin internalization [9].

Biological Activity of Cerulenin

IC_{50} for fatty acid synthetase of *E. coli*: 30 μM [2]
IC_{50} for β-ketoacyl-acyl carrier protein synthetase of *E. coli*: <5 μM [2]
IC_{50} for HIV-1 protease in vitro: 2.5 mM [8]

References

1. Sato Y, Nomura S, Kamio Y, Omura S, Hata T (1967) Studies on cerulenin, 3. Isolation and physico-chemical properties of cerulenin. J Antibiot (Tokyo) 20:344–348
2. D''Agnolo G, Rosenfeld IS, Awaya J, Omura S, Vagelos PR (1973) Inhibition of fatty acid synthesis by the antibiotic cerulenin. Specific inactivation of β-ketoacyl-acyl carrier protein synthetase. Biochim Biophys Acta 326:155–156
3. Vance D, Goldberg I, Mitsuhashi O, Bloch K (1972) Inhibition of fatty acid synthetases by the antibiotic cerulenin. Biochem Biophys Res Commun 48:649–656
4. Child CJ, Shoolingin-Jordan PM (1998) Inactivation of the polyketide synthase, 6-methylsalicylic acid synthase, by the specific modification of Cys-204 of the β-ketoacyl synthase by the fungal mycotoxin cerulenin. Biochem J 330:933–937
5. Funabashi H, Kawaguchi A, Tomoda H, Omura S, Okuda S, Iwasaki S (1989) Binding site of cerulenin in fatty acid synthetase. J Biochem 105:751–755
6. Moche M, Schneider G, Edwards P, Dehesh K, Lindqvist Y (1999) Structure of the complex between the antibiotic cerulenin and its target, β-ketoacyl-acyl carrier protein synthase. J Biol Chem 274:6031–6034
7. Falo LD, Jr., Benacerraf B, Rothstein L, Rock KL (1987) Cerulenin is a potent inhibitor of antigen processing by antigen-presenting cells. J Immunol 139:3918–3923
8. Moelling K, Schulze T, Knoop MT, Kay J, Jupp R, Nicolaou G, Pearl LH (1990) In vitro inhibition of HIV-1 proteinase by cerulenin. FEBS Lett 261:373–377
9. Jochen AL, Hays J, Mick G (1995) Inhibitory effects of cerulenin on protein palmitoylation and insulin internalization in rat adipocytes. Biochim Biophys Acta 1259:65–72

Chromeceptin*

Keywords: [Insulin-induced adipogenesis] [IGF2] [STAT6] [MFP-2] [ACC1]

Structure

	Chromeceptin
Molecular formula	$C_{19}H_{16}F_3N_3O$
Molecular weight	359.35
Solubility	DMSO, +++; H_2O, ±; MeOH, ++

Discovery/Isolation

Chromeceptin was identified as an inhibitor of insulin-induced adipogenesis of 3 T3-L1 cells by the morphology-based adipogenesis screen of a synthetic chemical library containing 10,000 divergent compounds [1].

Biological Studies

Chromeceptin exhibited cytotoxicity against IGF2-producing hepatocarcinoma cells with selectivity similar to that of a neutralizing antibody against IGF2 [1]. This compound blocks the autocrine loop of IGF2 in the hepatocarcinoma cells by activating the expression of IGFBP-1, a secreted IGF binding polypeptide, and STAT6 was identified as a chromeceptin-responsive element binding transcription factor in IGFBP-1 [1, 2]. Furthermore, chromoceptin induced the expression of SOCS-3 (suppressor of cytokine signaling-3) via STAT-6 activation. These results indicate that the expression of SOCS-3 and IGFBP-1, and perhaps that of the other genes that are controlled by STAT6, is up-regulated by chromeceptin and may work together to suppress the function of IGF2.

The chromeceptin binding protein was identified as MFP-2 (multifunctional protein-2) by affinity purification using biotinylated chromeceptin [2]. Furthermore, the chromeceptin-MFP-2 complex binds to and inhibits ACC1 (acetyl-CoA carboxylase 1), an enzyme important for the de novo synthesis of malonyl-CoA and fatty acids [3]. The formation of this ternary complex removes ACC1 from the cytosol and sequesters it in peroxisomes. As a result, chromeceptin impairs fatty acid synthesis.

Biological Activity of Chromeceptin

IC_{50} for IGF2-producing hepatocellular carcinoma cells (Hep-G2): 29 nM [1]
IC_{50} for hepatocellular carcinoma cells with low level of IGF2 (SK-Hep-1): ~1000 nM [1]
Inhibition of the promoter of IGF2 in hepatocellular carcinoma cells: 1 μM [1]
Induction of IGFBP-1 and SOCS-3: 1 μM [2]
Pex5p-dependent peroxisomal translocation of MPF1 and ACC1: 1 μM [3]

References

1. Choi Y, Kawazoe Y, Murakami K, Misawa H, Uesugi M (2003) Identification of bioactive molecules by adipogenesis profiling of organic compounds. J Biol Chem 278:7320–7324
2. Choi Y, Shimogawa H, Murakami K, Ramdas L, Zhang W, Qin J, Uesugi M (2006) Chemical genetic identification of the IGF-linked pathway that is mediated by STAT6 and MFP2. Chem Biol 13:241–249
3. Jung D, Abu-Elheiga L, Ayuzawa R, Gu Z, Shirakawa T, Fujiki Y, Nakatsuji N, Wakil SJ, Uesugi M (2012) Mislocalization and inhibition of acetyl-CoA carboxylase 1 by a synthetic small molecule. Biochem J 448:409–416

Compactin/ML-236B, Lovastatin

Keywords: [Hypolipidemic agent] [HMG-CoA reductase]

Structure

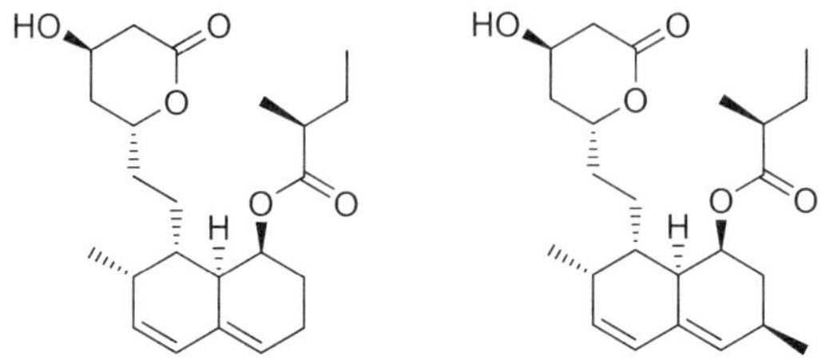

	Compactin	Lovastatin
Molecular formula	$C_{23}H_{34}O_5$	$C_{24}H_{36}O_5$
Molecular weight	390.51	404.54
Solubility	DMSO, +++; H_2O, ±; MeOH, ++	DMSO, +++; H_2O, ±; MeOH, ++

Discovery/Isolation

ML-236B was isolated from *Penicillum citrinum* as an inhibitor of cholesterol synthesis [1]. The same compound, compactin, was isolated as an antifungal metabolite from *Penicillium brevicompactum* [2]. Lovastatin (monacolin K, mevinolin) was isolated from *Monascus* sp. [3, 4] and *Asperugillus terreus* [5] as a hypocholesterolemic agents.

Biological Studies

Compactin, lovastatin, and other statins inhibit 3-hydroxy-3-methylglutaryl (HMG)–CoA reductase, the rate-limiting enzyme in the cholesterol synthetic pathway [6]. The inhibition is competitive with respect to the substrate HMG-CoA. The crystal structure of six statins with HMG-CoA reductase revealed that statins occupy a portion of the binding site of HMG-CoA, then blocking substrate access to the active site [7].

Statins not only decrease cholesterol content but also show several other phenotypes which are related or unrelated with HMG-CoA reductase inhibition. By HMG-CoA reductase inhibition, statins show antitumor effects on leukemia stem cells [8], antiviral activity by inhibiting the hepatitis C virus (HCV) RNA replication and assembly of the viral replication complex [9], and extracellular amyloid β-peptide degradation by microglia via stimulation of exosome-associated insulin-degrading enzyme secretion [10]. On the other hand, statins also show effects unrelated to HMG-CoA reductase inhibition, for example, cell cycle arrest at G1 phase by proteasome inhibition [11–13], and immunoresponse modulation via inhibition of LFA-1-mediated adhesion of leukocytes to ICAM-1 [14, 15] or via repression of MHC-II-mediated T-cell activation [16]. Other therapeutic effects on fragile X syndrome [17], *Listeria monocytogenes* infection [18], progressive multiple sclerosis [19], skeletal dysplasia [20], and Duchenne muscular dystrophy [21] have been reported.

Biological Activity of Compactin and Lovastatin
Compactin
IC_{50} for cholesterol synthesis in ^{14}C–acetate incorporation into digitonin-precipitable sterols in a rat liver enzyme system: 26 nM [1]
IC_{50} for HMG-CoA reductase: 100 nM [6]
Lovastatin
Decrease of HCV RNA levels to 10–20% in Huh7-K2040 cells: 50 μM, 24-hour treatment [9]
IC_{50} for LFA-1–ICAM-1 interaction: 2.1 ± 0.8 [15]
Inhibition of IFN-γ induced MHC-II expression on human endothelial cells and macrophages: 10 μM, 48-hour treatment [16]

References

1. Endo A, Kuroda M, Tsujita Y (1976) ML-236A, ML-236B, and ML-236C, new inhibitors of cholesterogenesis produced by *Penicillium citrinium*. J Antibiot (Tokyo) 29:1346–1348
2. Brown AG, Smale TC, King TJ, Hasenkamp R, Thompson RH (1976) Crystal and molecular structure of compactin, a new antifungal metabolite from *Penicillium brevicompactum*. J Chem Soc Perkin 1:1165–1170
3. Endo A. (1979) Monacolin K, a new hypocholesterolemic agent produced by a *Monascus* species. J Antibiot (Tokyo) 32:852–854
4. Endo A. (1980) Monacolin K, a new hypocholesterolemic agent that specifically inhibits 3-hydroxy-3-methylglutaryl coenzyme A reductase. J Antibiot (Tokyo) 33:334–336
5. Alberts AW, Chen J, Kuron G, Hunt V, Huff J, Hoffman C, Rothrock J, Lopez M, Joshua H, Harris E, Patchett A, Monaghan R, Currie S, Stapley E, Albers-Schonberg G, Hensens O, Hirshfield J, Hoogsteen K, Liesch J, Springer J. (1980) Mevinolin: a highly potent competitive inhibitor of hydroxymethylglutaryl-coenzyme A reductase and a cholesterol-lowering agent. Proc Natl Acad Sci U S A 77:3957–3961
6. Endo A, Kuroda M, Tanzawa K (1976) Competitive inhibition of 3-hydroxy-3-methylglutaryl coenzyme A reductase by ML-236A and ML-236B fungal metabolites, having hypocholesterolemic activity. FEBS Lett 72:323–326
7. Istvan ES, Deisenhofer J. (2001) Structural mechanism for statin inhibition of HMG-CoA reductase. Science 292:1160–1164
8. Hartwell KA, Miller PG, Mukherjee S, Kahn AR, Stewart AL, Logan DJ, Negri JM, Duvet M, Järås M, Puram R, Dancik V, Al-Shahrour F, Kindler T, Tothova Z, Chattopadhyay S, Hasaka T, Narayan R, Dai M, Huang C, Shterental S, Chu LP, Haydu JE, Shieh JH, Steensma DP, Munoz B, Bittker JA, Shamji AF, Clemons PA, Tolliday NJ, Carpenter AE, Gilliland DG, Stern

AM, Moore MA, Scadden DT, Schreiber SL, Ebert BL, Golub TR. (2013) Niche-based screening identifies small-molecule inhibitors of leukemia stem Cells. Nat Chem Biol 9:840–848
9. Ye J, Wang C, Sumpter R Jr., Brown MS, Goldstein JL, Gale M Jr. (2003) Disruption of hepatitis C virus RNA replication through inhibition of host protein geranylgeranylation. Proc Natl Acad Sci U S A 100:15865–15870
10. Tamboli IY, Barth E, Christian L, Siepmann M, Kumar S, Singh S, Tolksdorf K, Heneka MT, Lütjohann D, Wunderlich P, Walter J. (2010) Statins promote the degradation of extracellular amyloid β-peptide by microglia via stimulation of exosome-associated insulin-degrading enzyme secretion. J Biol Chem 285:37405–37414
11. Cornell RB, Horwitz AF. (1980) Apparent coordination of the biosynthesis of lipids in cultured cells: its relationship to the regulation of the membrane sterol: phospholipid ratio and cell cycling. J Cell Biol 86:810–819
12. Habenicht AJ, Glomset JA, Ross R. (1980) Relation of cholesterol and mevalonic acid to the cell cycle in smooth muscle and Swiss 3T3 cells stimulated to divide by platelet-derived growth factor. J Biol Chem 255:5134–5140
13. Rao S, Porter DC, Chen X, Herliczek T, Lowe M, Keyomarsi K. (1999) Lovastatin-mediated G1 arrest is through inhibition of the proteasome, independent of hydroxymethyl glutaryl-CoA reductase. Proc Natl Acad Sci U S A 96:7797–7802
14. Kallen J, Welzenbach K, Ramage P, Geyl D, Kriwacki R, Legge G, Cottens S, Weitz-Schmidt G, Hommel U. (1999) Structural basis for LFA-1 inhibition upon lovastatin binding to the CD11a I-domain.J Mol Biol 292:1–9
15. Weitz-Schmidt G, Welzenbach K, Brinkmann V, Kamata T, Kallen J, Bruns C, Cottens S, Takada Y, Hommel U. (2001) Statins selectively inhibit leukocyte function antigen-1 by binding to a novel regulatory integrin site. Nat Med 7:687–692
16. Kwak B, Mulhaupt F, Myit S, Mach F. (2000) Statins as a newly recognized type of immunomodulator. Nat Med 6:1399–402
17. Osterweil EK, Chuang SC, Chubykin AA, Sidorov M, Bianchi R, Wong RK, Bear MF. (2013) Lovastatin corrects excess protein synthesis and prevents epileptogenesis in a mouse model of fragile X syndrome. Neuron 77:243–250
18. Parihar SP, Guler R, Lang DM, Suzuki H, Marais AD, Brombacher F. (2013) Simvastatin enhances protection against Listeria monocytogenes infection in mice by counteracting *Listeria*-induced phagosomal escape. PLoS One 8: e75490
19. Palace J, Robertson N. (2014) Modifying disability in progressive multiple sclerosis. Lancet 383:2189–2191
20. Yamashita A, Morioka M, Kishi H, Kimura T, Yahara Y, Okada M, Fujita K, Sawai H, Ikegawa S, Tsumaki N. (2014) Statin treatment rescues FGFR3 skeletal dysplasia phenotypes. Nature 513:507–511
21. Whitehead NP, Kim MJ, Bible KL, Adams ME, Froehner SC. (2015) A new therapeutic effect of simvastatin revealed by functional improvement in muscular dystrophy. Proc Natl Acad Sci U S A 112:12864–12869

Concanamycin A

Keywords: [Antifungal] [V-ATPase inhibitor]

Structure

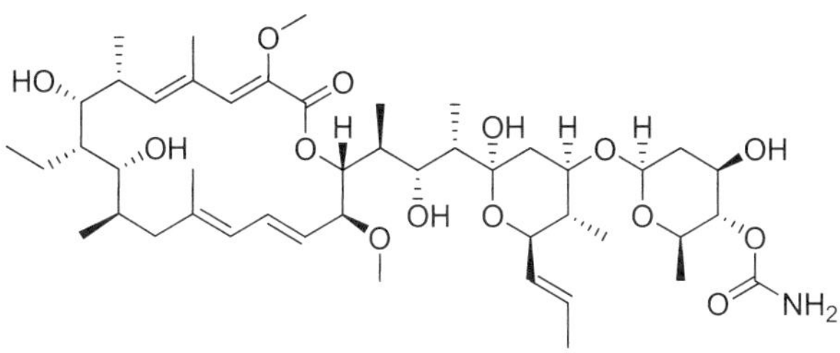

	Concanamycin A
Molecular formula	$C_{46}H_{75}NO_{14}$
Molecular weight	866.10
Solubility	DMSO, +++; H_2O, ±; MeOH, +++

Discovery/Isolation/Chemical Studies

The first concanamycin to be isolated was folimycin which shows antifungal activities [1]. Concanamycins A, B, and C were isolated from *Streptomyces diastatochromogenes* S-45 as effective inhibitors of the proliferation of mouse splenic lymphocytes stimulated by concanavalin A [2, 3]. Folimycin is identical to concanamycin A [3]. Concanamycins D, E, and F were also isolated from *Streptomyces* sp. [4]. The structures were determined by several spectroscopic analyses and X-ray crystallography [2, 5].

Biological Studies

Concanamycins were reisolated as inhibitors of the acidification of endosome/lysosomes and of oxidized LDL-induced accumulation of lipid droplets [4, 6]. These results and other features indicate that concanamycins are V-ATPase inhibitors.

Concanamycins block intracellular protein secretion before arrival at the Golgi apparatus [7, 8], perforin-/Fas-based cytolytic activity of CTL [9, 10], and bone resorption by osteoclasts [11] via V-ATPase inhibition.

Biological Activity of Concanamycins

MIC for *Sacchromyces cerevisiae*: <0.39 μg/ml [3]
IC_{50} for lysosome acidification of concanamycin A: 0.061 nM [4]
95% inhibition of the production of infectous VSV production: >10 ng/ml [8]
ID_{50} for cytolysis activity of CTL clone OE4: 1.2 ng/ml [10]

References

1. Yamamoto H, Nakazawa K, Horii S, Miyake A (1960) Studies on agricultural antibiotic. Folimycin, a new antibiotic produced by *Streptomyces neyagawaensis* nov. sp. Nippon Nogeikagaku Kaishi 34: 268–272
2. Kinashi H, Sakaguchi K, Higashijima T, Miyazawa T (1982) Structures of concanamycins B and C. J Antibiot (Tokyo) 35: 1618–1620
3. Kinashi H, Someno K, Sakaguchi K (1984) Isolation and characterization of concanamycins A, B and C. J Antibiot (Tokyo) 37: 1333–1343
4. Woo JT, Shinohara C, Sakai K, Hasumi K, Endo A (1992) Isolation, characterization and biological activities of concanamycins as inhibitors of lysosomal acidification. J Antibiot (Tokyo) 45: 1108–1116
5. Westley JW, Liu CM, Sello LH, Evans RH, Troupe N, Blount JF, Chiu AM, Todaro LJ, Miller PA (1984) The structure and absolute configuration of the 18-membered macrolide lactone antibiotic X-4357B (concanamycin A). J Antibiot (Tokyo) 37: 1738–1740
6. Woo JT, Shinohara C, Sakai K, Hasumi K, Endo A (1992) Inhibition of the acidification of endosomes and lysosomes by the antibiotic concanamycin B in macrophage J774. Eur J Biochem 207: 383–389
7. Muroi M, Shiragami N, Nagao K, Yamasaki M, Takatsuki A (1993) Folimycin (concanamycin A), a specific inhibitor of V-ATPase, blocks intracellular translocation of the glycoprotein of vesicular stomatitis virus before arrival to the Golgi apparatus. Cell Struct Funct 18: 139–149
8. Muroi M, Takasu A, Yamasaki M, Takatsuki A (1993) Folimycin (concanamycin A), an inhibitor of V-type H^+-ATPase, blocks cell-surface expression of virus-envelope glycoproteins. Biochem Biophys Res Commun 193: 999–1005
9. Kataoka T, Shinohara N, Takayama H, Takaku K, Kondo S, Yonehara S, Nagai K (1996) Concanamycin A, a powerful tool for characterization and estimation of contribution of perforin- and Fas-based lytic pathways in cell-mediated cytotoxicity. J Immunol 156: 3678–3686
10. Kataoka T, Takaku K, Magae J, Shinohara N, Takayama H, Kondo S, Nagai K (1994) Acidification is essential for maintaining the structure and function of lytic granules of CTL. Effect of concanamycin A, an inhibitor of vacuolar type H^+-ATPase, on CTL-mediated cytotoxicity. J Immunol 153: 3938–3947
11. Woo JT, Ohba Y, Tagami K, Sumitani K, Yamaguchi K, Tsuji T (1996) Concanamycin B, a vacuolar H^+-ATPase specific inhibitor suppresses bone resorption in vitro. Biol Pharm Bull 19: 297–299

Cyclosporine A

Keywords: [Immunosuppressant] [cyclophilin] [calcineurin]

Structure

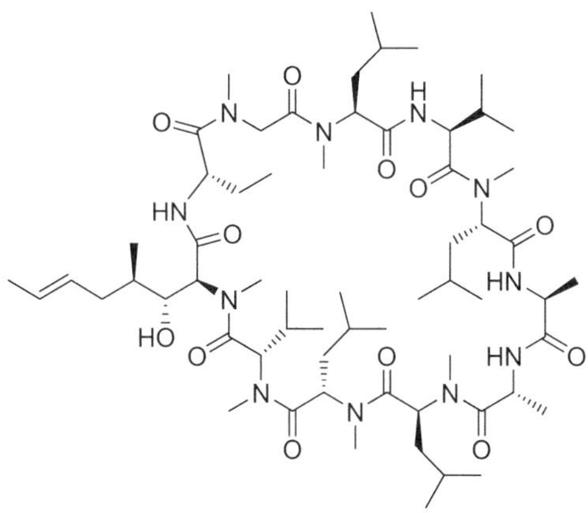

	Cyclosporine A
Molecular formula	$C_{62}H_{111}N_{11}O_{12}$
Molecular weight	1202.64
Solubility	DMSO, ++; H_2O, ±; MeOH, +++

Discovery/Isolation

Cyclosporine A, a cyclic undecapeptide, was found by researchers at Sandoz Ltd. as an antifungal antibiotic from a fungus (originally identified as *Cylindrocarpum lucidum*), collected in Wisconsin in the USA [1]. Later, the same compound was isolated from *Trichoderma inflatum* which was isolated from a soil sample from Hardangervidda in Norway [2].

Biological Studies

Cyclosporine A was originally isolated as an antifungal antibiotic; however, its remarkable immunosuppressive activity in several pharmacological models was discovered by Sandoz Ltd. in the early 1970s [2, 3]. Cyclosporine A inhibits DNA binding activity of NF-AT and the production of interleukin 2 (IL-2), and prevents the expression of IL-2 receptor, suggesting that cyclosporine A inhibits the initial steps of T-lymphocyte activation [4, 5].

A cyclosporine A binding protein, named cyclophlin, was isolated using tritium-labeled cyclosporine A [6], and identified as the peptidyl-prolyl *cis-trans* isomerase (PPIase) [7, 8]. Cyclosporine A specifically binds to cyclophilin and inhibits its enzyme activity; however, the inhibition of PPI activity is not sufficient for immunosuppressive activity [9]. Liu et al. found that the complex of cyclosporine

A and cyclophilin binds to calcineurin and inhibits its phosphatase activity [10]. The inhibition of calcineurin is sufficient for the inhibition of T-cell activation, and it was revealed that NF-AT dephosphoryation by calcineurin is required for T-cell activation [11].

Because cyclosporine A binds to cyclophilin D which is a component of mitochondrial permeability pore complex, cyclosporine A is also used as a bioprobe for investigating the function of cyclophilin D on mitochondria-mediated apoptosis [12–14].

Biological Activity of Cyclosporine A

IC_{90} of NF-AT binding in Jurkat nuclear extract: 10 ng/ml [5]
IC_{50} for swelling of mitochondria: 25–50 pmole/mg protein [12]
Inhibition between cyclophilin D and Bcl-2: 1 μM [14]

References

1. Svarstad H, Bugge HC, Dhillion SS (2000) From Norway to Novartis: cyclosporin from *Tolypocladium inflatum* in an open access bioprospecting regime. Biodivers Conserv 9:1521–1541
2. Stahelin HF (1996) The history of cyclosporin A (Sandimmune) revisited: another point of view. Experientia 52:5–13
3. Ruegger A, Kuhn M, Lichti H, Loosli HR, Huguenin R, Quiquerez C, Wartburg AV (1976) Cyclosporin-A, a peptide metabolite from *Trichoderma-Polysporum* (Link Ex Pers) *Rifai*, with a remarkable immunosuppressive activity. Helv Chim Acta 59:1075–1092
4. Elliott JF, Lin Y, Mizel SB, Bleackley RC, Harnish DG, Paetkau V (1984) Induction of interleukin 2 messenger RNA inhibited by cyclosporin A. Science 226:1439–1441
5. Emmel EA, Verweij CL, Durand DB, Higgins KM, Lacy E, Crabtree GR (1989) Cyclosporin A specifically inhibits function of nuclear proteins involved in T cell activation. Science 246:1617–1620
6. Handschumacher RE, Harding MW, Rice J, Drugge RJ, Speicher DW (1984) Cyclophilin: a specific cytosolic binding protein for cyclosporin A. Science 226:544–547
7. Fischer G, Wittmann-Liebold B, Lang K, Kiefhaber T, Schmid FX (1989) Cyclophilin and peptidyl-prolyl *cis-trans* isomerase are probably identical proteins. Nature 337:476–478
8. Takahashi N, Hayano T, Suzuki M (1989) Peptidyl-prolyl *cis-trans* isomerase is the cyclosporin A-binding protein cyclophilin. Nature 337:473–475
9. Clipstone NA, Crabtree GR (1992) Identification of calcineurin as a key signalling enzyme in T-lymphocyte activation. Nature 357:695–697
10. Liu J, Farmer JD, Jr., Lane WS, Friedman J, Weissman I, Schreiber SL (1991) Calcineurin is a common target of cyclophilin-cyclosporin A and FKBP-FK506 complexes. Cell 66:807–815

11. McCaffrey PG, Perrino BA, Soderling TR, Rao A (1993) NF-ATp, a T lymphocyte DNA-binding protein that is a target for calcineurin and immunosuppressive drugs. J Biol Chem 268:3747–3752
12. Broekemeier KM, Dempsey ME, Pfeiffer DR (1989) Cyclosporin A is a potent inhibitor of the inner membrane permeability transition in liver mitochondria. J Biol Chem 264:7826–7830
13. Marzo I, Brenner C, Zamzami N, Jurgensmeier JM, Susin SA, Vieira HL, Prevost MC, Xie Z, Matsuyama S, Reed JC, Kroemer G (1998) Bax and adenine nucleotide translocator cooperate in the mitochondrial control of apoptosis. Science 281:2027–2031
14. Eliseev RA, Malecki J, Lester T, Zhang Y, Humphrey J, Gunter TE (2009) Cyclophilin D interacts with Bcl2 and exerts an anti-apoptotic effect. J Biol Chem 284:9692–9699

Cytotrienin A

Keywords: [Apoptosis inducer] [Protein translation inhibitor]

Structure

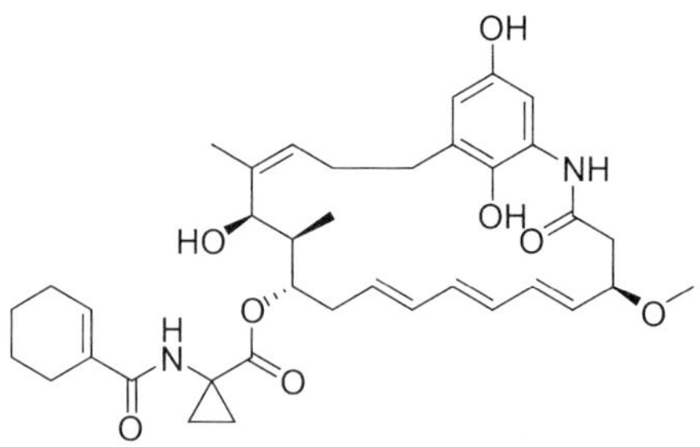

	Cytotrienin A
Molecular formula	$C_{37}H_{48}N_2O_8$
Molecular weight	648.80
Solubility	DMSO, +++; H_2O, ±; MeOH, +++

Discovery/Isolation/Chemical Studies

Cytotrienin A was isolated as an apoptosis inducer from *Streptomyces* sp. [1, 2]. This compound is the first ansamycins with a unique 1-aminocyclopropane-1-carboxylic acid unit.

Biological Studies

Cytotrienin A induces apoptosis in HL-60 cells accompanied with the activation of a 36 kDa protein kinase (p36 MBP (myelin basic protein) kinase) which phosphorylates MBP in a gel kinase assay [3]. The 36 kDa MBP kinase was identified as the active proteolytic fragment of MST1/Krs2 and MST2/Krs1 which was activated in response to anticancer drugs including camptothecin and etoposide.

During the course of a high-throughput screen to identify translation inhibitors, cytotrienin A was identified as an inhibitor of both cap- and IRES-dependent translation in vitro [4]. Detailed investigation shows that cytotrienin A inhibits eukaryotic protein synthesis by targeting translation elongation and interfering with eEF1A function. Cytotrienin A also triggers ribotoxic stress response and selectively inhibits TNF-α-induced ICAM-1 expression by inducing ectodomain shedding of TNF receptor 1 via the activation of ERK and p38 MAP kinase [5].

Biological Activity of Cytotrienin A

IC_{50} values for HL-60 and WI-38 cells: 5 and >1000 nM, respectively [2]
IC_{50} for protein synthesis in A549 cells: ~0.1 μM [5]

References

1. Zhang HP, Kakeya H, Osada H (1997) Novel triene-ansamycins, cytotrienins A and B, inducing apoptosis on human leukemia HL-60 cells. Tetrahedron Lett 38:1789–1792
2. Kakeya H, Zhang HP, Kobinata K, Onose R, Onozawa C, Kudo T, Osada H (1997) Cytotrienin A, a novel apoptosis inducer in human leukemia HL-60 cells. J Antibiot (Tokyo) 50:370–372
3. Kakeya H, Onose R, Osada H (1998) Caspase-mediated activation of a 36-kDa myelin basic protein kinase during anticancer drug-induced apoptosis. Cancer Res 58:4888–4894
4. Lindqvist L, Robert F, Merrick W, Kakeya H, Fraser C, Osada H, Pelletier J (2010) Inhibition of translation by cytotrienin A--a member of the ansamycin family. RNA 16:2404–2413
5. Yamada Y, Taketani S, Osada H, Kataoka T (2011) Cytotrienin A, a translation inhibitor that induces ectodomain shedding of TNF receptor 1 via activation of ERK and p38 MAP kinase. Eur J Pharmacol 667:113–119

15-Deoxyspergualin

Keywords: [Immunosuppressant] [Antiangiogenesis]

Structure

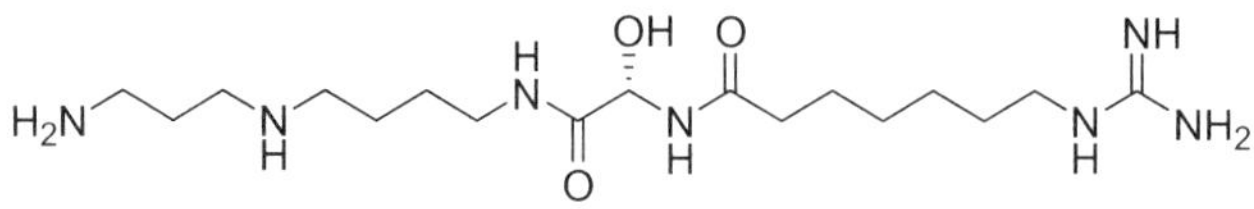

	15-Deoxyspergualin
Molecular formula	$C_{17}H_{37}N_7O_3$
Molecular weight	387.52
Solubility	DMSO, ++; H_2O, +++; MeOH, ++

Discovery/Isolation/Chemical Studies

15-Deoxyspergualin (DSG) is a synthetic derivative of spergualin, an antitumor natural product isolated from *Bacillus laterosporus* [1–3]. DSG has a spermidine moiety and a guanidine moiety in its structure, and the spermidine moiety has cell binding activity and the guanidine moiety has cytotoxic activity [4].

Biological Studies

DSG not only inhibits growth of various tumor cell lines in a xenograft model [5], but also has a potent immunosuppressive effect [6, 7]. Several biological activities of DSG have been reported, for example the indirect inhibition of phosphatidylcholine (PC) synthesis [8, 9] and the inactivation of eIF5A [10].

Nadler et al. identified an intracellular binding protein of DSG by using 11-methoxy-DSG-sepharose as Hsc70, the constitutive or cognate member of the Hsp70 protein family [11]. The members of the Hsp70 family of heat shock proteins are important for many cellular processes, including immune responses, and this finding suggests that heat shock proteins may represent a class of immunosuppressant binding proteins, or immunophilins. The binding site was determined as the EEVD regulatory domain at the extreme C-terminal of Hsc70 by 1-ethyl-3-(3-dimethylaminopropyl)carbodiimide (EDC)–cross-linking of [^{14}C]-DSG [12]. Because the members of the Hsp70 family are important for many cellular processes including immune responses, these findings suggest that the Hsp70/Hsc70 are a class of immunosuppressant binding proteins.

Biological Activity of 15-Deoxysperugualin

In vivo antitumor activity in L-1210: 6.25 mg/kg/day, intraperitoneally [1]

K_d values for DSK binding to Hsc70 and Hsp90: 4 and 5 μM, respectively [13]

References

1. Takeuchi T, Iinuma H, Kunimoto S, Masuda T, Ishizuka M, Takeuchi M, Hamada M, Naganawa H, Kondo S, Umezawa H (1981) A new antitumor antibiotic, spergualin: isolation and antitumor activity. J Antibiot (Tokyo) 34:1619–1621
2. Umezawa H, Kondo S, Iinuma H, Kunimoto S, Ikeda Y, Iwasawa H, Ikeda D, Takeuchi T (1981) Structure of an antitumor antibiotic, spergualin. J Antibiot (Tokyo) 34:1622–1624
3. Iwasawa H, Kondo S, Ikeda D, Takeuchi T, Umezawa H (1982) Synthesis of (−)-15-deoxyspergualin and (−)-spergualin-15-phosphate. J Antibiot (Tokyo) 35:1665–1669
4. Umeda Y, Moriguchi M, Kuroda H, Nakamura T, Fujii A, Iinuma H, Takeuchi T, Umezawa H (1987) Synthesis and antitumor activity of spergualin analogues. II. Chemical modification of the spermidine moiety. J Antibiot (Tokyo) 40:1303–1315
5. Plowman J, Harrison SD, Jr., Trader MW, Griswold DP, Jr., Chadwick M, McComish MF, Silveira DM, Zaharko D (1987) Preclinical antitumor activity and pharmacological properties of deoxyspergualin. Cancer Res 47:685–689
6. Masuda T, Mizutani S, Iijima M, Odai H, Suda H, Ishizuka M, Takeuchi T, Umezawa H (1987) Immunosuppressive activity of 15-deoxyspergualin and its effect on skin allografts in rats. J Antibiot (Tokyo) 40:1612–1618
7. Nemoto K, Hayashi M, Abe F, Nakamura T, Ishizuka M, Umezawa H (1987) Immunosuppressive activities of 15-deoxyspergualin in animals. J Antibiot (Tokyo) 40:561–562
8. Kawada M, Masuda T, Ishizuka M, Takeuchi T (2002) 15-Deoxyspergualin inhibits Akt kinase activation and phosphatidylcholine synthesis. J Biol Chem 277:27765–27771
9. Kawada M, Ishizuka M (2003) Inhibition of CTP: phosphocholine cytidylyltransferase activity by 15-deoxyspergualin. J Antibiot (Tokyo) 56:725–726
10. Nishimura K, Ohki Y, Fukuchi-Shimogori T, Sakata K, Saiga K, Beppu T, Shirahata A, Kashiwagi K, Igarashi K (2002) Inhibition of cell growth through inactivation of eukaryotic translation initiation factor 5A (eIF5A) by deoxyspergualin. Biochem J 363:761–768
11. Nadler SG, Tepper MA, Schacter B, Mazzucco CE (1992) Interaction of the immunosuppressant deoxyspergualin with a member of the Hsp70 family of heat shock proteins. Science 258:484–486
12. Nadler SG, Dischino DD, Malacko AR, Cleaveland JS, Fujihara SM, Marquardt H (1998) Identification of a binding site on Hsc70 for the immunosuppressant 15-deoxyspergualin. Biochem Biophys Res Commun 253:176–180
13. Nadeau K, Nadler SG, Saulnier M, Tepper MA, Walsh CT (1994) Quantitation of the interaction of the immunosuppressant deoxyspergualin and analogs with Hsc70 and Hsp90. Biochemistry 33:2561–2567

7-Desacetoxy-6,7-Dehydrogedunin

Keywords: [Antimalarial] [PKR inhibitor] [HSF1-dependent stress response]

Structure

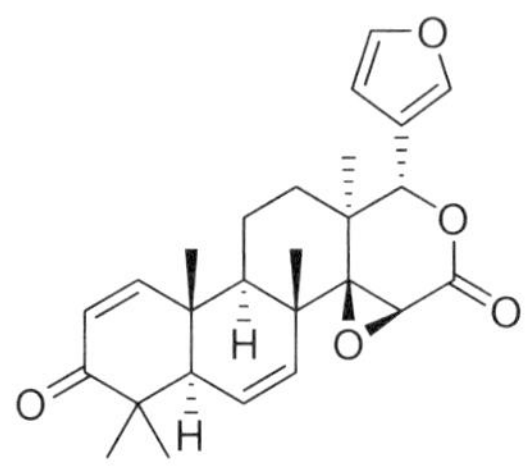

	7-Desacetoxy-6,7-dehydrogedunin
Molecular formula	$C_{26}H_{30}O_5$
Molecular weight	422.52
Solubility	DMSO, ++; H_2O, ±; MeOH, ?

Discovery/Isolation/Chemical Studies

7-Desacetoxy-6,7-dehydrogedunin is a derivative of gedunin, an antimalarial compound isolated from *Melia azedarach* [1] and *Azadirachta indica* [2]. 7-Desacetoxy-6,7-dehydrogedunin and gedunin contain an epoxide and α,β-unsaturated ketone.

Biological Studies

7-Desacetoxy-6,7-dehydrogedunin was identified as a compound that protects macrophages from death induced by anthrax lethal toxin [3]. SAR analysis revealed that epoxide and α,β-unsaturated ketone are not essential for the macrophage protection. Eleven interacting proteins were identified by biotinylated analog, suggesting that this compound is moderately selective. Further analysis using siRNA showed that PKR is needed for macrophage cell death and the phenotype of PKR knockdown resembles that of treatment with 7-desacetoxy-6,7-dehydrogedunin. Different from the other PKR inhibitors, 2-aminopurine and C16,7-desacetoxy-6,7-dehydrogedunin dose not bind the ATP catalytic domain and inhibit kinase activity of PKR, but prevents ASC assembly. Because the lethal toxin induces inflammasome formation and caspase-1 activation, these results suggest that PKR interactions, rather than PKR kinase activity, are involved in the broad mechanisms of inflammasome-mediated caspase-1 activation.

It was also reported that 7-desacetoxy-6,7-dehydrogedunin activates the heat shock factor (HSF) 1-dependent stress response [4].

Biological Activity of 7-Desacetoxy-6,7-Dehydrogedunin

IC_{50} for J774 macrophage protection from anthrax lethal toxin: 5 μM [3]

References

1. Khalid SA, Farouk A, Geary TG, Jensen JB (1986) Potential antimalarial candidates from African plants: and in vitro approach using *Plasmodium falciparum*. J Ethnopharmacol 15:201–209
2. Khalid SA, Duddeck H, Gonzalez-Sierra M (1989) Isolation and characterization of an antimalarial agent of the neem tree *Azadirachta indica*. J Nat Prod 52:922–926
3. Hett EC, Slater LH, Mark KG, Kawate T, Monks BG, Stutz A, Latz E, Hung DT (2013) Chemical genetics reveals a kinase-independent role for protein kinase R in pyroptosis. Nat Chem Biol 9:398–405
4. Santagata S, Xu YM, Wijeratne EM, Kontnik R, Rooney C, Perley CC, Kwon H, Clardy J, Kesari S, Whitesell L, Lindquist S, Gunatilaka AA (2012) Using the heat-shock response to discover anticancer compounds that target protein homeostasis. ACS Chem Biol 7:340–349

Destruxin E

Keywords: [V-ATPase inhibitor]

Structure

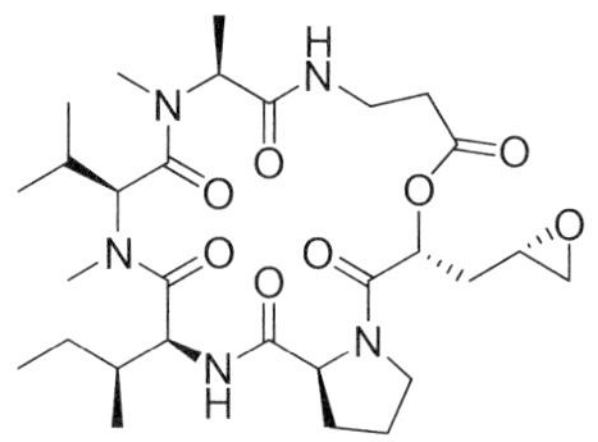

	Destruxin E
Molecular formula	$C_{29}H_{47}N_5O_8$
Molecular weight	593.72
Solubility	DMSO, +++; H_2O, +; MeOH, +++

Discovery/Isolation/Chemical Studies

Destruxin E was isolated from the culture medium of *Metarhizium anisopliae*, which is a fungi known to cause diseases in several kinds of insects [1]. The absolute structure was determined by total synthesis [2].

Biological Studies

Destruxin E shows cytotoxicity against murine lymphocytes and L1210 leukemia cells [3]. In the course of screening of microbial metabolites that inhibit the accumulation of cholesteryl ester in macrophages incubated with oxidized LDL, destruxin E was reisolated as an active compound [4]. Detailed biochemical investigations revealed that destruxin E inhibits V-ATPase-depedent acidification of endosomes and lysosomes [4]. Destruxin E inhibited perforin-dependent cytotoxicity mediated by $CD8^+$ CTL [5] and bone resorption by osteoclasts [6] through V-ATPase inhibition. Antitumor and antiangiogenic acitivities were also reported [7].

Biological Activity of Destruxin E

Inhibition of ATP-depedent acidification of endosomes and lysosomes in macrophage: 1 μM [4]

IC_{50} for perforin-dependent cytotoxicity mediated by CD^{8+} CTL: 4.0 μM [5]

Induction of morphological changes of osteoclast-like multinucleated cells: 10–50 nM [6]

References

1. Pais M, Das BC, Ferron P (1981) Depsipeptides from *Metarhizium anisopliae*. Phytochemistry 20:715–723
2. Yoshida M, Takeuchi H, Ishida Y, Yashiroda Y, Takagi M, Shin-ya K, Doi T (2010) Synthesis, structure determination, and biological evaluation of destruxin E. Org Lett 12:3792–3795
3. Morel E, Pais M, Turpin M, Guyot M (1983) Cytotoxicity of cyclodepsipeptides on murine lymphocytes and on L1210 leukemia cells. Biomed Pharmacother 37:184–185
4. Naganuma S, Kuzuya N, Sakai K, Hasumi K, Endo A (1992) Inhibition of the accumulation of lipid droplets in macrophage J774 by bafilomycin B1 and destruxin E. Biochim Biophys Acta 1126:41–48
5. Togashi K, Kataoka T, Nagai K (1997) Characterization of a series of vacuolar type H^+ –ATPase inhibitors on CTL-mediated cytotoxicity. Immunol Lett 55:139–144
6. Nakagawa H, Takami M, Udagawa N, Sawae Y, Suda K, Sasaki T, Takahashi N, Wachi M, Nagai K, Woo JT (2003) Destruxins, cyclodepsipeptides, block the formation of actin rings and prominent clear zones and ruffled borders in osteoclasts. Bone 33:443–455
7. Dornetshuber-Fleiss R, Heffeter P, Mohr T, Hazemi P, Kryeziu K, Seger C, Berger W, Lemmens-Gruber R (2013) Destruxins: fungal-derived cyclohexadepsipeptides with multifaceted anticancer and antiangiogenic activities. Biochem Pharmacol 86:361–377

Diazonamide A

Keywords: [Mitotic inhibitor] [Ornithine δ-aminotransferase]

Structure

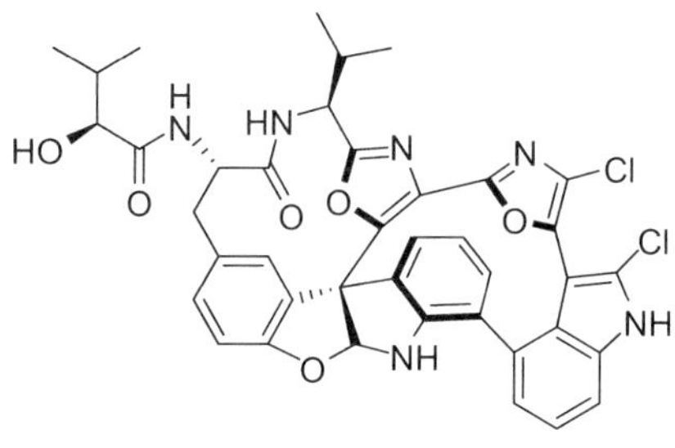

	Diazonamide A
Molecular formula	$C_{40}H_{34}Cl_2N_6O_6$
Molecular weight	765.65
Solubility	DMSO, ? ; H_2O, ? ; MeOH, ++

Discovery/Isolation/Chemical Studies

Diazonamide A was isolated as a cytotoxic compound from the colonial ascidian *Diazona chinensis* (order Phlebobranchia) collected from the ceilings of small caves along the northwest coast of Siquijor Island, Philippines [1]. The structure was determined except for the stereochemistry of C-37 by total synthesis [2, 3].

Biological Studies

Diazonamide A causes cells to arrest in mitosis, and treated cells lose both interphase and spindle microtubules [4]. This compound inhibits microtubule assembly by inhibition of tubulin-dependent GTP hydrolysis. Because diazonamide A does not inhibit the binding of [^{3}H]vinblastine, [^{3}H]dolastatin 10, or [8-^{14}C]GTP to tubulin, it was suggested that the binding site of diazonamide A on tubulin is different from that of the vinca alkaloid and dolastatin 10.

Wang et al. purified a mitochondria matrix protein, ornithine δ-aminotransferase (OAT), as a diazonamide A binding protein using biotinylated diazonamide A [5]. They showed that RanGTP disrupts the interaction between diazonamide A and OAT, and that OAT knockdown blocks cell division and causes cell death in cancer cell lines. These results revealed the new function of OAT in mitotic regulation.

Biological Activity of Diazonamide A

IC_{50} for HCT116 human colon carcinoma and B16 murine melanoma cancer cell lines: 15 ng/ml [1]

Inhibition of cell cycle progression of human ovarian carcinoma 2008 cells in M phase: 40–100 nM [4]

IC_{50} for in vitro microtubule polymerization stimulated by MAP or glutamate: 0.3 and 0.75 μM [4]

References

1. Lindquist N, Fenical W, Vanduyne GD, Clardy J (1991) Isolation and structure determination of diazonamide-A and diazonamide-B, unusual cytotoxic metabolites from the marine Ascidian *Diazona chinensis*. J Am Chem Soc 113:2303–2304
2. Nicolaou KC, Hao JL, Reddy MV, Rao PB, Rassias G, Snyder SA, Huang XH, Chen DYK, Brenzovich WE, Giuseppone N, Giannakakou P, O″Brate A (2004) Chemistry and biology of diazonamide A: second total synthesis and biological investigations. J Am Chem Soc 126:12897–12906
3. Nicolaou KC, Chen DYK, Huang XH, Ling TT, Bella M, Snyder SA (2004) Chemistry and biology of diazonamide A: first total synthesis and confirmation of the true structure. J Am Chem Soc 126:12888–12896
4. Cruz-Monserrate Z, Vervoort HC, Bai R, Newman DJ, Howell SB, Los G, Mullaney JT, Williams MD, Pettit GR, Fenical W, Hamel E (2003) Diazonamide A and a synthetic structural analog: disruptive effects on mitosis and cellular microtubules and analysis of their interactions with tubulin. Mol Pharmacol 63:1273–1280
5. Wang G, Shang L, Burgett AW, Harran PG, Wang X (2007) Diazonamide toxins reveal an unexpected function for ornithine δ-amino transferase in mitotic cell division. Proc Natl Acad Sci U S A 104:2068–2073

Didemnin B

Keywords: [Antitumor] [Antiviral] [EF-1α]

Structure

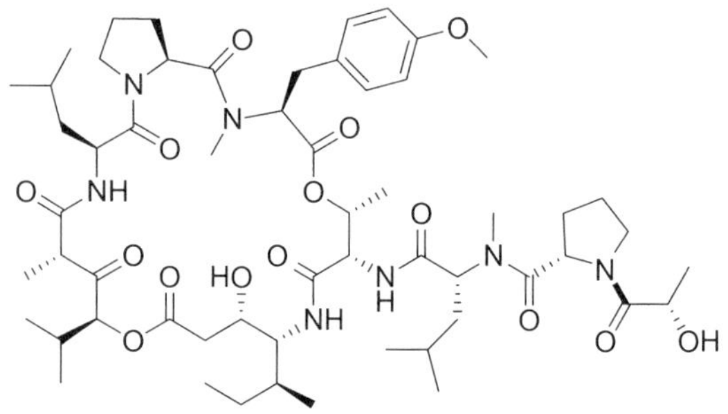

	Didemnin B
Molecular formula	$C_{57}H_{89}N_7O_{15}$
Molecular weight	1112.37
Solubility	DMSO, ++; H_2O, ? ; MeOH, ?

Discovery/Isolation

Cyclic depsipeptide didemnin B was isolated as a highly active antiviral and antitumor compound from Caribbean tunicate *Trididemnum solidum* [1].

Biological Studies

In addition to antiviral and antitumor activity, didemnin B inhibits polypeptide synthesis in both cellular and cell-free assay, and cell cycle progression in G1 phase [2]. Crews et al. purified the elongation factor 1α (EF-1α) as a target protein of didemnin by using biotinylated didemnin A [3]. EF-1α is an abundant guanine nucleotide binding protein that transports aminoacyl-tRNAs to the ribosomal A site in a GTP-dependent manner. [^{3}H] *N*-acetyl didemnin A bound to EF-1α in vitro in a GTP-dependent manner. Didemnin B stimulates EF-1α-dependent aminoacyl-tRNA binding to ribosome and inhibits translocation by stabilizing aminoacyl-tRNA bound to the ribosomal A site without prevention of peptide bond formation [4].

Biological Activity of Didemnin B

Antiviral activity against herpes simplex virus type I and II: 0.05 μM [1]
IC_{50} for L1210 cell line: 0.001 μg/ml [1]
IC_{50} for in vitro protein synthesis in rabbit reticulocyte lysates: 3 μM [4]

References

1. Rinehart KL, Jr., Gloer JB, Hughes RG, Jr., Renis HE, McGovren JP, Swynenberg EB, Stringfellow DA, Kuentzel SL, Li LH (1981) Didemnins: antiviral and antitumor depsipeptides from a caribbean tunicate. Science 212:933–935
2. Crampton SL, Adams EG, Kuentzel SL, Li LH, Badiner G, Bhuyan BK (1984) Biochemical and cellular effects of didemnins A and B. Cancer Res 44:1796–1801
3. Crews CM, Collins JL, Lane WS, Snapper ML, Schreiber SL (1994) GTP-dependent binding of the antiproliferative agent didemnin to elongation factor 1α. J Biol Chem 269:15411–15414
4. SirDeshpande BV, Toogood PL (1995) Mechanism of protein synthesis inhibition by didemnin B in vitro. Biochemistry 34:9177–9184

DX-52-1* (Quinocarcin/Quinocarmycin Derivative)

Keywords: [Antitumor] [Hypoxia-inducible factor 1α (HIF1α)] [Radixin] [Galectin-3]

Structure

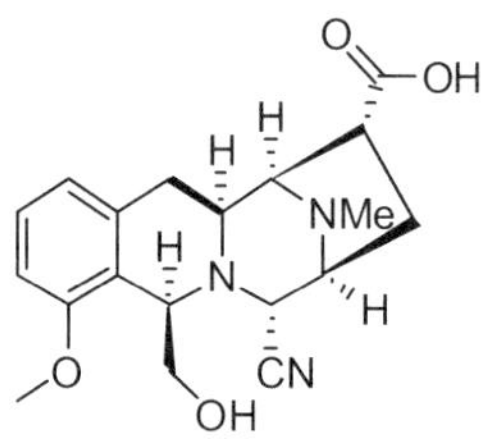

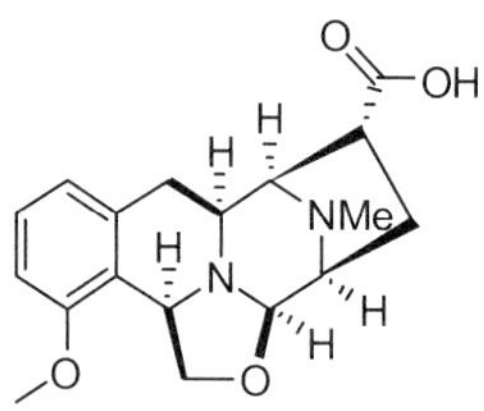

	DX-52-1	Quinocarcin/Quinocarmycin
Molecular formula	$C_{19}H_{23}N_3O_4$	$C_{18}H_{22}N_2O_4$
Molecular weight	357.41	330.38
Solubility	DMSO, ? ; H_2O, +; MeOH, ?	DMSO, ? ; H_2O, +; MeOH, ?

Discovery/Isolation

DX-52-1 is an analog of quinocarcin/quinocarmycin which was isolated from *Streptomyces melanovinaceus* nov. sp. as an antitumor compound [1–3].

Biological Studies

DX-52-1 was identified as an inhibitor of HIF1α transcriptional activation pathway under both deferoxamine mesylate-treatment and hypoxia conditions [4, 5].

Kahsai et al. identified DX-52-1 as an inhibitor of epithelial cell migration in the course of screening for new small-molecule modulators of cell motility [6]. They found four major DX-52-1 binding proteins using biotinylated DX-52-1, and identified radixin [6] and galactin-3 [7] as the primary and secondary targets of DX-52-1, respectively. Radixin is a member of the ezrin/radixin/moesin (ERM) family of membrane-actin cytoskeleton linker protein, and DX-52-1 is the first inhibitor of radixin which interferes with the interaction between radixin and both actin and the cell adhesion molecule CD44. Galectin-3 is a multifunctional carbohydrate binding protein, and DX-52-1 binds galectin-3 outside its β-galactoside binding site. Furthermore, it was recently reported that DX-52-1 also inhibits moesin–CD44 interaction and reduces the proliferation of glioblastoma [8]. Because moesin is a member of the ERM family, these results suggest that DX-52-1 inhibits cell migration and proliferation by not only disrupting the interaction of ERM proteins with actin and CD44 but also inhibiting a carbohydrate binding–independent function of galectin-3.

Biological Activity of DX-52-1

IC_{50} for HeLa S3 cell proliferation: 50 ng/ml [9]

Reduction of HIF1α protein level under hypoxia conditions: 20 μM [4]

IC_{50} for inhibition of wound closure at 24 hours in MDCK cell monolayer: 140 nM [6]

Inhibition of moesin–CD44 interaction in LN299 cells: 200 nM [8]

References

1. Saito H, Hirata T (1987) Synthetic Approach to quinocarcin. Tetrahedron Lett 28:4065–4068
2. Takahashi K, Tomita F (1983) DC-52, a novel antitumor antibiotic. 2. Isolation, physico-chemical characteristics and structure determination. J Antibiot (Tokyo) 36:468–470
3. Tomita F, Takahashi K, Shimizu K (1983) DC-52, a novel antitumor antibiotic. 1. Taxonomy, fermentation and biological activity. J Antibiot (Tokyo) 36:463–467
4. Chau NM, Rogers P, Aherne W, Carroll V, Collins I, McDonald E, Workman P, Ashcroft M (2005) Identification of novel small molecule inhibitors of hypoxia-inducible factor-1 that differentially block hypoxia-inducible factor-1 activity and hypoxia-inducible factor-1α induction in response to hypoxic stress and growth factors. Cancer Res 65:4918–4928
5. Rapisarda A, Uranchimeg B, Scudiero DA, Selby M, Sausville EA, Shoemaker RH, Melillo G (2002) Identification of small molecule inhibitors of hypoxia-inducible factor 1 transcriptional activation pathway. Cancer Res 62:4316–4324
6. Kahsai AW, Zhu S, Wardrop DJ, Lane WS, Fenteany G (2006) Quinocarmycin analog DX-52-1 inhibits cell migration and targets radixin, disrupting interactions of radixin with actin and CD44. Chem Biol 13:973–983
7. Kahsai AW, Cui J, Kaniskan HU, Garner PP, Fenteany G (2008) Analogs of tetrahydroisoquinoline natural products that inhibit cell migration and target galectin-3 outside of its carbohydrate-binding site. J Biol Chem 283:24534–24545
8. Zhu X, Morales FC, Agarwal NK, Dogruluk T, Gagea M, Georgescu MM (2013) Moesin is a glioma progression marker that induces proliferation and Wnt/β-catenin pathway activation via interaction with CD44. Cancer Res 73:1142–1155
9. Saito H, Kobayashi S, Uosaki Y, Sato A, Fujimoto K, Miyoshi K, Ashizawa T, Morimoto M, Hirata T (1990) Synthesis and biological evaluation of quinocarcin derivatives. Chem Pharm Bull 38:1278–1285

Eeyarestatin I* and II*

Keywords: [ER-associated protein degradation (ERAD) inhibitor] [Sec61 complex inhibitor]

Structure

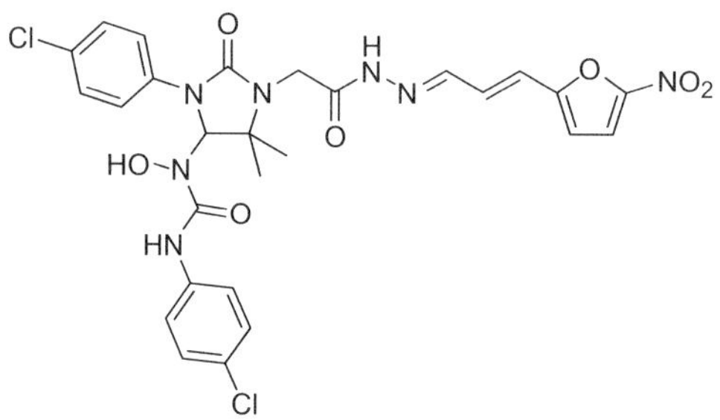

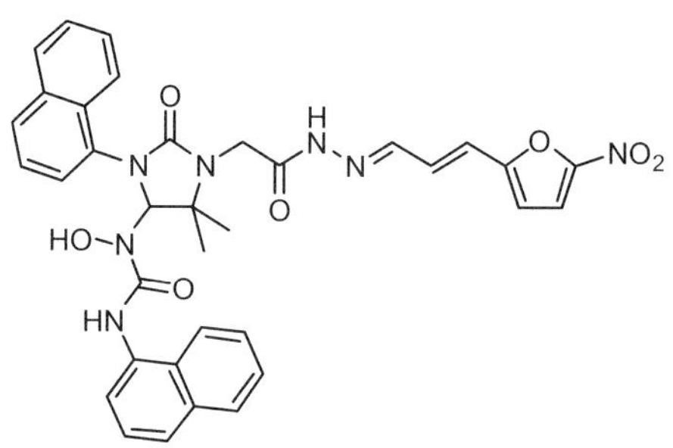

	Eeyarestatin I	Eeyarestatin II
Molecular formula	$C_{27}H_{25}Cl_2N_7O_7$	$C_{35}H_{31}N_7O_7$
Molecular weight	630.44	661.68
Solubility	DMSO, ++; H_2O, –; MeOH, +++	DMSO, ++; H_2O, –; MeOH, +++

Discovery/Isolation

Eeyarestatin I and II were found from a chemical library of 16,320 compounds by high-throughput screen of the inhibitor which prevents the turnover of EGFP-fused class I MHC heavy chain (EGFP-HC) [1].

Biological Studies

Eeyarestatin I and II inhibited the degradation of three substrates: EGFP-HC, wild-type class I HC, and T-cell receptor α, by retaining them in the ER [1]. Therefore, it is thought that these compounds inhibit ER-associated protein degradation (ERAD), which is the system eliminating the misfolded protein from the

ER. Several biochemical analyses revealed that eeyarestatin I accumulated the polyubiquitinated proteins by inhibiting p97 ATPase complex–mediated deubiquitination [2]. p97 ATPase is an essential component of the ERAD machinery, and eeyarestatin I directly binds p97 via the nitrofuran-containing group [3]. Inhibition of ERAD by eeyarestatin I further evokes the ER stress response- and histone H2A–mediated up-regulation of BH3-only protein NOXA, leading to apoptosis in cancer cells [4].

It has also been reported that eeyarestatin I inhibits nascent polypeptide cotranslational translocation into the ER by targeting Sec61 complex which forms the ER translocon [5] and intercellular vesicle transport [6].

Biological Activity of Eeyarestatin I
Inhibition of class I HC degradation in U373: 8 μM [1]
Accumulation of polyubiquitinated proteins in A9 cells: 10 μM [2]
IC_{50} values for cell growth of Jurkat, MINO, JEKO, KMS-12, BJAB cells: 6.43, 6.29, 4.1, 3.07, and 5.21 μM, respectively [4]
NOXA induction in cancer cells: 10 μM [4]
IC_{50} for cotranslational translcation of the P2X2 purinergic receptor in vitro:70 μM [5]

References

1. Fiebiger E, Hirsch C, Vyas JM, Gordon E, Ploegh HL, Tortorella D (2004) Dissection of the dislocation pathway for type I membrane proteins with a new small molecule inhibitor, eeyarestatin. Mol Biol Cell 15:1635–1646
2. Wang Q, Li L, Ye Y (2008) Inhibition of p97-dependent protein degradation by eeyarestatin I. J Biol Chem 283:7445–7454
3. Wang Q, Shinkre BA, Lee JG, Weniger MA, Liu Y, Chen W, Wiestner A, Trenkle WC, Ye Y (2010) The ERAD inhibitor eeyarestatin I is a bifunctional compound with a membrane-binding domain and a p97/VCP inhibitory group. PLoS One 5:e15479
4. Wang Q, Mora-Jensen H, Weniger MA, Perez-Galan P, Wolford C, Hai T, Ron D, Chen W, Trenkle W, Wiestner A, Ye Y (2009) ERAD inhibitors integrate ER stress with an epigenetic mechanism to activate BH3-only protein NOXA in cancer cells. Proc Natl Acad Sci U S A 106:2200–2205
5. Cross BC, McKibbin C, Callan AC, Roboti P, Piacenti M, Rabu C, Wilson CM, Whitehead R, Flitsch SL, Pool MR, High S, Swanton E (2009) Eeyarestatin I inhibits Sec61-mediated protein translocation at the endoplasmic reticulum. J Cell Sci 122:4393–4400
6. Aletrari MO, McKibbin C, Williams H, Pawar V, Pietroni P, Lord JM, Flitsch SL, Whitehead R, Swanton E, High S, Spooner RA (2011) Eeyarestatin 1 interferes with both retrograde and anterograde intracellular trafficking pathways. PLoS One 6:e22713

Epigallocatechin 3-Gallate (EGCG)

Keywords: [Tumor prevention] [Antiangiogenesis] [67-kDa laminin receptor]

Structure

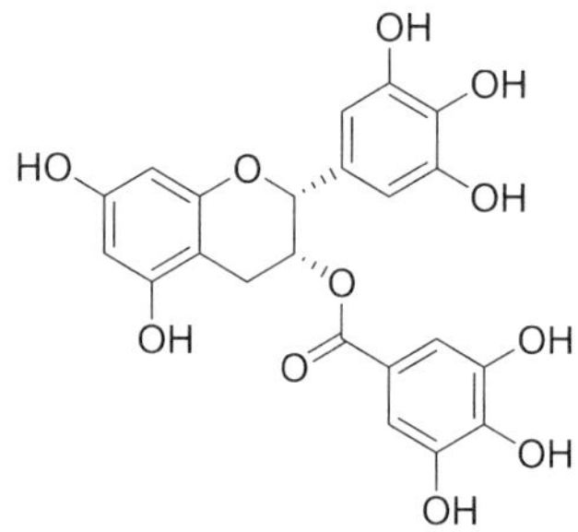

	(−)-Epicallocatechin 3-gallate
Molecular formula	$C_{22}H_{18}O_{11}$
Molecular weight	458.38
Solubility	DMSO, ++; H_2O, +; MeOH, ++

Discovery/Isolation

Epigallochatechin gallate (EGCG) was isolated from green tea as a member of the catechins [1].

Biological Studies

EGCG, a bioactive constituent of green tea, shows cancer prevention, antimatrix metalloproteinase, and antiangiogenesis activities [2–4]. The binding protein of EGCG was identified as the 67 kDa laminin receptor by a subtraction cloning strategy from cells treated or untreated with all-*trans*-retinoic acid, which enhances the binding of EGCG to the cell surface of cancer cells [5]. Further analyses revealed that EGCG signaling is mediated by the 67 kDa laminin receptor, eukaryotic translation elongation factor 1A (eEF1A), and myosin phosphatase-targeting subunit (MYPT1), and inhibits tumor growth [6, 7]. Because the 67 kDa laminin receptor is expressed on a variety of tumor cells, and the expression level is strongly correlated with the risk of tumor invasion and metastasis, EGCG might be a useful chemopreventing agent for cancer.

Biological Activity of Epigallocatechin 3-Gallate

IC_{50} values for MMP-2 and MMP-9: 20 and 50 μM, respectively [4]
K_d for binding to 67 kDa laminin receptor: 39.9 nM [5]

References

1. Bradfield AE, Penney M (1948) 456. The catechins of green tea. Part II. J Chem Soc:2249–2254
2. Yamane T, Takahashi T, Kuwata K, Oya K, Inagake M, Kitao Y, Suganuma M, Fujiki H (1995) Inhibition of *N*-methyl-*N*′-nitro-*N*-nitrosoguanidine-induced carcinogenesis by (−)-epigallocatechin gallate in the rat glandular stomach. Cancer Res 55:2081–2084
3. Cao Y, Cao R (1999) Angiogenesis inhibited by drinking tea. Nature 398:381
4. Garbisa S, Biggin S, Cavallarin N, Sartor L, Benelli R, Albini A (1999) Tumor invasion: molecular shears blunted by green tea. Nat Med 5:1216
5. Tachibana H, Koga K, Fujimura Y, Yamada K (2004) A receptor for green tea polyphenol EGCG. Nat Struct Mol Biol 11:380–381
6. Umeda D, Yano S, Yamada K, Tachibana H (2008) Involvement of 67-kDa laminin receptor-mediated myosin phosphatase activation in antiproliferative effect of epigallocatechin-3-*O*-gallate at a physiological concentration on Caco-2 colon cancer cells. Biochem Biophys Res Commun 371:172–176
7. Umeda D, Yano S, Yamada K, Tachibana H (2008) Green tea polyphenol epigallocatechin-3-gallate signaling pathway through 67-kDa laminin receptor. J Biol Chem 283:3050–3058

Epolactaene, ETB*

Keywords: [HSP60] [DNA polymerase] [Topoisomerase]

Structure

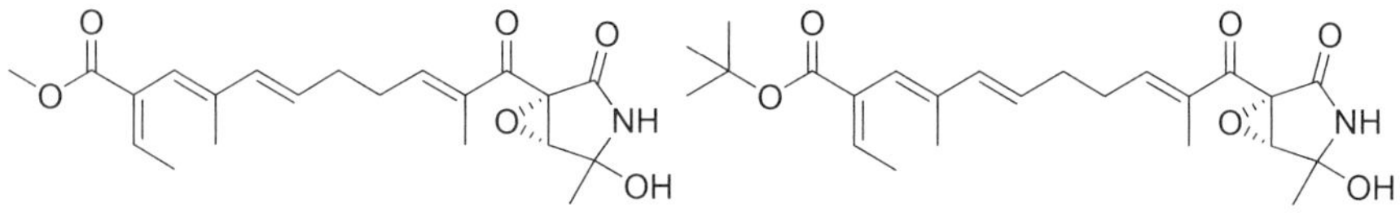

	Epolactaene	ETB
Molecular formula	$C_{21}H_{27}NO_6$	$C_{24}H_{33}NO_6$
Molecular weight	389.45	431.52
Solubility	DMSO, ++; H_2O, +; MeOH, ++	DMSO, ++; H_2O, +; MeOH, ++

Discovery/Isolation/Chemical Studies

Epolactaene was isolated from the fungal strain *Penicillium* sp. BM 1689-P as a neuritogenic compound which induced neurite outgrowth in the human neuroblastoma cell line SH-SY5Y that lacks significant TRK family mRNAs [1]. This compound possesses a highly functionalized α,β-epoxy-γ-lactam and a conjugated triene moiety in the side chain. Epolactaene is present in a 5:1 diasteromeric mixture derived from epimerization at the hemiaminal moiety.

Biological Studies

Epolactaene promoted neurite outgrowth in a neuroblastoma cell line [1] and induced apoptosis in a human leukemia B-cell line BALL-1 in a dose- and time-dependent manner [2]. At the molecular level, epolactaene inhibits DNA polymerases α and β, and DNA topoisomerase II [3].

Nagumo et al. found that the methyl ester moiety could be substituted by tertiary butyl ester, and synthesized ETB (epolactaene tertiary butyl ester) [4]. Using a biotinylated compound, human HSP60 was identified as a binding protein of epolactaene. Epolactaene inhibits chaperone activity of HSP60 by covalent binding on Cys442 [5]. The binding is reversible; therefore, it was speculated that epolactaene reacts with the α,β-unsaturated ketone via Michael addition. Recently, Sun et al. reported that ETB inhibits Mtf1-dependent mitochondrial transcription but not nuclear transcription by disrupting the binding between Hsp60 and Mtf1 in *Schizossacharomyces pombe* [6]. The specificity of ETB as an inhibitor of mitochondrial transcription suggests a novel chemical approach to the study of mitochondria.

Interestingly, it has been reported that epolactaene derivatives lacking α,β-unsaturated ketone also showed cytotoxicity [2, 4, 7]. Kuramochi et al. prepared a biotinylated derivative, Bio-Epo-C12, that lacks α,β-unsaturated ketone but possesses α,β-epoxy-γ-lactam moiety, and identified several binding proteins including

HSP60 [8]. They further investigated the binding mechanism and proposed the model that epolactaene induces intramolecular or intermolecular disulfide formation between protein cysteines via *retro*-Claisen reaction [8, 9].

Biological Activity of Epolactaene and Its Derivatives
Epolactaene
Promotion of neurite outgrowth in SH-SY5Y: 2.5–10 μg/ml [1]
IC_{50} for BALL-1 cells: 3.82 μM [2]
ETB
Inhibition of Mtf1-dependent mitochondrial transcription: 1 μM [6]

References

1. Kakeya H, Takahashi I, Okada G, Isono K, Osada H (1995) Epolactaene, a novel neuritogenic compound in human neuroblastoma cells, produced by a marine fungus. J Antibiot (Tokyo) 48:733–735
2. Nakai J, Kawada K, Nagata S, Kuramochi K, Uchiro H, Kobayashi S, Ikekita M (2002) A novel lipid compound, epolactaene, induces apoptosis: its action is modulated by its side chain structure. Biochim Biophys Acta 1581:1–10
3. Mizushina Y, Kobayashi S, Kuramochi K, Nagata S, Sugawara F, Sakaguchi K (2000) Epolactaene, a novel neuritogenic compound in human neuroblastoma cells, selectively inhibits the activities of mammalian DNA polymerases and human DNA topoisomerase II. Biochem Biophys Res Commun 273:784–788
4. Nagumo Y, Kakeya H, Yamaguchi J, Uno T, Shoji M, Hayashi Y, Osada H (2004) Structure-activity relationships of epolactaene derivatives: structural requirements for inhibition of Hsp60 chaperone activity. Bioorg Med Chem Lett 14:4425–4429
5. Nagumo Y, Kakeya H, Shoji M, Hayashi Y, Dohmae N, Osada H (2005) Epolactaene binds human Hsp60 Cys442 resulting in the inhibition of chaperone activity. Biochem J 387:835–840
6. Sun W, Wang L, Jiang H, Chen D, Murchie AI (2012) Targeting mitochondrial transcription in fission yeast with ETB, an inhibitor of HSP60, the chaperone that binds to the mitochondrial transcription factor Mtf1. Genes Cells 17:122–131
7. Kuramochi K, Matsui R, Matsubara Y, Nakai J, Sunoki T, Arai S, Nagata S, Nagahara Y, Mizushina Y, Ikekita M, Kobayashi S (2006) Apoptosis-inducing effect of epolactaene derivatives on BALL-1 cells. Bioorg Med Chem 14:2151–2161
8. Kuramochi K, Yukizawa S, Ikeda S, Sunoki T, Arai S, Matsui R, Morita A, Mizushina Y, Sakaguchi K, Sugawara F, Ikekita M, Kobayashi S (2008) Syntheses and applications of fluorescent and biotinylated epolactaene derivatives: epolactaene and its derivative induce disulfide formation. Bioorg Med Chem 16:5039–5049
9. Kuramochi K, Sunoki T, Tsubaki K, Mizushina Y, Sakaguchi K, Sugawara F, Ikekita M, Kobayashi S (2011) Transformation of thiols to disulfides by epolactaene and its derivatives. Bioorg Med Chem 19:4162–4172

Epoxomicin

Keywords: [Proteasome inhibitor] [Chymotrypsin]

Structure

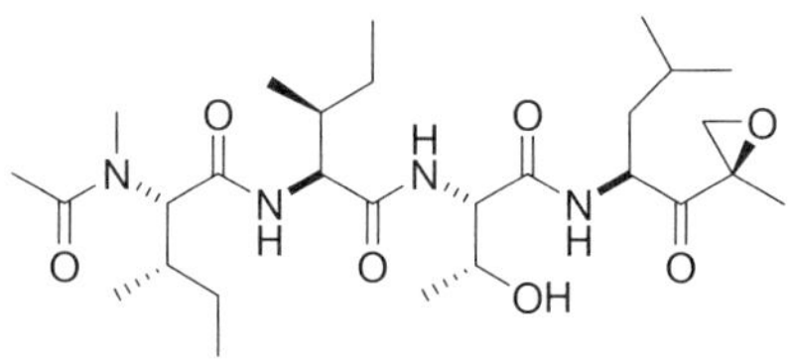

	Epoxomicin
Molecular formula	$C_{28}H_{50}N_4O_7$
Molecular weight	554.73
Solubility	DMSO, ++; H_2O, −; MeOH, ++

Discovery/Isolation/Chemical Studies

In the course of screening for antitumor agents showing specific activity against B16 murine melanoma, epoxomicin was isolated from an unidentified actinomycete strain No. Q996-17 [1]. Epoxomicin has an epoxy-β-aminoketone moiety in the structure, suggesting that this compound covalently binds to its target protein.

Biological Studies

Epoxomicin exhibits strong in vitro cytotoxicities against various tumor cell lines, and showed strong in vivo inhibitory activity against B16 melanoma but moderate activity against P388 mouse leukemia [1]. Meng et al. identified the target protein of epoxomicin by using a biotinylated epoxomicin analog [2]. They found that the epoxomicin probe covalently binds to two major (23 and 30 kDa) and one minor (28 kDa) proteins. All these proteins are proteasome components, and it was found that epoxomicin inhibits proteolytic activity, especially chymotrypsin-like activity, of 20S proteasome. Furthermore, epoxomicin induces accumulation of ubiquitinated proteins and inhibits activation of NFκB DNA binding activity by preventing IκB degradation.

Biological Activity of Epoxomicin

IC_{50} for several tumor cells: 2–44 ng/ml [1]
Accumulation of p53 in HUVECs: 100 nM [2]
Accumulation of ubiquitinated proteins in HeLa cells: 10 μM [2]
Inhibition of NFκB activation in HeLa cells: 10 μM [2]

References

1. Hanada M, Sugawara K, Kaneta K, Toda S, Nishiyama Y, Tomita K, Yamamoto H, Konishi M, Oki T (1992) Epoxomicin, a new antitumor agent of microbial origin. J Antibiot (Tokyo) 45:1746–1752
2. Meng L, Mohan R, Kwok BH, Elofsson M, Sin N, Crews CM (1999) Epoxomicin, a potent and selective proteasome inhibitor, exhibits in vivo antiinflammatory activity. Proc Natl Acad Sci U S A 96:10403–10408

EPZ004777*

Keywords: [Methyltransferase] [DOT1L]

Structure

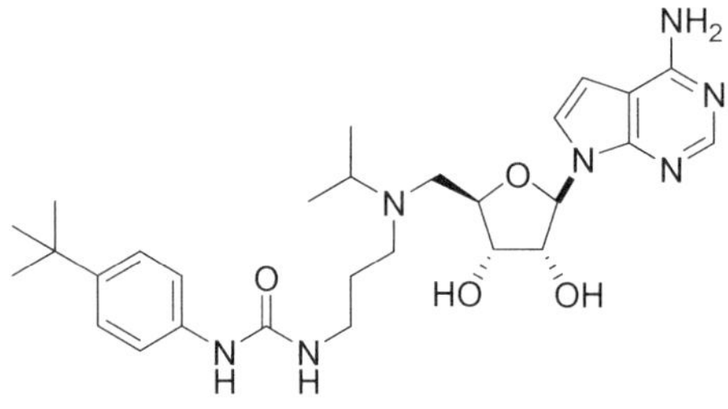

	EPZ004777
Molecular formula	$C_{28}H_{41}N_7O_4$
Molecular weight	539.68
Solubility	DMSO, +++; H_2O, −; MeOH, +++

Discovery/Isolation/Chemical Studies
EPZ004777 was designed as a potent and highly selective DOT1L inhibitor based on the structures of *S*-adenosylmethionine and *S*-adenosylhomocysteine, which are a DOT1L substrate and product, respectively, and the crystal structure of the DOT1L active site [1].

Biological Studies
DOT1L catalyzes the methylation of lysine 79 of histone H3 (H3K79), a chromatin modification associated with actively transcribed genes. EPZ004777 decreases the cellular methylation level of H3K79 [1]. DOT1L mislocation by MLL (mixed-lineage leukemia)–fusion protein is assumed to be a driver for leukemogenesis in MLL. EPZ004777 blocks MLL-fusion target gene expression and selectively inhibits proliferation of MLL-rearranged cells. Furthermore, leukemic transformation by MLL-fusion oncoprotein is inhibited by EPZ004777, suggesting that MLL patients may benefit from therapeutic agents targeting aberrant H3K79 methylation [2, 3].

Biological Activity of EPZ004777
IC_{50} for DOT1L in vitro: 400 ± 100 pM [1]
Inhibition of cellular H3K79 methylation in MLL- and non-MLL cell lines: 3 μM [1]
Apoptosis induction in MLL-rearranged cells: 3 μM [1]

References

1. Daigle SR, Olhava EJ, Therkelsen CA, Majer CR, Sneeringer CJ, Song J, Johnston LD, Scott MP, Smith JJ, Xiao Y, Jin L, Kuntz KW, Chesworth R, Moyer MP, Bernt KM, Tseng JC, Kung AL, Armstrong SA, Copeland RA, Richon VM, Pollock RM (2011) Selective killing of mixed lineage leukemia cells by a potent small-molecule DOT1L inhibitor. Cancer Cell 20:53–65
2. Chen L, Deshpande AJ, Banka D, Bernt KM, Dias S, Buske C, Olhava EJ, Daigle SR, Richon VM, Pollock RM, Armstrong SA (2013) Abrogation of MLL-AF10 and CALM-AF10-mediated transformation through genetic inactivation or pharmacological inhibition of the H3K79 methyltransferase Dot1l. Leukemia 27:813–822
3. Deshpande AJ, Chen L, Fazio M, Sinha AU, Bernt KM, Banka D, Dias S, Chang J, Olhava EJ, Daigle SR, Richon VM, Pollock RM, Armstrong SA (2013) Leukemic transformation by the MLL-AF6 fusion oncogene requires the H3K79 methyltransferase Dot1l. Blood 121:2533–2541

EX-527*

Keywords: [SIRT1 inhibitor]

Structure

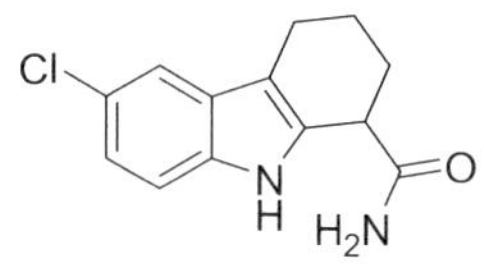

	EX-527
Molecular formula	$C_{13}H_{13}ClN_2O$
Molecular weight	248.71
Solubility	DMSO, +++; H_2O, ? ; MeOH, ?

Discovery/Isolation

EX-527 is an indole-containing potent and selective synthetic SIRT1 inhibitor, which was identified by high-throughput screening [1].

Biological Studies

Treatment with EX-527 increases acetylation of a number of substrates of SIRT1 including p53 [2] and tau [3]. EX-527 suppresses pathology of Huntington's disease in both mammalian cells and animal models [4]. Administration of EX-527 either intraperitoneally or directly into the brain in mice decreases food intake through the melanocortin system [5]. These reports indicate that EX-527 is an effective SIRT1 inhibitor in vivo.

Biological Activity of EX-527

In vitro IC_{50} for SIRT1: 98 nM [1]
In vitro IC_{50} values for SIRT2 and SIRT3: 19.6 μM and 48.7 μM, respectively [1]
Increase of p53 acetylation in response to DNA damage in cells: 1 μM [2]
Increase of tau acetylation in cells: 10–50 μM [3]

References

1. Napper AD, Hixon J, McDonagh T, Keavey K, Pons JF, Barker J, Yau WT, Amouzegh P, Flegg A, Hamelin E, Thomas RJ, Kates M, Jones S, Navia MA, Saunders JO, DiStefano PS, Curtis R (2005) Discovery of indoles as potent and selective inhibitors of the deacetylase SIRT1. J Med Chem 48:8045–8054
2. Solomon JM, Pasupuleti R, Xu L, McDonagh T, Curtis R, DiStefano PS, Huber LJ (2006) Inhibition of SIRT1 catalytic activity increases p53 acetylation but does not alter cell survival following DNA damage. Mol Cell Biol 26:28–38

3. Min SW, Cho SH, Zhou Y, Schroeder S, Haroutunian V, Seeley WW, Huang EJ, Shen Y, Masliah E, Mukherjee C, Meyers D, Cole PA, Ott M, Gan L. (2010) Acetylation of tau inhibits its degradation and contributes to tauopathy. Neuron 6:953–966
4. Smith MR, Syed A, Lukacsovich T, Purcell J, Barbaro BA, Worthge SA, Wei SR, Pollio G, Magnoni L, Scali C, Massai L, Franceschini D, Camarri M, Gianfriddo M, Diodato E, Thomas R, Gokce O, Tabrizi SJ, Caricasole A, Landwehrmeyer B, Menalled L, Murphy C, Ramboz S, Luthi-Carter R, Westerberg G, Marsh JL (2014) A potent and selective Sirtuin 1 inhibitor alleviates pathology in multiple animal and cell models of Huntington's disease. Hum Mol Genet 23:2995–3007
5. Dietrich MO, Antunes C, Geliang G, Liu ZW, Borok E, Nie Y, Xu AW, Souza DO, Gao Q, Diano S, Gao XB, Horvath TL (2010) Agrp neurons mediate Sirt1's action on the melanocortin system and energy balance: roles for Sirt1 in neuronal firing and synaptic plasticity. J Neurosci 30:11815–11825

FK228 (Romidepsin, FR901228)

Keywords: [Antitumor] [Histone deacetylase (HDAC) inhibitor]

Structure

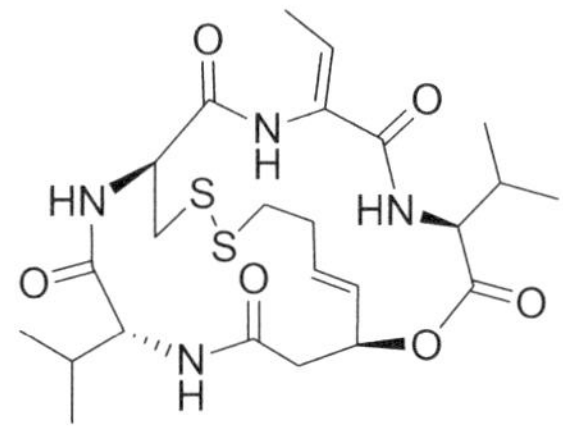

	FK228
Molecular formula	$C_{24}H_{36}N_4O_6S_2$
Molecular weight	540.69
Solubility	DMSO, ++; H_2O, –; MeOH, –

Discovery/Isolation/Chemical Studies

FK228 (romidepsin, FR901228) was isolated from a broth culture of *Chromobacterium violaceum* as an antitumor bicyclic depsipeptide [1].

Biological Studies

FK228 reversed the transformed morphology of Ha-*ras* transformant, Ras-1 cells, and inhibited their cell cycle progression at G0/G1 phases [2]. FK228 also reduced c-myc and Fas ligand expression, then effectively blocked activation-induced apoptosis in T-cell hybridomas [3].

FK228 was reisolated as a transcriptional activator of the SV40 promoter. This compound caused cell cycle arrest at both G1 and G2/M phases, induction of internucleosomal breakdown of chromatin, and inhibition of intracellular histone deacetylase activity [4].

Biological Activity of FK228

Antitumor activity against human tumor A549 implanted under the kidney capsule of immunosuppressed BDF1 mice: 0.56–3.2 mg/kg [1]
Morphological reversion of Ras1, Ha-*ras*-transformed NIH/3T3: 2.5 ng/ml [1]
Cell cycle arrest at G1 and G2/M phases in M-8 cells: 10 ng/ml [4]
IC_{50} for histone deacetylase activity in vitro: 1.1 nM [4]

References

1. Ueda H, Nakajima H, Hori Y, Fujita T, Nishimura M, Goto T, Okuhara M (1994) FR901228, a novel antitumor bicyclic depsipeptide produced by *Chromobacterium violaceum* No. 968. I. Taxonomy, fermentation, isolation, physico-chemical and biological properties, and antitumor activity. J Antibiot (Tokyo) 47:301–310
2. Shigematsu N, Ueda H, Takase S, Tanaka H, Yamamoto K, Tada T (1994) FR901228, a novel antitumor bicyclic depsipeptide produced by *Chromobacterium violaceum* No. 968. II. Structure determination. J Antibiot (Tokyo) 47:311–314
3. Wang R, Brunner T, Zhang L, Shi Y (1998) Fungal metabolite FR901228 inhibits c-Myc and Fas ligand expression. Oncogene 17:1503–1508
4. Nakajima H, Kim YB, Terano H, Yoshida M, Horinouchi S (1998) FR901228, a potent antitumor antibiotic, is a novel histone deacetylase inhibitor. Exp Cell Res 241:126–133

FK506 (Tacrolimus)

Keywords: [Immunosuppressant]

Structure

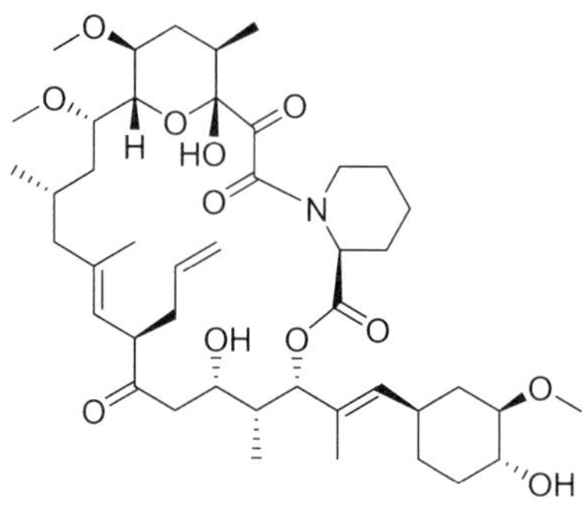

	FK506
Molecular formula	$C_{44}H_{69}NO_{12}$
Molecular weight	804.02
Solubility	DMSO, +++; H_2O, –; MeOH, +++

Discovery/Isolation

FK506 was isolated in the course of the search for an immunosuppressant from *Streptomyces tsukubaensis* [1–3].

Biological Studies

FK506 inhibited mixed lymphocyte reaction, cytotoxic T-cell generation, and expression of early T-cell activation genes [4]. An FK506 binding protein named FKBP12 was isolated by an affinity purification method [5] and was identified as peptidyl-prolyl *cis-trans* isomerase (PPIase) [6, 7]. Therefore, it was speculated that FKBP regulates T-cell activation by the recognition of proline-containing epitopes in target proteins. However, Bierer et al. found that the inhibition of peptidyl-prolyl isomerase activity is insufficient for immunosuppressive effects by using an unnatural immunophilin ligand, 506BD, which contains only the common structural element of FK506 and rapamycin [8]. Liu et al. finally identified calcineurin as the target protein of the FKBP-FK506 complex [9]. They showed that the FKBP-FK506 complex, but not FKBP-506BD, competitively binds to and inhibits phosphatase activity of calcineurin. Inhibition of calcineurin phosphatase activity is strongly correlated with the inhibition of transcriptional activation by NF-AT, a T-cell-specific transcription factor that regulates IL-2 gene synthesis in human T cells [10].

Biological Activity of FK506
IC_{50} for mouse mixed lymphocyte reaction in vitro: 0.32 nM [1]
ED_{50} for humoral and cellular immunity in mice: 4.4 mg/kg and 14 mg/kg [1]
IC_{50} values for production of IL-2, IL-3, IFN-γ, and expression of IL-2 receptor in MLR in vitro: 0.1, 0.3, 1, and 10 nM, respectively [2]
K_d for FKBP12 by Scatchard analysis: 0.4 nM [7]

References

1. Kino T, Hatanaka H, Hashimoto M, Nishiyama M, Goto T, Okuhara M, Kohsaka M, Aoki H, Imanaka H (1987) FK-506, a novel immunosuppressant isolated from a *Streptomyces*. I. Fermentation, isolation, and physico-chemical and biological characteristics. J Antibiot (Tokyo) 40:1249–1255
2. Kino T, Hatanaka H, Miyata S, Inamura N, Nishiyama M, Yajima T, Goto T, Okuhara M, Kohsaka M, Aoki H, et al. (1987) FK-506, a novel immunosuppressant isolated from a *Streptomyces*. II. Immunosuppressive effect of FK-506 in vitro. J Antibiot (Tokyo) 40:1256–1265
3. Tanaka H, Kuroda A, Marusawa H, Hashimoto M, Hatanaka H, Kino T, Goto T, Okuhara M (1987) Physicochemical properties of FK-506, a novel immunosuppressant isolated from *Streptomyces tsukubaensis*. Transplant Proc 19:11–16
4. Kino T, Inamura N, Sakai F, Nakahara K, Goto T, Okuhara M, Kohsaka M, Aoki H, Ochiai T (1987) Effect of FK-506 on human mixed lymphocyte reaction in vitro. Transplant Proc 19:36–39
5. Harding MW, Galat A, Uehling DE, Schreiber SL (1989) A receptor for the immunosuppressant FK506 is a *cis-trans* peptidyl-prolyl isomerase. Nature 34:758–760
6. Rosen MK, Standaert RF, Galat A, Nakatsuka M, Schreiber SL (1990) Inhibition of FKBP rotamase activity by immunosuppressant FK506: twisted amide surrogate. Science 248:863–866
7. Siekierka JJ, Hung SH, Poe M, Lin CS, Sigal NH (1989) A cytosolic binding protein for the immunosuppressant FK506 has peptidyl-prolyl isomerase activity but is distinct from cyclophilin. Nature 341:755–757
8. Bierer BE, Somers PK, Wandless TJ, Burakoff SJ, Schreiber SL (1990) Probing immunosuppressant action with a nonnatural immunophilin ligand. Science 250:556–559
9. Liu J, Farmer JD, Jr., Lane WS, Friedman J, Weissman I, Schreiber SL (1991) Calcineurin is a common target of cyclophilin-cyclosporin A and FKBP-FK506 complexes. Cell 66:807–815
10. Liu J, Albers MW, Wandless TJ, Luan S, Alberg DG, Belshaw PJ, Cohen P, MacKintosh C, Klee CB, Schreiber SL (1992) Inhibition of T cell signaling by immunophilin-ligand complexes correlates with loss of calcineurin phosphatase activity. Biochemistry 31:3896–3901

FR177391

Keywords: [Anti-hyperlipidemic] [Protein phosphatase 2A (PP2A)]

Structure

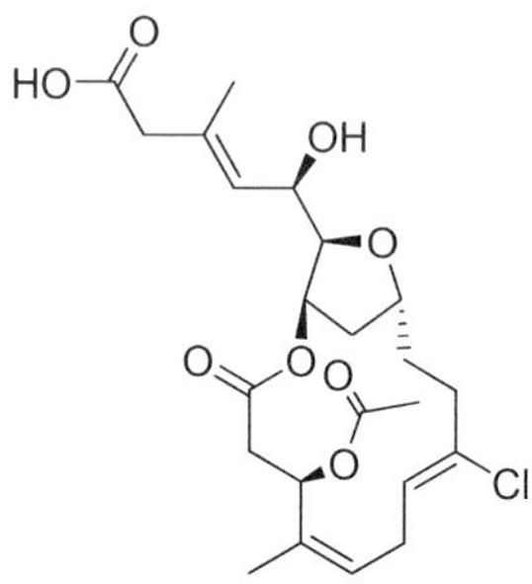

	FR177391
Molecular formula	$C_{23}H_{31}ClO_8$
Molecular weight	470.94
Solubility	DMSO, ++; H_2O, +; MeOH, +++

Discovery/Isolation

In the course of the screening for new antihyperlipidemia drugs, FR177391 was isolated from *Serratia liquefaciens* [1].

Biological Studies

FR177391 potently enhanced the differentiation of mouse 3Y3-L1 fibroblasts to mature adipocytes with oil droplets. The oral administration of FR177391 to C57BL/KsJ-db/db non-insulin-dependent diabetes mellitus (NIDDM) model mice also reduced the blood levels of triglyceride [1].

The target protein for FR177391 was identified to be protein phosphatase 2A (PP2A) by affinity chromatography using biotinylated FR177391 [2, 3]. FR177391 potently inhibited PP2A activity and shared its binding pocket with okadaic acid [3].

In the FR177391-treated 3 T3-L1 cells, the phosphorylation/activation of ERK was prolonged by the inhibition of its dephosphorylation by PP2A, suggesting that PP2A inhibition plays an important role in adipocyte maturation and regulation of blood levels of lipids [3].

Biological Activity of FR177391

Minimum effective concentration for adipogenesis-enhancing effect: 0.1 μM [1]
Enhancement of ERK1/2 phosphorylation in 3 T3-L1 adipocyte: 10 μM [3]
K_i values for PP2A and PP1: 2.9 and 7853 nM, respectively [3]

References

1. Sato B, Nakajima H, Fujita T, Takase S, Yoshimura S, Kinoshita T, Terano H (2005) FR177391, a new anti-hyperlipidemic agent from *Serratia*. I. Taxonomy, fermentation, isolation, physico-chemical properties, structure elucidation and biological activities. J Antibiot (Tokyo) 58:634–639
2. Kobayashi M, Sato K, Yoshimura S, Yamaoka M, Takase S, Ohkubo M, Fujii T, Nakajima H (2005) FR177391, a new anti-hyperlipidemic agent from *Serratia*. III. Microbial conversion of FR177391 and synthesis of FR177391 derivatives for its target protein screening by chemical genetic approaches. J Antibiot (Tokyo) 58:648–653
3. Yamaoka M, Sato K, Kobayashi M, Nishio N, Ohkubo M, Fujii T, Nakajima H (2005) FR177391, a new anti-hyperlipidemic agent from *Serratia*. IV. Target identification and validation by chemical genetic approaches. J Antibiot (Tokyo) 58:654–662

Fumagillin

Keywords: [Antiangiogenesis] [Methionine aminopeptidase inhibitor]

Structure

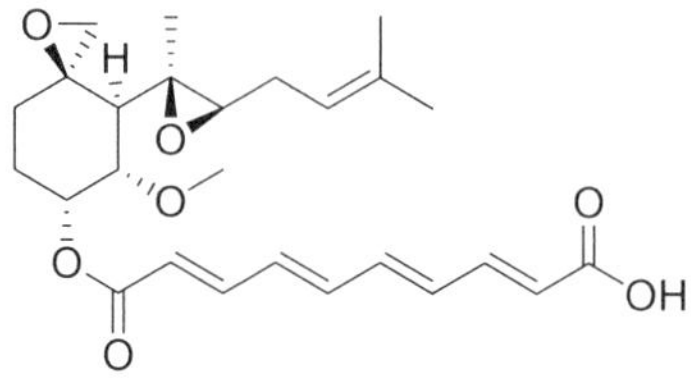

	Fumagillin
Molecular formula	$C_{26}H_{34}O_7$
Molecular weight	458.55
Solubility	DMSO, +++; H_2O, –; MeOH, +

Discovery/Isolation/Chemical Studies

Fumagillin was isolated from a culture broth of *Aspergillus* sp. as active concentrates that were capable of inhibiting *Staphylococcus aureus* 209 bacteriophage [1]. Total synthesis was done by Corey and Snider [2].

Biological Studies

Ingber et al. reisolated fumagillin from *Aspergillus fumigatus*, as a potent inhibitor of angiogenesis [3]. The compound completely inhibited endothelial cell proliferation and suppressed tumor-induced neovascularization in the mouse dorsal air sac. However, severe side effects limited the in vivo effectiveness.

Fumagillin contains two epoxides, suggesting that this compound covalently binds its target protein. Sin et al. [4]. and Griffith et al. [5] synthesized biotinylated fumagillin or fumagillin derivative, and showed that fumagillin covalently binds and inhibits type II methionine aminopeptidase (MetAP2). Fumagillin covalently modified histidine231 in the active site of MetAP2, and irreversibly inhibited the aminopeptidase activity of MetAP2 [6–8].

The reason why MetAP2 inhibition by fumagillin resulted in antiangiogenic activity is largely unknown. It was reported that MetAP2 binds and down-regulates ERK1/2 in a fumagillin-dependent manner, resulting in the inhibition of cell proliferation [9]. On the other hand, Kim et al. reported that depletion of MetAP2 did not change the fumagillin sensitivity of endothelial cells, suggesting that MetAP2 function is not required for endothelial cell proliferation [10]. Further investigation is required for elucidation of fumagillin functions.

Biological Activity of Fumagillin

Inhibition of tumor-induced angiogenesis in the subcutaneous dorsal air sac: 100 mg/kg [3]

Growth inhibition of MetAP-1 deleted *Sacchromyces cerevisiae*: 0.5 pmol/filter disk [4]

IC_{50} for cell growth of C2C12 myoblasts: 2.6 nM [9]

References

1. Hanson FR, Eble TE (1949) An antiphage agent Isolated from *Aspergillus* sp. J Bacteriol 58:527–529
2. Corey EJ, Snider BB (1972) Total synthesis of (+/−)-fumagillin. J Am Chem Soc 94:2549–255
3. Ingber D, Fujita T, Kishimoto S, Sudo K, Kanamaru T, Brem H, Folkman J (1990) Synthetic analogues of fumagillin that inhibit angiogenesis and suppress tumour growth. Nature 348:555–557
4. Sin N, Meng LH, Wang MQW, Wen JJ, Bornmann WG, Crews CM (1997) The anti-angiogenic agent fumagillin covalently binds and inhibits the methionine aminopeptidase, MetAP-2. Proc Natl Acad Sci U S A 94:6099–6103
5. Griffith EC, Su Z, Turk BE, Chen S, Chang YH, Wu Z, Biemann K, Liu JO (1997) Methionine aminopeptidase (type 2) is the common target for angiogenesis inhibitors AGM-1470 and ovalicin. Chem Biol 4:461–471
6. Griffith EC, Su Z, Niwayama S, Ramsay CA, Chang YH, Liu JO (1998) Molecular recognition of angiogenesis inhibitors fumagillin and ovalicin by methionine aminopeptidase 2. Proc Natl Acad Sci U S A 95:15183–15188
7. Liu SP, Widom J, Kemp CW, Crews CM, Clardy J (1998) Structure of human methionine aminopeptidase-2 complexed with fumagillin. Science 282:1324–1327
8. Lowther WT, McMillen DA, Orville AM, Matthews BW (1998) The anti-angiogenic agent fumagillin covalently modifies a conserved active-site histidine in the *Escherichia coli* methionine aminopeptidase. Proc Natl Acad Sci U S A 95:12153–12157
9. Datta B, Majumdar A, Datta R, Balusu R (2004) Treatment of cells with the angiogenic inhibitor fumagillin results in increased stability of eukaryotic initiation factor 2-associated glycoprotein, p67, and reduced phosphorylation of extracellular signal-regulated kinases. Biochemistry 43:14821–14831
10. Kim S, LaMontagne K, Sabio M, Sharma S, Versace RW, Yusuff N, Phillips PE (2004) Depletion of methionine aminopeptidase 2 does not alter cell response to fumagillin or bengamides. Cancer Res 64:2984–2987

Fusicoccin, Cotylenin A

Keywords: [14-3-3] [Plasma membrane H^+-ATPase] [Protein-protein interaction inhibitor]

Structure

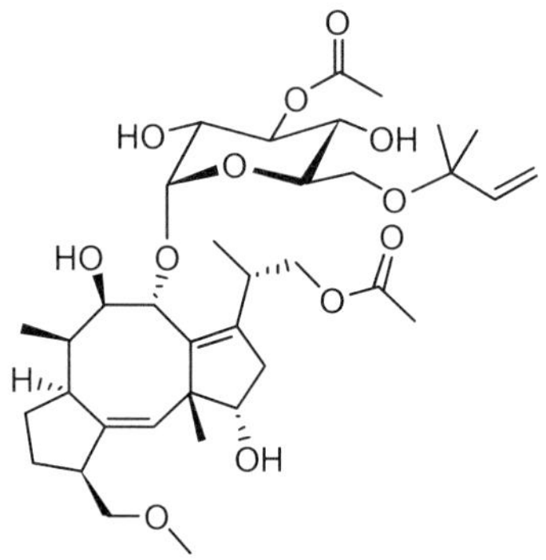

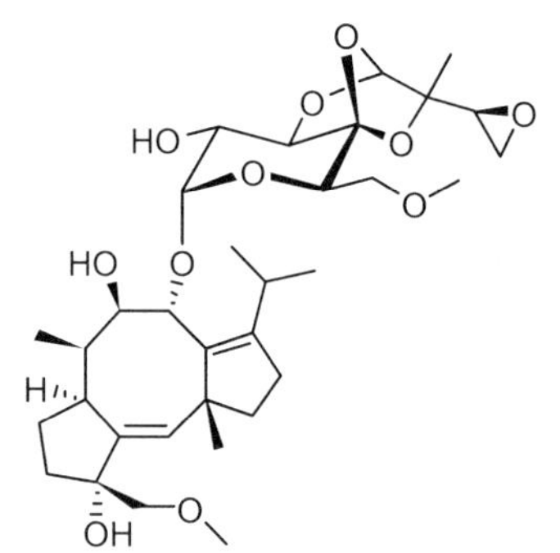

	Fusicoccin	Cotylenin A
Molecular formula	$C_{36}H_{56}O_{12}$	$C_{33}H_{50}O_{11}$
Molecular weight	680.83	622.75
Solubility	DMSO, ++; H_2O, ±; MeOH, +++	DMSO, ++; H_2O, ±; MeOH, +++

Discovery/Isolation

Fusicoccin was isolated as the main phytotoxin produced by the fungus *Fusicoccum amygdali* Del [1]. Cotylenin A, the major metabolite of novel fusicocane-diterpene glycosides from *Cladosporium* sp. 501-7 W, was isolated as a potent plant growth stimulant possessing phytohormone-like activity [2].

Biological Studies

Fusicoccin stimulated stomatal opening of *Commelina communis* [3] and rapid cell elongation in *Avena coleoptiles* [4], which were accompanied by potassium entry and proton excretion, respectively. Further in vitro biochemical investigations suggested that the ultimate target of fusicoccin action is H^+-APTase [5], but the molecular mechanism of this activation remained to be revealed.

The two fusicoccin binding proteins were purified by affinity chromatography using FC-linked adipic acid dihydrazide agarose from *Avena sativa* [6] or by photoaffinity labeling from *Commelina communis* [7], and these proteins were identified as 14-3-3 family proteins [8–10]. In vitro binding assay revealed that 14-3-3 protein binds to H^+-ATPase, and that fusicoccin strengthens the interaction between 14-3-3 protein and H^+-ATPase [11]. Furthermore, fusicoccin activates plant plasma membrane H^+-APTase through the formation of a ternary complex with plant 14-3-3 [12, 13], suggesting that fusicoccin is a protein–protein interaction stabilizer.

Different from fusicoccin, studies of cotylenin A has been focused on its antitumor activities [14]. Because 14-3-3 family proteins play a critical role in serine/threonine kinase–dependent signaling pathways through protein–protein interactions with multiple phosphorylated ligands, fusicoccin and cotylenin A derivatives have been developed as antitumor agents [15, 16].

Biological Activity of Fusicoccin and Cotylenin A

Fusicoccin

Enhancement of $NaHCO_3$ uptake in *Avena sativa*: 20 μM [17]

Cotylenin A

Morphological differentiation in M1 and HL60 cells: 20 μM [14]

EC_{50} for 14-3-3/C-RAFpS233pS259 interaction in vitro: 60 μM [15]

References

1. Ballio A, Chain EB, De Leo P, Erlanger BF, Mauri M, Tonolo A (1964) Fusicoccin: a new wilting toxin produced by *Fusicoccum amygdali* Del. Nature 203:297
2. Sassa T (1971) Cotylenins, leaf growth substances produced by a fungus: isolation and characterization of cotylenins A and B. Agric Biol Chem 35:1415–1418
3. Squire GR, Mansfield TA (1972) Studies of the mechanism of action of fusicoccin, the fungal toxin that induces wilting, and its interaction with abscisic acid. Planta 105:71–78
4. Cleland RE (1976) Fusicoccin-induced growth and hydrogen ion excretion of *Avena coleoptiles*: relation to auxin responses. Planta 128:201–206
5. Rasi-Caldogno F, De Michelis MI, Pugliarello MC, Marre E (1986) H^+-pumping driven by the plasma membrane ATPase in membrane vesicles from radish: stimulation by fusicoccin. Plant Physiol 82:121–125

6. de Boer AH, Watson BA, Cleland RE (1989) Purification and identification of the fusicoccin binding protein from oat root plasma membrane. Plant Physiol 89:250–259
7. Oecking C, Weiler EW (1991) Characterization and purification of the fusicoccin-binding complex from plasma membranes of *Commelina communis*. Eur J Biochem 199:685–689
8. Korthout HA, de Boer AH (1994) A fusicoccin binding protein belongs to the family of 14-3-3 brain protein homologs. Plant Cell 6: 1681–1692
9. Marra M, Fullone MR, Fogliano V, Pen J, Mattei M, Masi S, Aducci P (1994) The 30-kilodalton protein present in purified fusicoccin receptor preparations is a 14-3-3-like protein. Plant Physiol 106:1497–1501
10. Oecking C, Eckerskorn C, Weiler EW (1994) The fusicoccin receptor of plants is a member of the 14-3-3 superfamily of eukaryotic regulatory proteins. FEBS Lett 352:163–166
11. Fullone MR, Visconti S, Marra M, Fogliano V, Aducci P (1998) Fusicoccin effect on the in vitro interaction between plant 14-3-3 proteins and plasma membrane H^+ –ATPase. J Biol Chem 273:7698–7702
12. Fuglsang AT, Visconti S, Drumm K, Jahn T, Stensballe A, Mattei B, Jensen ON, Aducci P, Palmgren MG (1999) Binding of 14-3-3 protein to the plasma membrane H^+-ATPase AHA2 involves the three C-terminal residues Tyr946-Thr-Val and requires phosphorylation of Thr947. J Biol Chem 274:36774–36780
13. Svennelid F, Olsson A, Piotrowski M, Rosenquist M, Ottman C, Larsson C, Oecking C, Sommarin M (1999) Phosphorylation of Thr-948 at the C terminus of the plasma membrane H^+-ATPase creates a binding site for the regulatory 14-3-3 protein. Plant Cell 11:2379–2391
14. Asahi K, Honma Y, Hazeki K, Sassa T, Kubohara Y, Sakurai A, Takahashi N (1997) Cotylenin A, a plant-growth regulator, induces the differentiation in murine and human myeloid leukemia cells. Biochem Biophys Res Commun 238:758–763
15. Molzan M, Kasper S, Röglin L, Skwarczynska M, Sassa T, Inoue T, Breitenbuecher F, Ohkanda J, Kato N, Schuler M, Ottmann C (2013) Stabilization of physical RAF/14-3-3 interaction by cotylenin A as treatment strategy for RAS mutant cancers. ACS Chem Biol 8:1869–1875
16. Takahashi M, Kawamura A, Kato N, Nishi T, Hamachi I, Ohkanda J (2012) Phosphopeptide-dependent labeling of 14-3-3ζ proteins by fusicoccin-based fluorescent probes. Angew Chem Int Ed Eng 51:509–512
17. Johnson KD, Rayle DL (1976) Enhancement of CO_2 uptake in *Avena coleoptiles* by fusicoccin. Plant Physiol 57:806–811

Geldanamycin

Keywords: [HSP90 inhibitor]

Structure

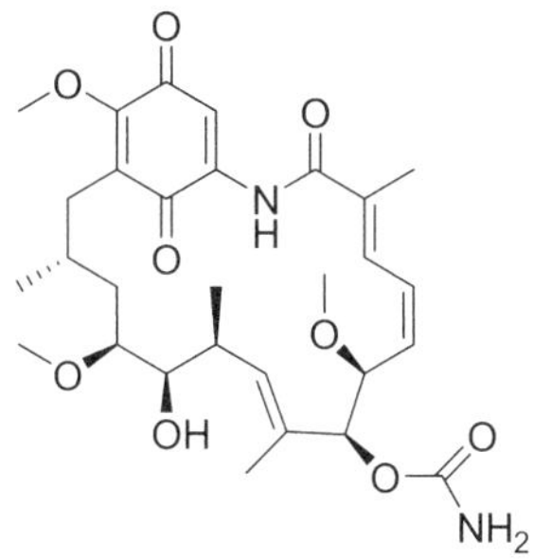

	Geldanamycin
Molecular formula	$C_{29}H_{40}N_2O_9$
Molecular weight	560.64
Solubility	DMSO, +++; H_2O, –; MeOH, –

Discovery/Isolation

Geldanamycin was isolated from culture broth of *Streptomyces hygroscopicus* as a compound showing moderate inhibitory activity against protozoa, bacteria, fungi, and tumor cells [1].

Biological Studies

Uehara et al. found that geldanamycin and other benzoquinonoid ansamycins reverse the morphology of *src*-transformed cells from the transformed to the normal phenotype by inhibiting the autophosphorylation of $p60^{src}$ [2]. Because these compound did not inhibit the autophosphorylation of *src* kinase in vitro, it was thought that geldanamycins are not kinase inhibitors. The target protein of geldanamycin was determined as HSP90 by using affinity resin which immobilizes the diamine derivative of geldanamycin [3].

Geldanamycin binds the *N*-terminal ATP/ADP binding domain of HSP90 [4], and the inhibition of ATP/ADP binding leads the inhibition of the Hsp90-mediated conformational maturation/refolding reaction, and the degradation of HSP90 substrates [5]. Because HSP90 client proteins include several oncogenic proteins, such as mutated p53, Bcr-Abl, Raf-1, Akt, ErbB2, and HIF-1α, geldanamycin is a promising antitumor compound in the preclinical model system [6].

Biological Activity of Geldanamycin

Morphological changes from the transformed to the normal phenotype: 0.1 μg/ml [2]
Disruption of multimolecular complexes containing Raf-1: 2 μM [7]

References

1. DeBoer C, Meulman PA, Wnuk RJ, Peterson DH (1970) Geldanamycin, a new antibiotic. J Antibiot (Tokyo) 23:442–447
2. Uehara Y, Hori M, Takeuchi T, Umezawa H (1986) Phenotypic change from transformed to normal induced by benzoquinonoid ansamycins accompanies inactivation of $p60^{src}$ in rat kidney cells infected with Rous sarcoma virus. Mol Cell Biol 6:2198–2206
3. Whitesell L, Mimnaugh EG, De Costa B, Myers CE, Neckers LM (1994) Inhibition of heat shock protein HSP90-$pp60^{v\text{-}src}$ heteroprotein complex formation by benzoquinone ansamycins: essential role for stress proteins in oncogenic transformation. Proc Natl Acad Sci U S A 91:8324–8328
4. Stebbins CE, Russo AA, Schneider C, Rosen N, Hartl FU, Pavletich NP (1997) Crystal structure of an Hsp90-geldanamycin complex: targeting of a protein chaperone by an antitumor agent. Cell 89:239–250
5. Schneider C, Sepp-Lorenzino L, Nimmesgern E, Ouerfelli O, Danishefsky S, Rosen N, Hartl FU (1996) Pharmacologic shifting of a balance between protein refolding and degradation mediated by Hsp90. Proc Natl Acad Sci U S A 93:14536–14541
6. Neckers L (2002) Hsp90 inhibitors as novel cancer chemotherapeutic agents. Trends Mol Med 8:S55–61
7. Schulte TW, Blagosklonny MV, Ingui C, Neckers L (1995) Disruption of the Raf-1-Hsp90 molecular complex results in destabilization of Raf-1 and loss of Raf-1-Ras association. J Biol Chem 270:24585–24588

Gerfelin, Methyl Gerfelin*

Keywords: [Geranylgeranyl diphosphate (GGPP) synthase inhibitor] [Osteoclastogenesis inhibitor] [Glyoxalase inhibitor]

Structure

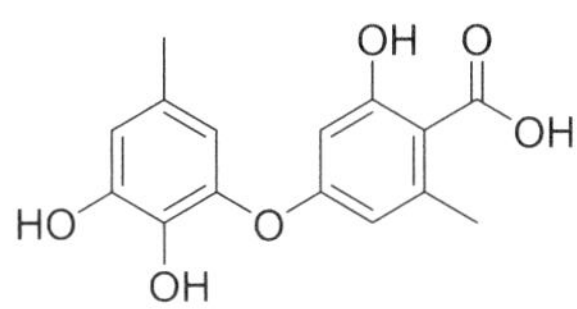

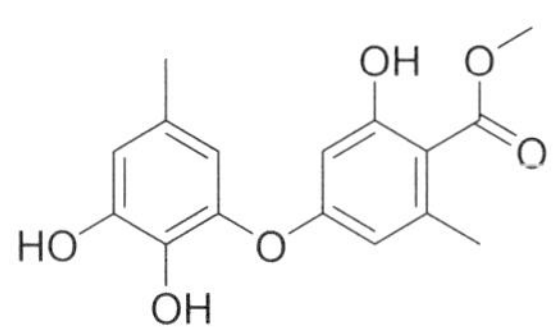

	Gerfelin	Methyl gerfelin
Molecular formula	$C_{15}H_{14}O_6$	$C_{16}H_{16}O_6$
Molecular weight	209.27	304.29
Solubility	DMSO, +++; H_2O, ? ; MeOH, ++	DMSO, +++; H_2O, ±; MeOH, ++

Discovery/Isolation/Chemical Studies

Gerfelin was isolated from *Beauveria felina* QN22047 as an inhibitor of geranylgeranyl diphosphate (GGPP) synthase [1, 2]. Structure–activity relationship studies using synthetic gerfelin derivatives revealed that the 2-hydroxybenzoic acid moiety and the catechol moiety are important for GGPP synthase inhibition [3]. Methyl gerfelin, a methyl ester of gerfelin, was identified as an osteoclastogenesis inhibitor in cellular phenotype-based screening of osteoclast inhibitors [4].

Biological Studies

Gerfelin inhibits GGPP synthase noncompetitively against isopentenyl diphosphate and uncompetitively against farnesyl diphosphate [1].

Gerfelin and methyl gerfelin suppressed the formation of tartrate-resistant acid phosphatase–positive multinucleated osteoclasts induced by RANKL and M-CSF [4]. However, this osteoclastogenesis inhibition was not dependent on the GGPP synthase inhibition of methyl gerfelin, suggesting that this compound could interact with another protein. Kawatani et al. made a methyl gerfelin affinity matrix and isolated glyoxalase I (GLO1) as a binding protein [4]. Further analysis revealed that GLO1 activity is required for osteoclastogenesis. SAR analysis using methyl gerfelin derivatives, each having an amine-terminated tri(ethylene glycol) linker at the peripheral position, revealed that derivatization at the C6-methyl group would be suitable for the further development of selective GLO1 inhibitors [5].

Methyl gerfelin also inhibited *Plasmodium falciparum* GLO1 with a complicated mechanism [6]. Because it has been reported that glycolytic flux increases in malaria parasites [7], methyl gerfelin might be a good lead compound for developing antimalarial medicines that increase harmful methylglyoxal concentrations in *P. falciparum*.

Biological Activity of Gerfelin
IC_{50} for GGPP synthase in vitro: 3.5 μg/ml [1]
K_i for GGPP synthase: 5.5 μg/ml [1] and 19.0 μM [4]
IC_{50} for osteoclastogenesis: 61 μM [4]

Biological Activity of Methyl Gerfelin
IC_{50} for osteoclastogenesis: 2.8 μM [4]
IC_{50} for GGPP synthase in vitro: 2.5 μM [4]
K_i for GLO1 activity in vitro: 0.23 μM, competitively [4]

References

1. Zenitani S, Tashiro S, Shindo K, Nagai K, Suzuki K, Imoto M (2003) Gerfelin, a novel inhibitor of geranylgeranyl diphosphate synthase from *Beauveria felina* QN22047. I. Taxonomy, fermentation, isolation, and biological activities. J Antibiot (Tokyo) 56:617–621
2. Zenitani S, Shindo K, Tashiro S, Sekiguchi M, Nishimori M, Suzuki K, Imoto M (2003) Gerfelin, a novel inhibitor of geranylgeranyl diphosphate synthase from *Beauveria felina* QN22047. II. Structural elucidation. J Antibiot (Tokyo) 56:658–660
3. Islam MS, Kitagawa M, Imoto M, Kitahara T, Watanabe H. (2006) Synthesis of gerfelin and related analogous compounds. Biosci Biotechnol Biochem 70:2523–2528
4. Kawatani M, Okumura H, Honda K, Kanoh N, Muroi M, Dohmae N, Takami M, Kitagawa M, Futamura Y, Imoto M, Osada H (2008) The identification of an osteoclastogenesis inhibitor through the inhibition of glyoxalase I. Proc Natl Acad Sci U S A 105:11691–11696
5. Kanoh N, Suzuki T, Kawatani M, Katou Y, Osada H, Iwabuchi Y. (2013) Dual structure-activity relationship of osteoclastogenesis inhibitor methyl gerfelin based on TEG scanning. Bioconjug Chem 24:44–52
6. Urscher M, Przyborski JM, Imoto M, Deponte M. (2010) Distinct subcellular localization in the cytosol and apicoplast, unexpected dimerization and inhibition of *Plasmodium falciparum* glyoxalases. Mol Microbiol 76:92–103
7. Akoachere M, Iozef R, Rahlfs S, Deponte M, Mannervik B, Creighton DJ, Schirmer H, Becker K. (2005) Characterization of the glyoxalases of the malarial parasite *Plasmodium falciparum* and comparison with their human counterparts. Biol Chem 386:41–52

Glaziovianin A, Gatastatin*

Keywords: [Tubulin binder] [Microtubule dynamics inhibitor] [γ-tubulin inhibitor]

Structure

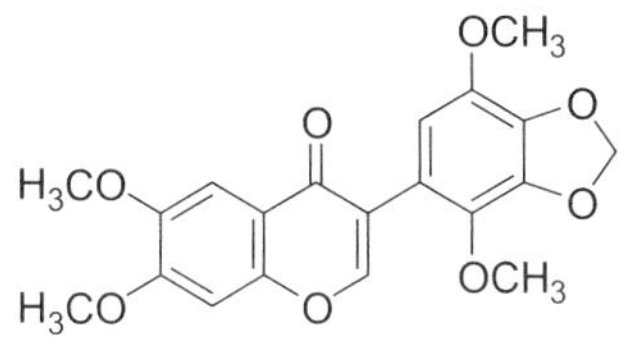

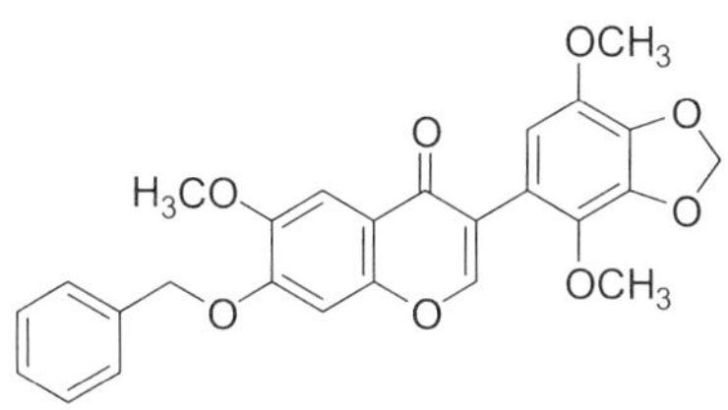

	Glaziovianin A	Gatastatin
Molecular formula	$C_{20}H_{18}O_8$	$C_{26}H_{22}O_8$
Molecular weight	386.36	462.45
Solubility	DMSO, +++; H_2O, –; MeOH, +	DMSO, +++; H_2O, –; MeOH, +

Discovery/Isolation/Chemical Studies

Glaziovianin A was isolated from the leaves of *Ateleia glazioviana* (Legminosae) as a cytotoxic compound against human leukemia cell line HL-60 [1]. SAR studies using synthetic glaziovianin A derivatives revealed that O^7-alkylated glaziovianin A analogs have strong cytotoxicity against HeLa cells [2, 3]. Gatastatin, O^7-demethyl benzyl glaziovianin A, was identified as a γ-tubulin-specific inhibitor in the course of screening the γ-tubulin binder [4].

Biological Studies

Glaziovianin A induced a multipole spindle strucuture in 3Y1 cells, suggesting that this compound targets microtubules [1]. Detailed biochemical investigations revealed that glaziovianin A and O^7-demethyl propargyl glaziovianin A bind the colchicine binding site of tubulin and suppresses microtubule dynamics in vitro and in cells [5]. The microtubule dynamic inhibition causes the delay of endosome maturation and the down-regulation of growth factor–dependent signaling.

Gatastatin inhibited γ-tubulin-dependent microtubule nucleation and minus end capping of microtubules by a γ-tubulin complex in vitro [4]. Gatastatin also inhibited microtubule nucleation from the centrosome, and prolonged not only metaphase but also anaphase duration by affecting anaphase spindle structure. These results indicate that the microtubule nucleation activity of γ-tubulin is continuously required throughout mitosis.

Biological Activity of Glaziovianin A

IC_{50} for glaziovianin A on cell growth of HL-60 cells: 0.29 μM [1]
Inhibition of microtubule dynamics in PtK2 cells: 1 μM [2]
Inhibition of endosome transport and maturation in HeLa cells: 2 μM [2]

Biological Activity of Gatastatin

K_d for γ-tubulin determined by tryptophan fluorescence–based drug binding assay: 3.6 μM [4]

References

1. Yokosuka A, Haraguchi M, Usui T, Kazami S, Osada H, Yamori T, Mimaki Y (2007) Glaziovianin A, a new isoflavone, from the leaves of *Ateleia glazioviana* and its cytotoxic activity against human cancer cells. Bioorg Med Chem Lett 17: 3091–3094
2. Ikedo A, Hayakawa I, Usui T, Kazami S, Osada H, Kigoshi H. (2010) Structure-activity relationship study of glaziovianin A against cell cycle progression and spindle formation of HeLa S3 cells. Bioorg Med Chem Lett 20:5402–5404
3. Hayakawa I, Ikedo A, Chinen T, Usui T, Kigoshi H. (2012) Design, synthesis, and biological evaluation of the analogues of glaziovianin A, a potent antitumor isoflavone. Bioorg Med Chem 20:5745–5756
4. Chinen T, Liu P, Shioda S, Pagel J, Cerikan B, Lin T, Gruss O, Hayashi Y, Takeno H, Shima T, Okada T, Hayakawa I, Hayashi Y, Kigoshi H, Usui T, Schiebel E. (2015) The γ-tubulin specific inhibitor gatastatin reveals temporal requirements of microtubule nucleation during the cell cycle. Nat Commun 6:8722
5. Chinen T, Kazami S, Nagumo Y, Hayakawa I, Ikedo A, Takagi M, Yokosuka A, Imamoto N, Mimaki Y, Kigoshi H, Osada H, Usui T (2013) Glaziovianin A prevents endosome maturation via inhibiting microtubule dynamics. ACS Chem Biol 8: 884–889

Gleevec*

Keywords: [Antitumor] [Bcr-Abl inhibitor]

Structure

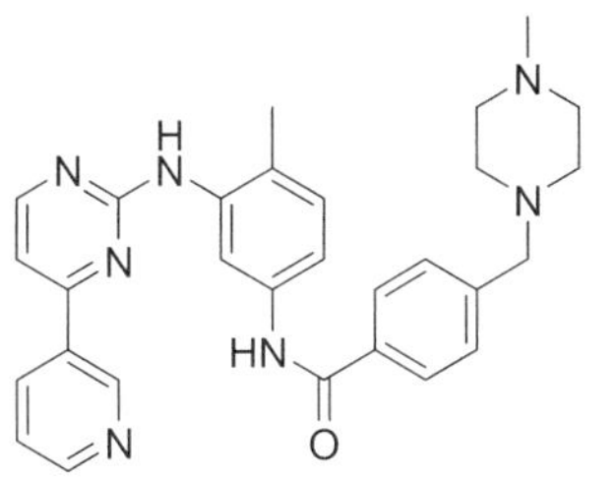

	Gleevec
Molecular formula	$C_{29}H_{31}N_7O$
Molecular weight	493.60 (589.71: mesylate)
Solubility	DMSO, +; H_2O, +++ (mesilate); MeOH, +

Discovery/Isolation/Chemical Studies

As the protein kinase activity of Bcr-Abl is essential for its transformation-inducing activity, a research program began to develop low molecular weight inhibitors of Bcr-Abl. The high-throughput screening study identified 2-phenylaminopyrimidine as a lead compound of the inhibitor of Bcr-Abl. To improve the binding properties of this molecule to Bcr-Abl, methyl and benzamide groups were introduced and, as a result, a drug named Gleevec (CGP-57148, imatinib mesylate, STI-571, and Glivec) emerged [1]. This compound inhibits the Abl protein tyrosine kinases from several sources with an IC_{50} of 0.025–0.038 μM in vitro. This is a very selective inhibitor of Abl kinase and no significant inhibition of other protein kinases except c-kit and PDGF-R has been reported.

Biological Studies

In Philadelphia chromosome (Ph)–positive CML cells, a tyrosine kinase, Bcr-Abl, is kept active, and it is shown that this causes the CML properties. Gleevec inhibits this Bcr-Abl kinase by binding to the catalytic cleft between the two major structural lobes of the kinase domain of Bcr-Abl. As a result, these cells stop growing and even die by a process of apoptosis. Because the fusion gene product from Ph, Bcr-Abl tyrosine kinase, exists only in leukemia cells and not in normal cells, Gleevec kills cancer cells selectively. Gleevec was one of the first cancer treatments inhibiting a specific target molecule and is often cited as a paradigm for research in cancer therapeutics [2].

Biological Activity of Gleevec

IC_{50} values for v-Abl, Bcr-Abl, and c-Abl protein kinases; 0.038, 0.025, and 0.025 μM, respectively [1]

Growth of cells with expression of Bcr-Abl (MO7p210); 1 μM [1].

References

1. Druker BJ, Tamura S, Buchdunger E, Ohno S, Segal GM, Fanning S, Zimmermann J, Lydon NB (1996) Effects of a selective inhibitor of the Abl tyrosine kinase on the growth of Bcr-Abl positive cells. Nat Med 2:561–566
2. Goldman JM, Melo JV (2003) Chronic myeloid leukemia–advances in biology and new approaches to treatment. N Engl J Med 349:1451–1464

Glucopiericidin A

Keywords: [Filopodia protrusion inhibitor] [Glucose transporter inhibitor]

Structure

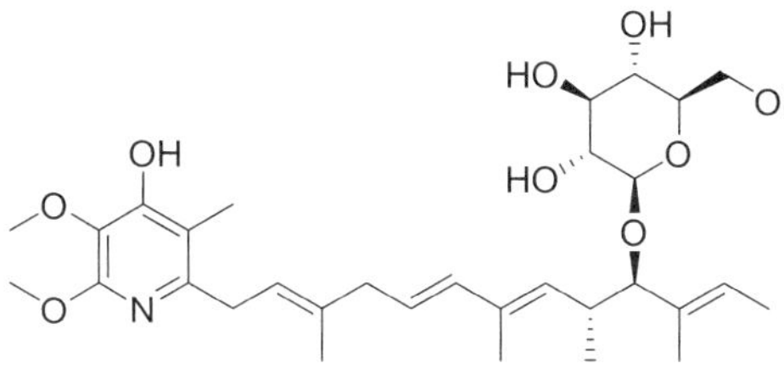

	Glucopiericidin A
Molecular formula	$C_{31}H_{47}NO_9$
Molecular weight	577.71
Solubility	DMSO, ++; H_2O, –; MeOH, +++

Discovery/Isolation/Chemical Studies

Glucopiericidin A was isolated as a new piericidin group antibiotic from a culture broth of *Streptomyces pactum* S48727 (FERM P-8117) [1]. Because glucopiericidin showed more potent inhibition of antibody formation in mouse spleen cell cultures and antimicrobial activities than piericidin A1 in vitro [1], it was suggested that D-glucose in glucopiericidin is important in modulating their physiological activities.

Biological Studies

Glucopiericidin A inhibited platelet-derived growth factor (PDGF)–induced activation of phospholipase γ1 (PLC-γ1) by reducing the tyrosine kinase activity of the PDGF receptor and PI turnover induced by PDGF [2].

Recently, in the course of screening microbial samples for a filopodia protrusion inhibitor, glucopiericidin A was reisolated [3]. Interestingly, glucopiericidin A alone did not inhibit filopodia protrusion, but synergistically inhibited protrusion with piericidin A, the mitochondria respiration inhibitor, suggesting that glucopiericidin A might be a glycolysis inhibitor. Furthermore, metabolomic analysis using capillary electrophoresis TOF-MS strongly suggested that glucopiericidin targets glucose transporters.

Biological Activity of Glucopiericidin

Inhibition of antibody formation: 0.1 ng/ml [1]

Inhibition of filopodia protrusion in combination with piericidin A (0.68 nM): 17 nM [3]

References

1. Matsumoto M, Mogi K, Nagaoka K, Ishizeki S, Kawahara R, Nakashima T (1987) New piericidin glucosides, glucopiericidins A and B. J Antibiot (Tokyo) 40:149–156
2. Ahn SC, Kim BY, Park CS, Lee HS, Suh PG, Ryu SH, Rho HM, Rhee JS, Mheen TI, Ahn JS (1995) Inhibition of PDGF-induced phosphoinositide-turnover by glucopiericidin A. Biochem Mol Biol Int 37:125–132
3. Kitagawa M, Ikeda S, Tashiro E, Soga T, Imoto M (2010) Metabolomic identification of the target of the filopodia protrusion inhibitor glucopiericidin A. Chem Biol 17:989–998

Gossypol

Keywords: [Apoptosis inducer] [Bcl-2 family inhibitor] [BH3 mimetics]

Structure

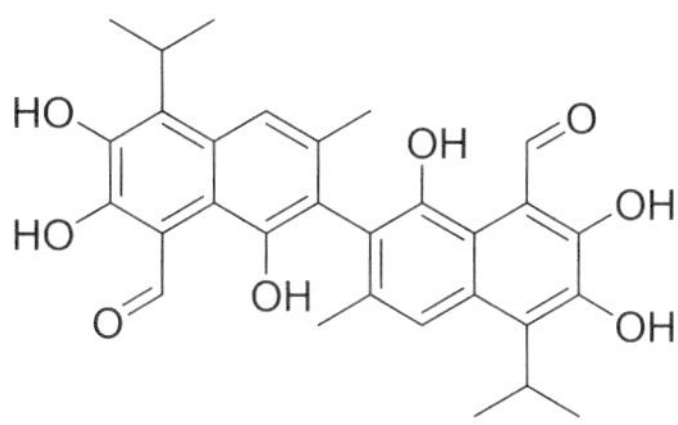

	Gossypol
Molecular formula	$C_{30}H_{30}O_8$
Molecular weight	518.56
Solubility	DMSO, +++; H_2O, –; MeOH, ++

Discovery/Isolation/Chemical Studies

Gossypol was isolated from crude cottonseed oil [1]. Withers and Carruth found that gossypol is quickly oxidized in an alcoholic solution of sodium hydroxide, suggesting that gossypol is a toxic substance but its oxidation by an alcoholic alkali renders its toxicity [2]. Chemical reaction with free ε–amino groups of lysine in proteins has been reported [3].

Biological Studies

Gossypol shows antitumor activities in vivo [4] and in vitro [5]. Gossypol induced cell cycle arrest in G0/G1 phase by up-regulation of p21, and apoptosis by both down-regulation of antiapoptosis Bcl-2 family proteins and up-regulation of the proapoptosis Bcl-2 member in HT-29 human colon carcinoma cells [6]. Furthermore, gossypol inhibited the binding of the BH3 domain of Bak to Bcl-xL in a dose-dependent manner. Detailed NMR analyses revealed that gossypol binds on the BH3 binding pocket of Bcl-xL. These results strongly suggest that gossypol is a BH3 mimetic and functions as a Bcl-2/Bcl-xL antagonist [7].

Biological Activity of Gossypol

IC_{50} for cell proliferation against several cancer cell lines: 5–35 μM [5]
IC_{50} for BH3 peptide binding to Bcl-X_L and Bcl-2: 0.4 and 10 μM [6]

References

1. Marchlewski L (1899) Gossypol, ein Bestandtheil der Baumwollsamen. J Prakt Chem 60:84–90
2. Withers WA, Carruth FE (1915) Gossypol–a toxic substance in cottonseed. A preliminary note. Science 41:324
3. Conkerton EJ, Frampton VL (1959) Reaction of gossypol with free epsilon-amino groups of lysine in proteins. Arch Biochem Biophys 81:130–134
4. Vermel EM, Krugliak SA (1963) Antineoplastic activity of gossypol in experimental transplanted tumors. Vopr Onkol 21:39–43
5. Tuszynski GP, Cossu G (1984) Differential cytotoxic effect of gossypol on human melanoma, colon carcinoma, and other tissue culture cell lines. Cancer Res 44:768–771
6. Zhang M, Liu H, Guo R, Ling Y, Wu X, Li B, Roller PP, Wang S, Yang D (2003) Molecular mechanism of gossypol-induced cell growth inhibition and cell death of HT-29 human colon carcinoma cells. Biochem Pharmacol 66:93–103
7. Kitada S, Leone M, Sareth S, Zhai D, Reed JC, Pellecchia M (2003) Discovery, characterization, and structure-activity relationships studies of proapoptotic polyphenols targeting B-cell lymphocyte/leukemia-2 proteins. J Med Chem 46:4259–4264

GSK126*, EPZ005687*

Keywords: [Enhancer of zeste 2 (EZH2) inhibitor] [Methylation of histone H3 on lysine 27]

Structure

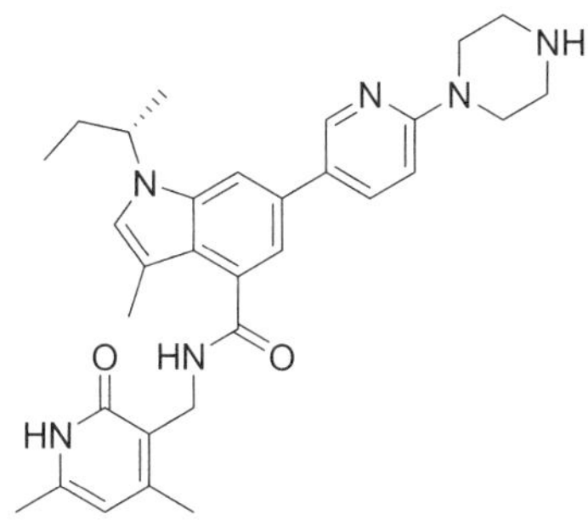

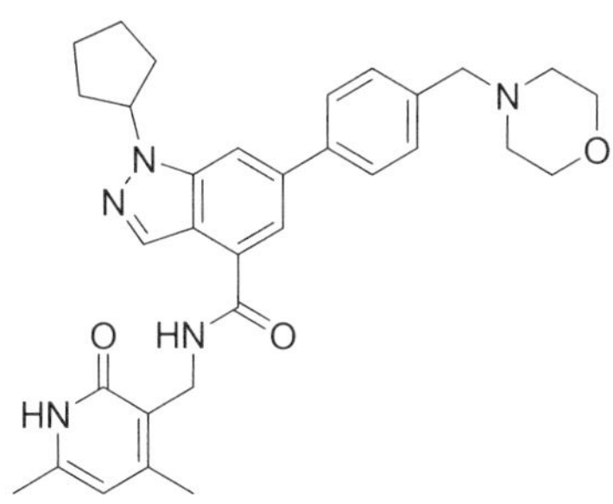

	GSK126	EPZ005687
Molecular formula	$C_{31}H_{38}N_6O_2$	$C_{32}H_{37}N_5O_3$
Molecular weight	526.67	539.68
Solubility	DMSO, ++; H_2O, ±; MeOH, ±	DMSO, ++; H_2O, ±; MeOH, ±

Discovery/Isolation/Chemical Studies

Methylation of histone H3 lysine 27 is catalyzed by polycomb repressive complex 2 (PRC2), containing the enzymatic subunit EZH2 or EZH1. GSK126 [1, 2] and EPZ005687 [3] were synthesized by optimization of the EZH2 inhibitor which was previously identified by a high-throughput biochemical screen with PRC2 protein.

Biological Studies

GSK126 and EPZ005687 are potent and selective inhibitors of PRC2 activity in a competitive manner against PRC2 substrate *S*-adenosylmethionine [1, 3].

GSK126 is more than 1000-fold more selective for EZH2 than 20 other human methyltransferases. Furthermore, this compound is a slow dissociation inhibitor from the activated enzyme and has a significantly longer residence time of inhibition on the activated form of EZH2 as compared to unactivated EZH2, suggesting that GSK126 may be beneficial in vivo and result in durable target inhibition after drug systemic clearance [4].

EPZ005687 shows ~50-fold selectivity for EZH2 over EZH1-containing PRC2. This compound decreased the methylation level of H3K27 in lymphoma cell line OCI-LY19 (wild-type EZH2) in a dose-dependent manner [3]. Similar results were obtained in EZH2 Tyr641 and Ala677 mutant lymphoma cell lines as well as cell lines of other cancer types. Furthermore, EPZ005687 decreases proliferation in mutant but not wild-type EZH2 lymphoma cells by inhibiting the cell cycle progression in G1 phase.

Biological Activity of GSK126 and EPZ004777:
GSK126

Apparent K_i values for EZH2 and EZH1 in vitro: 0.5–3 and 89 nM, respectively [1]
IC_{50} for EZH2 mutant DLBCL cell lines: 28–861 nM [1]
IC_{50} for wild-type EZH2 cell line (HT cells): 516 nM [1]

EPZ004777

IC_{50} for PRC2 in vitro: 54 ± 5 nM [3]
IC_{50} for the level of methylated H3K27 in OCI-LY19 cells: 80 ± 30 nM [3]

References

1. McCabe MT, Ott HM, Ganji G, Korenchuk S, Thompson C, Van Aller GS, Liu Y, Graves AP, Della Pietra A, 3rd, Diaz E, LaFrance LV, Mellinger M, Duquenne C, Tian X, Kruger RG, McHugh CF, Brandt M, Miller WH, Dhanak D, Verma SK, Tummino PJ, Creasy CL (2012) EZH2 inhibition as a therapeutic strategy for lymphoma with EZH2-activating mutations. Nature 492:108–112
2. Diaz E, Machutta CA, Chen S, Jiang Y, Nixon C, Hofmann G, Key D, Sweitzer S, Patel M, Wu Z, Creasy CL, Kruger RG, LaFrance L, Verma SK, Pappalardi MB, Le B, Van Aller GS, McCabe MT, Tummino PJ, Pope AJ, Thrall SH, Schwartz B, Brandt M (2012) Development and validation of reagents and assays for EZH2 peptide and nucleosome high-throughput screens. J Biomol Screen 17:1279–1292
3. Knutson SK, Wigle TJ, Warholic NM, Sneeringer CJ, Allain CJ, Klaus CR, Sacks JD, Raimondi A, Majer CR, Song J, Scott MP, Jin L, Smith JJ, Olhava EJ, Chesworth R, Moyer MP, Richon VM, Copeland RA, Keilhack H, Pollock RM, Kuntz KW (2012) A selective inhibitor of EZH2 blocks H3K27 methylation and kills mutant lymphoma cells. Nat Chem Biol 8:890–896
4. Van Aller GS, Pappalardi MB, Ott HM, Diaz E, Brandt M, Schwartz BJ, Miller WH, Dhanak D, McCabe MT, Verma SK, Creasy CL, Tummino PJ, Kruger RG (2014) Long residence time inhibition of EZH2 in activated polycomb repressive complex 2. ACS Chem Biol 9:622–629

GSK-J4*

Keywords: [JMJD3/UTX inhibitor]

Structure

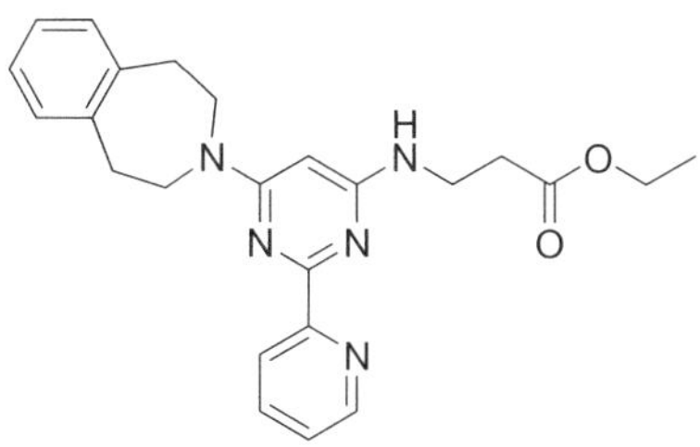

	GSK-J4
Molecular formula	$C_{24}H_{27}N_5O_2$
Molecular weight	417.50
Solubility	DMSO, +++; H_2O, ? ; MeOH, ?

Discovery/Isolation

GSK-J4 is the ethyl ester analog of GSK-J1, a potent and selective inhibitor of Jumonji domain–containing histone H3K27 demethylases including JMJD3 and UTX. GSK-J1 was identified by optimizing weakly active hit compounds discovered by high-throughput screening [1].

Biological Studies

GSK-J4 is a cell permeable prodrug which is rapidly hydrolyzed in cells by esterases to GSK-J1, a potent selective JMJD3/UTX inhibitor [1]. GSK-J4 inhibits TNF-α production in human primary macrophages accompanied by an increase in histone H3K27 trimethylation [1]. GSK-J4 reduces proliferation of T cells in acute lymphoblastic leukemia [2] and glioma cells harboring a histone K27M mutation in both in vitro and in vivo xenografts [3]. An inhibitory effect of GSK-J4 on reactivation of herpes simplex virus 1 in sensory neurons was also reported [4].

Biological Activity of GSK-J4

In vitro IC_{50} for GSK-J1 against JMJD3: 60 nM [1]
Increase in histone H3K27 trimetylation: 10–50 μM [1]
IC_{50} for proliferation of T cells in acute lymphoblastic leukemia: 2 μM [2]
IC_{50} for proliferation of K27 M-expressing glioma cells: 1.3–3.0 2 μM [3]

References

1. Kruidenier L, Chung CW, Cheng Z, Liddle J, Che K, Joberty G, Bantscheff M, Bountra C, Bridges A, Diallo H, Eberhard D, Hutchinson S, Jones E, Katso R, Leveridge M, Mander PK, Mosley J, Ramirez-Molina C, Rowland P, Schofield CJ, Sheppard RJ, Smith JE, Swales C, Tanner R, Thomas P, Tumber A, Drewes G, Oppermann U, Patel DJ, Lee K, Wilson DM (2012) A selective jumonji H3K27 demethylase inhibitor modulates the proinflammatory macrophage response. Nature 488:404–408
2. Ntziachristos P, Tsirigos A, Welstead GG, Trimarchi T, Bakogianni S, Xu L, Loizou E, Holmfeldt L, Strikoudis A, King B, Mullenders J, Becksfort J, Nedjic J, Paietta E, Tallman MS, Rowe JM, Tonon G, Satoh T, Kruidenier L, Prinjha R, Akira S, Van Vlierberghe P, Ferrando AA, Jaenisch R, Mullighan CG, Aifantis I (2014) Contrasting roles of histone 3 lysine 27 demethylases in acute lymphoblastic leukaemia. Nature 514:513–517
3. Hashizume R, Andor N, Ihara Y, Lerner R, Gan H, Chen X, Fang D, Huang X, Tom MW, Ngo V, Solomon D, Mueller S, Paris PL, Zhang Z, Petritsch C, Gupta N, Waldman TA, James CD (2014) Pharmacologic inhibition of histone demethylation as a therapy for pediatric brainstem glioma. Nat Med 20:1394–1396
4. Messer HG, Jacobs D, Dhummakupt A, Bloom DC (2015) Inhibition of H3K27me3-specific histone demethylases JMJD3 and UTX blocks reactivation of herpes simplex virus 1 in trigeminal ganglion neurons. J Virol 89:3417–3420

Halichondrin B, Eribulin*

Keywords: [Antitumor] [Tubulin polymerization inhibitor]

Structure

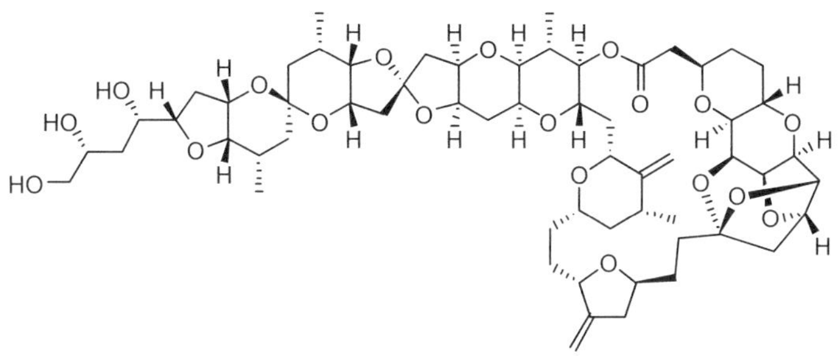

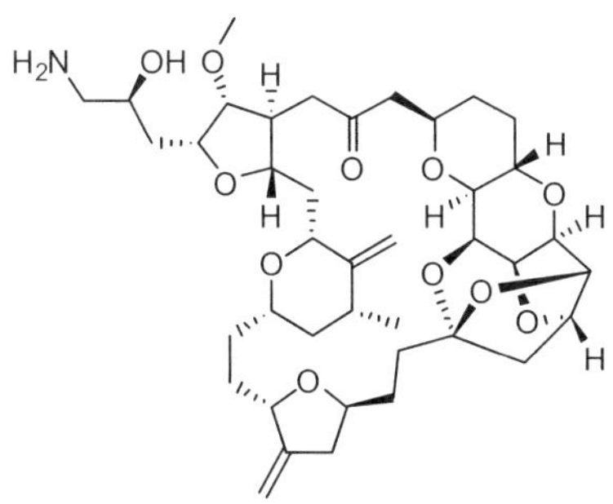

	Halichondrin B	Eribulin
Molecular formula	$C_{60}H_{86}O_{19}$	$C_{40}H_{59}NO_{11}$
Molecular weight	1111.33	729.91
Solubility	DMSO, ++; H_2O, ±; MeOH, +++	DMSO, ++; H_2O, ++; MeOH, +++

Discovery/Isolation

Halichondrin B was isolated as an antitumor compound from the marine sponge *Halichondria okadai* Kadota [1].

Biological Studies

Halichondrin B showed potent antitumor activity in vitro and in vivo [2, 1]. Data generated in an NCI drug evaluation program (60 human cell line panel) suggested that halichondrin B is a antimitotic agent [2]. Halichondrin B inhibited the cell cycle progression in mitosis and the polymerization of purified tubulin in vitro. Halichondrin B inhibited vinblastine and dolastatin 15 binding to tubulin [2, 3].

Based on the structure of halichondrin B, eribulin was synthesized and has been used as an antitumor agent clinically [4].

Biological Activity of Halichondrin B
IC_{50} for cytotoxicity against B16 melanoma in vitro: 0.093 ng/ml [1]
Antitumor activity against B16 melanoma in vivo: 2.5 μg/kg intraperitoneally, daily administration [1]
Apparent K_i for binding of vinblastine to tubulin: 5 μM, noncompetitive manner [2]

References

1. Hirata Y, Uemura D (1986) Halichondrins – antitumor polyether macrolides from a marine sponge. Pure Appl Chem 58:701–710
2. Bai RL, Paull KD, Herald CL, Malspeis L, Pettit GR, Hamel E (1991) Halichondrin B and homohalichondrin B, marine natural products binding in the vinca domain of tubulin. Discovery of tubulin-based mechanism of action by analysis of differential cytotoxicity data. J Biol Chem 266:15882–15889
3. Cruz-Monserrate Z, Mullaney JT, Harran PG, Pettit GR, Hamel E (2003) Dolastatin 15 binds in the *vinca* domain of tubulin as demonstrated by Hummel-Dreyer chromatography. Eur J Biochem 270:3822–3828
4. Towle MJ, Salvato KA, Budrow J, Wels BF, Kuznetsov G, Aalfs KK, Welsh S, Zheng W, Seletsky BM, Palme MH, Habgood GJ, Singer LA, Dipietro LV, Wang Y, Chen JJ, Quincy DA, Davis A, Yoshimatsu K, Kishi Y, Yu MJ, Littlefield BA (2001) In vitro and in vivo anticancer activities of synthetic macrocyclic ketone analogues of halichondrin B. Cancer Res 61:1013–1021

Hydrazinobenzoylcurcumin* (HBC, CTK7A)

Keywords: [Antitumor] [Ca^{2+}/calmodulin] [Histone acetyltransferase]

Structure

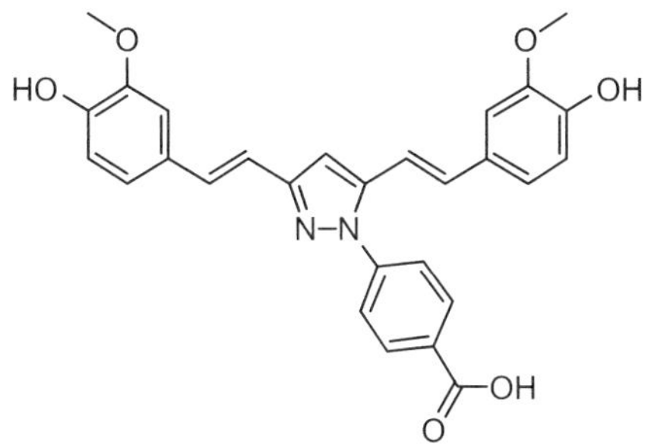

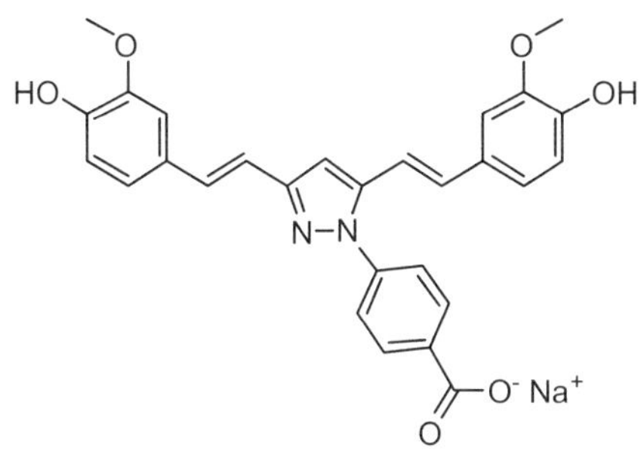

	HBC	CTK7A
Molecular formula	$C_{28}H_{24}N_2O_6$	$C_{28}H_{23}N_2NaO_6$
Molecular weight	484.51	506.49
Solubility	DMSO, +++; H_2O, ±; MeOH, +	DMSO, +++; H_2O, +; MeOH, +

Discovery/Isolation

HBC was synthesized as an antitumor compound from curcumin, a natural chemopreventive agent found in the rhizome of *Curcuma longa* [1]. CTK7A is a water-soluble sodium salt of HBC [2].

Biological Studies

HBC induced prolonged phosphorylation of ERK1/2 and activated p21WAF1 expression, resulting in the induction of G0/G1 cell cycle arrest in HCT15 colon cancer cells [3]. HBC also blocked angiogenesis by inhibiting the expression of HIF1α at the translational level [4].

The target molecule was identified as Ca^{2+}/calmodulin by phage-display biopanning [3]. More recently, it was reported that CTK7A, a water-soluble derivative of HBC, is a histone acetyltransferase inhibitor [2]. Therefore, antiproliferative activity of HBC/CTK7A might be dependent on the inhibition of both Ca^{2+}/calmodulin and histone acetlytransferase.

Biological Activity of HBC and CTK7A:

HBC

IC_{50} for the proliferation of HCT15: 20 μM [3]
IC_{50} for the proliferation of HUVECs: 20 μM [4]
Inhibition of HIF1α expression: 10 μM [4]

CTK7A

K_i values for acetyl-CoA and core histone in histone acetyltransfease assay: 13.8 and 18.6 μM, respectively [2]
Inhibition of wound-healing, induction of polyploidal cells, and senescence-like growth arrest: 100 μM [2]

References

1. Shim JS, Kim DH, Jung HJ, Kim JH, Lim D, Lee SK, Kim KW, Ahn JW, Yoo JS, Rho JR, Shin J, Kwon HJ (2002) Hydrazinocurcumin, a novel synthetic curcumin derivative, is a potent inhibitor of endothelial cell proliferation. Bioorg Med Chem 10:2987–2992
2. Arif M, Vedamurthy BM, Choudhari R, Ostwal YB, Mantelingu K, Kodaganur GS, Kundu TK (2010) Nitric oxide-mediated histone hyperacetylation in oral cancer: target for a water-soluble HAT inhibitor, CTK7A. Chem Biol 17:903–913
3. Shim JS, Lee J, Park HJ, Park SJ, Kwon HJ (2004) A new curcumin derivative, HBC, interferes with the cell cycle progression of colon cancer cells via antagonization of the Ca^{2+}/calmodulin function. Chem Biol 11:1455–1463
4. Jung HJ, Kim JH, Shim JS, Kwon HJ (2010) A novel Ca^{2+}/calmodulin antagonist HBC inhibits angiogenesis and down-regulates hypoxia-inducible factor. J Biol Chem 285:25867–25874

Iejimalides A–D

Keywords: [Antitumor] [V-ATPase inhibitor]

Structure

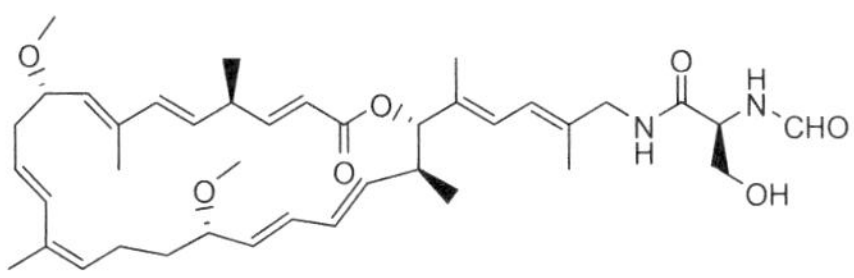

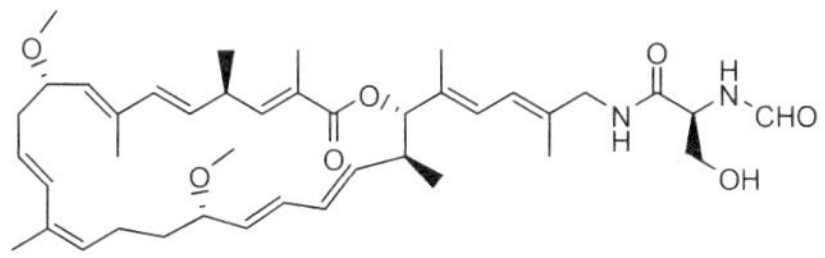

	Iejimalide A	Iejimalide B
Molecular formula	$C_{40}H_{58}N_2O_7$	$C_{41}H_{60}N_2O_7$
Molecular weight	678.90	692.94
Solubility	DMSO, ++; H_2O, ±; MeOH, +	DMSO, ++; H_2O, ±; MeOH, +

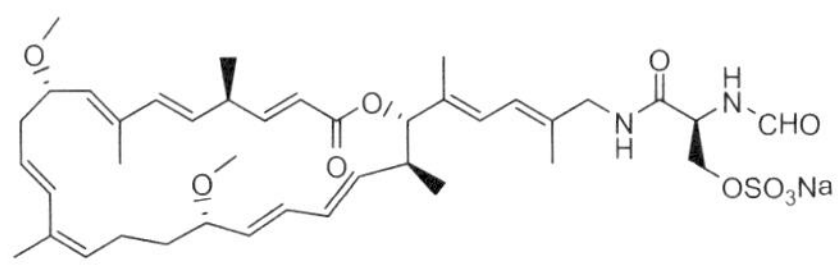

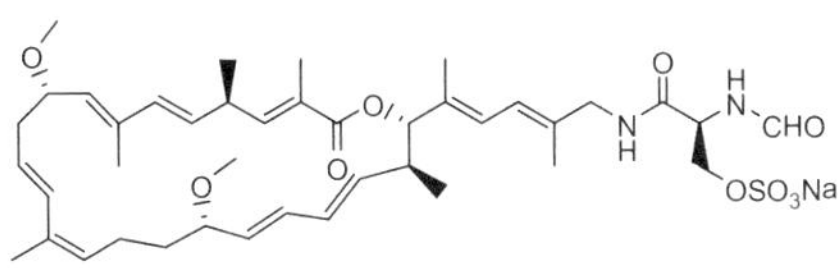

	Iejimalide C	Iejimalide D
Molecular formula	$C_{40}H_{57}N_2NaO_{10}S$	$C_{41}H_{59}N_2NaO_{10}S$
Molecular weight	780.95	794.98
Solubility	DMSO, ++; H_2O, +; MeOH, +	DMSO, ++; H_2O, +; MeOH, +

Discovery/Isolation/Chemical Studies
Iejimalides A–D, isolated from a marine tunicate *Eudistoma* cf. *rigida*, exhibit potent cytotoxic activity in vitro [1, 2] and antitumor activity in vivo [3]. The absolute structures were identified on the basis of detailed analysis of NMR data and chemical means [3]. Total synthesis was reported by Fürstner et al. [4, 5].

Biological Studies
Iejimalides A–D exerted antitumor and antiosteoporotic activity. The phenotypic assay revealed that these compounds inhibit vacuolar H^+-ATPases (V-ATPases) by binding with the *c* subunit in V_o domain of V-ATPases [6, 7]. It was also reported that iejimalide B and its analogs showed cytotoxicity via disorganization of the actin cytoskeleton [5]. However, different from actin inhibitors latrunculin A and jasplakinolide which inhibit cytokinesis, iejimalide B induces cell cycle arrest G0/G1 followed by S-phase arrest [8]. Furthermore, it has been confirmed that iejimalides inhibit V-ATPase activity in the context of epithelial tumor cells, and that this inhibition may lead to a lysosome-initiated cell death process [9].

Biological Activity of Iejimalides A–D
IC_{50} values for cytotoxicity of iejimalides A and B against L1210: 62 and 32 ng/ml, respectively [1]
IC_{50} values for cytotoxicity of iejimalides C and D against L1210: 10 and 0.58 μg/ml, respectively [2]
IC_{50} values for cytotoxicity of iejimalides A and B against TRAP-positive multinucleated cells: 5 and 6 nM, respectively [6]
K_i values for iejimalides A and B against yeast V-ATPase in vitro: 6.7 and 8.7 nM, respectively [6].

References

1. Kobayashi J, Cheng J-f, Ohta T, Nakamura H, Nozoe S, Hirata Y, Ohizumi Y, Sasaki T (1988) Iejimalides A and B, novel 24-membered macrolides with potent antileukemic activity from the Okinawan tunicate *Eudistoma* cf. *rigida*. J Org Chem 53:6147–6150
2. Kikuchi Y, Ishibashi M, Sasaki T, Kobayashi J (1991) Iejimalides C and D, new antineoplastic 24-membered macrolide sulfates from the Okinawan marine tunicate *Eudistoma* cf. *rigida*. Tetrahedron Lett 32:797–798
3. Nozawa K, Tsuda M, Ishiyama H, Sasaki T, Tsuruo T, Kobayashi J (2006) Absolute stereochemistry and antitumor activity of iejimalides. Bioorg Med Chem 14:1063–1067
4. Fürstner A, Nevado C, Tremblay M, Chevrier C, Teplý F, Aïssa C, Waser M (2006) Total synthesis of iejimalide B. Angew Chem Int Ed Eng 45:5837–5842
5. Fürstner A, Nevado C, Waser M, Tremblay M, Chevrier C, Teplý F, Aïssa C, Moulin E, Müller O (2007) Total synthesis of iejimalide A-D and assessment of the remarkable actin-depolymerizing capacity of these polyene macrolides. J Am Chem Soc 129:9150–9161

6. Kazami S, Muroi M, Kawatani M, Kubota T, Usui T, Kobayashi J, Osada H (2006) Iejimalides show anti-osteoclast activity *via* V-ATPase inhibition. Biosci Biotechnol Biochem 70:1364–1370
7. Kazami S, Takaine M, Itoh H, Kubota T, Kobayashi J, and Usui T. (2014) Iejimalide C is a potent V-ATPase inhibitor, and induces actin disorganization. Biol Pharm Bull, 37: 1944–1947
8. Wang WW, McHenry P, Jeffrey R, Schweitzer D, Helquist P, Tenniswood M (2008) Effects of Iejimalide B, a marine macrolide, on growth and apoptosis in prostate cancer cell lines. J Cell Biochem 105:998–1007
9. McHenry P, Wang WW, Devitt E, Kluesner N, Davisson VJ, McKee E, Schweitzer D, Helquist P, Tenniswood M (2010) Iejimalides A and B inhibit lysosomal vacuolar H^+ –ATPase (V-ATPase) activity and induce S-phase arrest and apoptosis in MCF-7 cells. J Cell Biochem 109:634–642

(+)-JQ1*

Keywords: [Bromodomain] [Signal transduction]

Structure

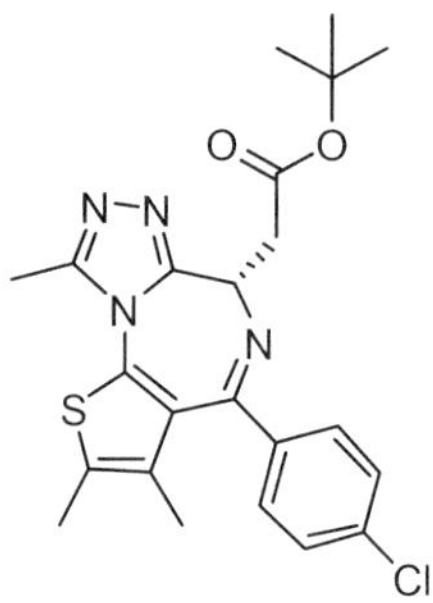

	(+)-JQ1
Molecular formula	$C_{23}H_{25}ClN_4O_2S$
Molecular weight	456.99
Solubility	DMSO, ++; H_2O, –; MeOH, ++

Discovery/Isolation/Chemical Studies

JQ1 was synthesized based on the patent that simple theinodiazepines possessed binding activity for BRD4, one of the BET (bromodomain and extraterminal) family, and the structure–activity relationship (SAR) data derived from molecular modeling of candidate ligands within the binding pocket of the apo crystal structure of the bromodomain of BRD4 [1].

Biological Studies

JQ1 increased the thermal stability of all bromodomains of the BET family in an (+)-enantiomer-specific manner [1]. Furthermore, (+)-JQ1 strongly inhibited the binding of a tetraacetlyated histone H4 peptide to BRD4. The determined cocrystal structure showed that (+)-JQ1 bound directly to the acetyl-lysine binding site of BRD4. Fluorescence recovery after photobleaching experiments using GFP-BRD4 showed that (+)-JQ1 displaced BRD4 from nuclear chromatin in cells.

(+)-JQ1 induced the differentiation and growth arrest of NMC (nuclear protein in testis (NUT) midline carcinoma) cells expressing BRD4-NUT fusion oncoprotein, and showed antitumor efficacy in an NMC xenograft model [1].

BET inhibition by JQ1 down-regulates MYC transcription, followed by genome-wide down-regulation of Myc-dependent target genes. In experimental models of multiple myeloma, an Myc-dependent hematologic malignancy, JQ1 produces a potent antiproliferative effect associated with cell cycle arrest and cellular senescence [2].

JQ1 reduced murine leukemia virus (MLV) DNA integration into the transcription start site of the host genome by inhibiting the interactions of BET proteins with modified histone sites. Therefore, JQ1 or the other BET inhibitors will facilitate the development of safer MLV-based vectors for human gene therapy by minimizing the risks of insertional activation of protooncogenes [3].

Biological Activity of (+)-JQ1

IC_{50} values for binding of a tetraacetylated histone H4 peptide to BRD4 for the first and second bromodomain: 77 nM and 33 nM, respectively [1]

Induction of a differentiation phenotype in a 797 NMC cell line: 500 nM (48 hours) [1]

G1 arrest of 797 NMC cells: 250 nM (48-hours) [1]

Antitumor activity in NMC 797 xenograft: 50 mg/kg daily intraperitoneal injection [1]

Down-regulation of MYC and cancer-related gene expression in MM cells: 500 nM [2]

IC_{50} for TGF-β1-mediated secretion of IL-6: 3.2 nM for JQ1 and 220.8 nM for iBET [4]

References

1. Filippakopoulos P, Qi J, Picaud S, Shen Y, Smith WB, Fedorov O, Morse EM, Keates T, Hickman TT, Felletar I, Philpott M, Munro S, McKeown MR, Wang Y, Christie AL, West N, Cameron MJ, Schwartz B, Heightman TD, La Thangue N, French CA, Wiest O, Kung AL, Knapp S, Bradner JE (2010) Selective inhibition of BET bromodomains. Nature 468:1067–1073
2. Delmore JE, Issa GC, Lemieux ME, Rahl PB, Shi J, Jacobs HM, Kastritis E, Gilpatrick T, Paranal RM, Qi J, Chesi M, Schinzel AC, McKeown MR, Heffernan TP, Vakoc CR, Bergsagel PL, Ghobrial IM, Richardson PG, Young RA, Hahn WC, Anderson KC, Kung AL, Bradner JE, Mitsiades CS (2011) BET bromodomain inhibition as a therapeutic strategy to target c-Myc. Cell 146: 904–917
3. Sharma A, Larue RC, Plumb MR, Malani N, Male F, Slaughter A, Kessl JJ, Shkriabai N, Coward E, Aiyer SS, Green PL, Wu L, Roth MJ, Bushman FD, Kvaratskhelia M (2013) BET proteins promote efficient murine leukemia virus integration at transcription start sites. Proc Natl Acad Sci U S A 110:12036–12041
4. Tang X, Peng R, Ren Y, Apparsundaram S, Deguzman J, Bauer CM, Hoffman AF, Hamilton S, Liang Z, Zeng H, Fuentes ME, Demartino JA, Kitson C, Stevenson CS, Budd DC (2013) BET bromodomain proteins mediate downstream signaling events following growth factor stimulation in human lung fibroblasts and are involved in bleomycin-induced pulmonary fibrosis. Mol Pharmacol 83: 283–293

KU0058684*, KU0058948*

Keywords: [Poly (ADP-ribose) polymerase (PARP) inhibitor]

Structure

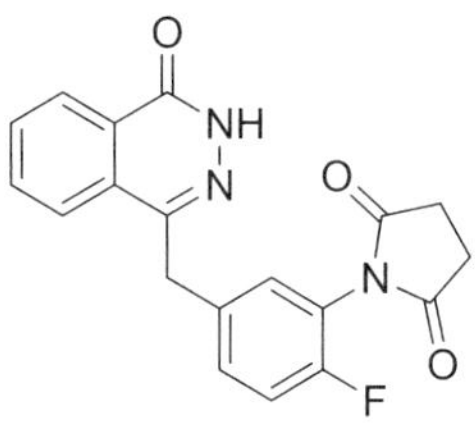

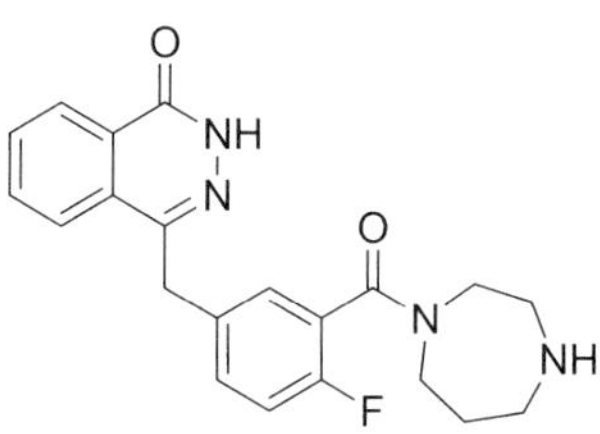

	KU0058684	KU0058948
Molecular formula	$C_{19}H_{14}FN_3O_3$	$C_{21}H_{21}FN_4O_2$
Molecular weight	351.34	380.42
Solubility	DMSO, +++; H_2O, –; MeOH, ++	DMSO, +++; H_2O, –; MeOH, ++

Discovery/Isolation

Several 4-aryl-2*H*-phthalazinones were identified as poly(ADP-ribose) polymerase-1 (PARP-1) inhibitors by high-throughput screening of a chemical library using a FlashPlate scintillation proximity assay. Based on these structures, both KU0058684 and KU0058948 were synthesized as specific and very potent PARP inhibitors [1, 2].

Biological Studies

KU0058684 and KU0058948 potently and irreversibly inhibit the cell proliferation of BRCA1- or BRCA2-deficient cells compared with heterogygous mutant or wild-type cells [3]. These compounds induce chromosomal instability, cell cycle arrest, and subsequent apoptosis. It is known that PARPs are required for efficient repair of DNA single-strand breaks and both BRCA1 and BRCA2 is involved in RAD51-dependent homologous recombination. Therefore, PARP inhibition results in the lack of a DNA repair pathway in BRCA1- or BRCA2-deficient cells, and would be a specific and less toxic therapy for cancer.

Biological Activity of KU0058684 and KU0058948
IC_{50} values for KU0058684 against PAR activity of PARP-1, -2, -3, and tankylase: 3.2, 1.5, 30, and 1600 nM, respectively [3]
IC_{50} values for KU0058948 against PAR activity of PARP-1, -2, -3, and tankylase: 3.4, 1.5, 40, and >10,000 nM, respectively [3]
IC_{50} values for KU0058684 on cell proliferation of BRCA1-, BRCA2-deficient and wild-type cells: 35, 15, and ~2000 nM, respectively [3]
IC_{50} values for KU0058684 and KU0058948 on cell proliferation of CAPAN-1. a BRCA2 mutant cell line: 3.2, and 3.4 nM, respectively [4]

References

1. Loh VM, Jr., Cockcroft XL, Dillon KJ, Dixon L, Drzewiecki J, Eversley PJ, Gomez S, Hoare J, Kerrigan F, Matthews IT, Menear KA, Martin NM, Newton RF, Paul J, Smith GC, Vile J, Whittle AJ (2005) Phthalazinones. Part 1: the design and synthesis of a novel series of potent inhibitors of poly(ADP-ribose) polymerase. Bioorg Med Chem Lett 15:2235–2238
2. Cockcroft XL, Dillon KJ, Dixon L, Drzewiecki J, Kerrigan F, Loh VM, Jr., Martin NM, Menear KA, Smith GC (2006) Phthalazinones 2: optimisation and synthesis of novel potent inhibitors of poly(ADP-ribose)polymerase. Bioorg Med Chem Lett 16:1040–1044
3. Farmer H, McCabe N, Lord CJ, Tutt AN, Johnson DA, Richardson TB, Santarosa M, Dillon KJ, Hickson I, Knights C, Martin NM, Jackson SP, Smith GC, Ashworth A (2005) Targeting the DNA repair defect in BRCA mutant cells as a therapeutic strategy. Nature 434:917–921
4. McCabe N, Lord CJ, Tutt AN, Martin NM, Smith GC, Ashworth A (2005) BRCA2-deficient CAPAN-1 cells are extremely sensitive to the inhibition of Poly (ADP-Ribose) polymerase: an issue of potency. Cancer Biol Ther 4:934–936

Lactacystin

Keywords: [Neurite outgrowth inducer] [Proteasome inhibitor]

Structure

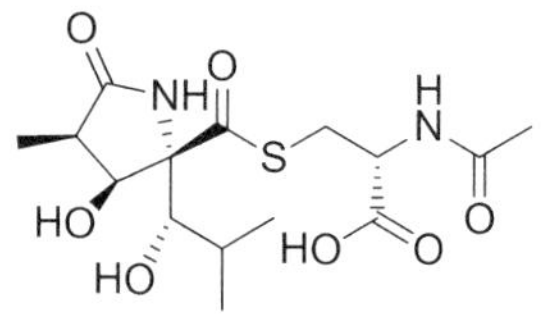

	Lactacystin
Molecular formula	$C_{15}H_{24}N_2O_7S$
Molecular weight	376.43
Solubility	DMSO, ++; H_2O, –, MeOH, ++

Discovery/Isolation

Lactacystin was isolated as the differentiation inducer of the mouse neuroblastoma Neuro 2A from the culture broth of a *Streptomyces* sp. OM-6519 [1, 2].

Biological Studies

Lactacystin induced neurite-like structure generation, increased the intracellular cAMP level transiently [2], and arrested cell cycle progression at both G0/G1 and G2 phases in Neuro 2A cells [3, 4].

[^{3}H]-labeled lactacystin covalently modified the highly conserved amino-terminal threonine of the mammalian proteasome subunit X, a component of 20S proteasome [5]. The ability of lactacystin analogs to inhibit cell cycle progression and induce neurite outgrowth correlated with their ability to inhibit the proteasome, strongly suggesting that 20S proteasome is a lactacystin-specific cellular target.

Because the ubiquitin–proteasome system plays important roles in several cellular processes including down-regulation of signal transduction by degradation of ligand–receptor complex, denaturated proteins, and apoptosis, there are many reports investigating the cellular function using lactacystin [6].

Biological Activity of Lactacystin

Neurite outgrowth in Neuro 2A: 1.3 μM [2]
Inhibition of cell cycle progression at both G0/G1 and G2 phases: 1.3 μM [4]
Accumulation of ubiquitinated proteins and apoptosis induction: 5 μM [6]

References

1. Omura S, Matsuzaki K, Fujimoto T, Kosuge K, Furuya T, Fujita S, Nakagawa A (1991) Structure of lactacystin, a new microbial metabolite which induces differentiation of neuroblastoma cells. J Antibiot (Tokyo) 44:117–118
2. Omura S, Fujimoto T, Otoguro K, Matsuzaki K, Moriguchi R, Tanaka H, Sasaki Y (1991) Lactacystin, a novel microbial metabolite, induces neuritogenesis of neuroblastoma cells. J Antibiot (Tokyo) 44:113–116
3. Fenteany G, Standaert RF, Reichard GA, Corey EJ, Schreiber SL (1994) A β-lactone related to lactacystin induces neurite outgrowth in a neuroblastoma cell line and inhibits cell cycle progression in an osteosarcoma cell line. Proc Natl Acad Sci U S A 91:3358–3362
4. Katagiri M, Hayashi M, Matsuzaki K, Tanaka H, Omura S (1995) The neuritogenesis inducer lactacystin arrests cell cycle at both G0/G1 and G2 phases in Neuro2A cells. J Antibiot (Tokyo) 48:344–346
5. Fenteany G, Standaert RF, Lane WS, Choi S, Corey EJ, Schreiber SL (1995) Inhibition of proteasome activities and subunit-specific amino-terminal threonine modification by lactacystin. Science 268:726–731
6. Soldatenkov VA, Dritschilo A (1997) Apoptosis of Ewing's sarcoma cells is accompanied by accumulation of ubiquitinated proteins. Cancer Res 57:3881–3885

Leptomycin B

Keywords: [Antitumor] [Nuclear export inhibitor]

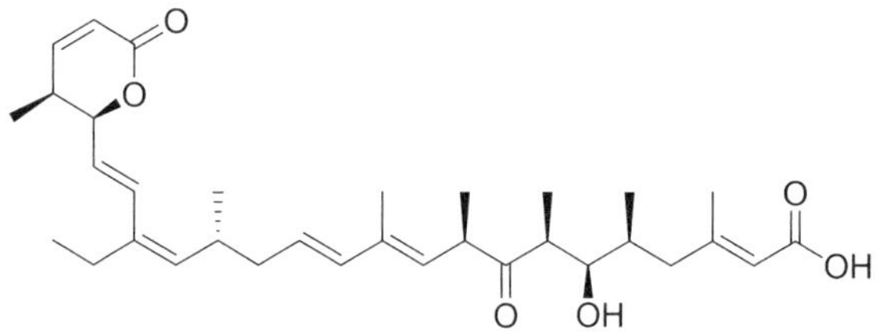

	Leptomycin B
Molecular formula	$C_{33}H_{48}O_6$
Molecular weight	540.74
Solubility	DMSO, +; H_2O, –; MeOH, +++

Discovery/Isolation/Chemical Studies

Leptomycin B was isolated from *Streptomyces* sp. ATS1287 as a strong antifungal antibiotic, which causes inhibition of cell division, elongated cells with morphologically altered nuclei, and several cell plates in *Schizosaccharomyces pombe* [1, 2]. There are several related compounds, anguinomycins A–D [3, 4], kazusamycin A and B [5, 6], leptolstatin [7, 8], and leptofuranins A–D [9, 10]. All these compounds have the structure characteristics of an unsaturated, branched-chain fatty acid with a terminal δ-lactone ring.

Biological Studies

Leptomycin B reversibly blocked cell cycle progression at both G1 and G2 phases in mammalian cells and *S. pombe* [11]. Interestingly, after removal of leptomycin B, G2-arrested cells started DNA replication without passage through mitosis, then proliferative tetraploid cells were produced. The *crm1*$^+$ gene, which was previously identified as the gene required for maintenance of higher-order chromosome structures, was identified as the molecular target of leptomycin B by genetic analyses using a leptomycin B–specific resistant mutant of *S. pombe* [12]. At first, the function of *crm1*$^+$ was unknown, but the discovery of leptomycin B in the course of screening of the compounds inhibiting the nuclear export of HIV-1 protein Rev revealed that leptomycin B is a potent and specific inhibitor of protein export from nuclei. Leptomycin B covalently binds with CRM1 and inhibits the binding of CRM1 to proteins containing nuclear export signal (NES) [13–17]. Mutation analysis of leptomycin B strongly suggests that leptomycin B binds covalently via its α,β-unsaturated δ-lactone to the sulfhydryl group of Cys529 in the central conserved region. Now leptomycin B has been widely used for investigation of biological function mediated by nuclear protein export.

The other leptomycins also shows antitumor activity, cell cycle proliferation inhibition, and nuclear export inhibition [3–6, 8, 10].

Biological Activity of Leptomycin B
Minimum inhibitory concentrations against *S. pombe* and *Mucor* sp.: 12–250 ng/ml [2]
Reversible cell cycle arrest at G1 and G2 phases: 1–200 ng/ml [11]
Inhibition of nuclear export in vitro: 40 nM [13]

References

1. Hamamoto T, Gunji S, Tsuji H, Beppu T (1983) Leptomycins A and B, new antifungal antibiotics. I. Taxonomy of the producing strain and their fermentation, purification and characterization. J Antibiot (Tokyo) 36:639–645
2. Hamamoto T, Seto H, Beppu T (1983) Leptomycins A and B, new antifungal antibiotics II Structure elucidation. J Antibiot (Tokyo) 36:646–650
3. Hayakawa Y, Adachi K, Komeshima N (1987) New antitumor antibiotics, anguinomycins A and B. J Antibiot (Tokyo) 40:1349–1352
4. Hayakawa Y, Sohda KY, Shin-Ya K, Hidaka T, Seto H (1995) Anguinomycins C and D, new antitumor antibiotics with selective cytotoxicity against transformed cells. J Antibiot (Tokyo) 48:954–961
5. Umezawa I, Komiyama K, Oka H, Okada K, Tomisaka S, Miyano T, Takano S (1984) A new antitumor antibiotic, kazusamycin. J Antibiot (Tokyo) 37:706–711
6. Funaishi K, Kawamura K, Sugiura Y, Nakahori N, Yoshida E, Okanishi M, Umezawa I, Funayama S, Komiyama K (1987) Kazusamycin B, a novel antitumor antibiotic. J Antibiot (Tokyo) 40:778–785
7. Abe K, Yoshida M, Horinouchi S, Beppu T (1993) Leptolstatin from *Streptomyces* sp. SAM1595, a new gap phase-specific inhibitor of the mammalian cell cycle. I. Screening, taxonomy, purification and biological activities. J Antibiot (Tokyo) 46:728–734
8. Abe K, Yoshida M, Naoki H, Horinouchi S, Beppu T (1993) Leptolstatin from *Streptomyces* sp. SAM1595, a new gap phase-specific inhibitor of the mammalian cell cycle. II. Physico-chemical properties and structure. J Antibiot (Tokyo) 46:735–740
9. Hayakawa Y, Sohda K, Seto H (1996) Studies on new antitumor antibiotics, leptofuranins A, B, C and D II Physiocochemical properties and structure elucidation. J Antibiot (Tokyo) 49:980–984
10. Hayakawa Y, Sohda K, Furihata K, Kuzuyama T, Shin-ya K, Seto H (1996) Studies on new antitumor antibiotics, leptofuranins A, B, C and D.I. Taxonomy, fermentation, isolation and biological activities. J Antibiot (Tokyo) 49:974–979
11. Yoshida M, Nishikawa M, Nishi K, Abe K, Horinouchi S, Beppu T (1990) Effects of leptomycin B on the cell cycle of fibroblasts and fission yeast cells. Exp Cell Res 187:150–156

12. Nishi K, Yoshida M, Fujiwara D, Nishikawa M, Horinouchi S, Beppu T (1994) Leptomycin B targets a regulatory cascade of crm1, a fission yeast nuclear protein, involved in control of higher order chromosome structure and gene expression. J Biol Chem 269:6320–6324
13. Fornerod M, Ohno M, Yoshida M, Mattaj IW (1997) CRM1 is an export receptor for leucine-rich nuclear export signals. Cell 90:1051–1060
14. Fukuda M, Asano S, Nakamura T, Adachi M, Yoshida M, Yanagida M, Nishida E (1997) CRM1 is responsible for intracellular transport mediated by the nuclear export signal. Nature 390:308–311
15. Kudo N, Khochbin S, Nishi K, Kitano K, Yanagida M, Yoshida M, Horinouchi S (1997) Molecular cloning and cell cycle-dependent expression of mammalian CRM1, a protein involved in nuclear export of proteins. J Biol Chem 272:29742–29751
16. Ossareh-Nazari B, Bachelerie F, Dargemont C (1997) Evidence for a role of CRM1 in signal-mediated nuclear protein export. Science 278:141–144
17. Watanabe M, Fukuda M, Yoshida M, Yanagida M, Nishida E (1999) Involvement of CRM1, a nuclear export receptor, in mRNA export in mammalian cells and fission yeast. Genes Cells 4:291–297

Locostatin* (UIC-1005)

Keywords: [Antitumor] [Protein-protein interaction inhibitor] [RKIP inhibitor]

Structure

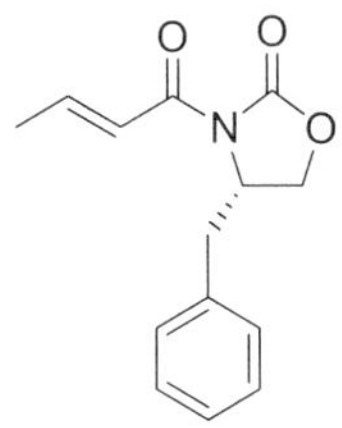

	Locostatin
Molecular formula	$C_{14}H_{15}NO_3$
Molecular weight	245.27
Solubility	DMSO, +; H_2O,-; MeOH, +++

Discovery/Isolation/Chemical Studies
Locostatin (former name: UIC-1005) was a 3,4-disubstituted oxazolidinone identified by a high-throughput assay for screening chemical libraries for compounds that affect cell sheet migration during wound closure in epithelial cell monolayers [1]. This compound bears an electrophilic α,β-unsaturated *N*-acyl group required for activity.

Biological Studies
Locostatin inhibited not only cell sheet migration during wound closure but also cell growth of epithelial cells [1]. Zhu et al. found that Raf kinase inhibitor protein (RKIP) is the relevant cellular target of locostatin using a [^{3}H]-labeled compound [2]. RKIP is an endogenous inhibitor of cell signaling by binding with multiple kinases, and locostatin covalently binds with RKIP and abrogates the binding of RKIP to and inhibition of kinase activity of Raf-1 kinase [2]. Locostatin also inhibits the binding of RKIP to G-protein coupled receptor kinase 2 (GRK2), but not to IKKα and TAK1, implying that there are at least two regions of RKIP interacting with different proteins [3]. The binding site of locostatin was identified as the residue alkylated by locostatin as His86, a highly conserved residue in RKIP's ligand binding pocket [3]. A detailed binding process between RKIP and locostatin was proposed [4].

Overexpression of RKIP converts epithelial cells into a highly migratory fibroblast-like phenotype, with dramatic reduction in the sensitivity of cells to locostatin, suggesting that RKIP is therefore the compound's valid target and a key regulator of cell motility [2].

Biological Activity of Locostatin

IC_{50} for wound closure in MDCK cells: 17.9 μM [2]

References

1. Mc Henry KT, Ankala SV, Ghosh AK, Fenteany G (2002) A non-antibacterial oxazolidinone derivative that inhibits epithelial cell sheet migration. ChemBioChem 3:1105–1111
2. Zhu S, Mc Henry KT, Lane WS, Fenteany G (2005) A chemical inhibitor reveals the role of Raf kinase inhibitor protein in cell migration. Chem Biol 12:981–991
3. Beshir AB, Argueta CE, Menikarachchi LC, Gascon JA, Fenteany G (2011) Locostatin disrupts association of raf kinase inhibitor protein with binding proteins by modifying a conserved histidine residue in the ligand-binding pocket. For Immunopathol Dis Therap 2:47–58
4. Rudnitskaya AN, Eddy NA, Fenteany G, Gascon JA (2012) Recognition and reactivity in the binding between Raf kinase inhibitor protein and its small-molecule inhibitor locostatin. J Phys Chem B 116:10176–10181

Miuraenamide A

Keywords: [Actin stabilizer] [NADH oxidase inhibitor]

Structure

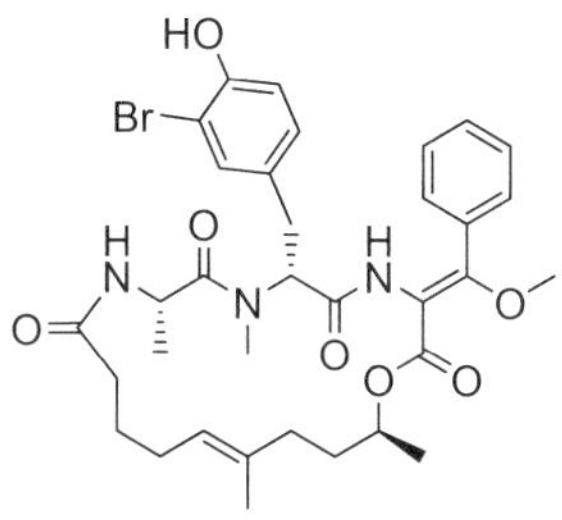

	Miuraenamide A
Molecular formula	$C_{34}H_{42}BrN_3O_7$
Molecular weight	684.63
Solubility	DMSO, ? ; H_2O, ? ; MeOH, ++

Discovery/Isolation

Miuraenamide A was produced by a slightly halophilic myxobacterial strain, SMH-27-4, which was a new myxobacterium genus isolated from nearshore soil [1].

Biological Studies

Miuraenamide A exhibited potent and selective inhibition against a phytopathogenic microorganism, *Phytophthora* sp., and moderate inhibition against some fungi and yeasts, but was ineffective against bacteria [1]. This compound inhibited NADH oxidase, suggesting the electron transfer system of the mitochondrial respiratory chain as the cellular target.

By cell morphological profiling of a natural product library using automated high-content image analysis, it was found that miuraenamide A induces morphological changes in HeLa cells similar to those induced by known actin-stabilizing compounds [2]. Miuraenamide A stabilized actin filaments in vitro, suggesting that this compound is an actin filament stabilizer.

Biological Activity of Miuraenamide A

IC_{50} for NADH oxidase: 50 μM [1]
Stabilization of actin in HeLa cell: 100 nM [2]
Stabilization of actin filament in vitro: 10 nM [2]

References

1. Iizuka T, Fudou R, Jojima Y, Ogawa S, Yamanaka S, Inukai Y, Ojika M (2006) Miuraenamides A and B, novel antimicrobial cyclic depsipeptides from a new slightly halophilic myxobacterium: taxonomy, production, and biological properties. J Antibiot (Tokyo) 59:385–391
2. Sumiya E, Shimogawa H, Sasaki H, Tsutsumi M, Yoshita K, Ojika M, Suenaga K, Uesugi M (2011) Cell-morphology profiling of a natural product library identifies bisebromoamide and miuraenamide A as actin filament stabilizers. ACS Chem Biol 6:425–431

MJE3*

Keywords: [Antitumor] [Phosphoglycerate mutase 1 inhibitor]

Structure

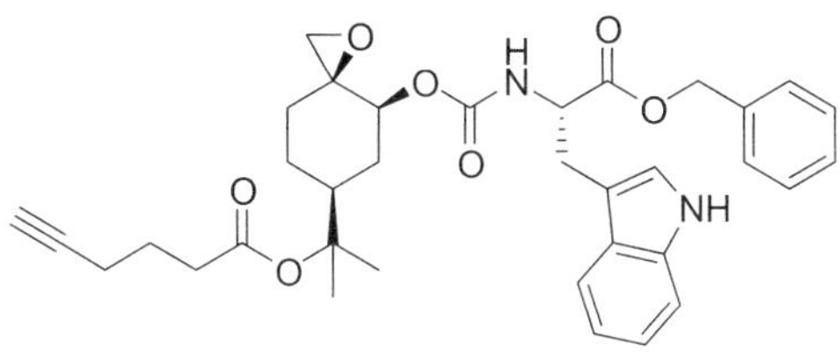

	MJE3
Molecular formula	$C_{35}H_{40}N_2O_7$
Molecular weight	600.70
Solubility	DMSO, ++; H_2O, ? ; MeOH, ?

Discovery/Isolation/Chemical Studies
MJE3 was found to block proliferation of invasive human breast cancer cell line MDA-MB-231 from a ~50-member probe library [1]. An analog of MJE3 in which the spiroepoxide was replaced by an exocyclic alkene did not exhibit antiproliferative effects, indicating that the reactivity of MJE3 was essential for biological activity.

Biological Studies
To identify the molecular target of MJE3, Evans et al. compared the in situ proteome reactivity profiles of library members, and found that MJE3 covalently labeled the glycolytic enzyme phosphoglycerate mutase 1 (PGAM1) [1]. Furthermore, treatment with MJE3, but not other derivatives, resulted in the inhibition of PGAM1 activity in situ. Tandem orthogonal proteolysis analysis suggested that MJE3 is hydrolyzed in situ to produce an acid product that covalently modifies the PGAM1 active site, resulting in enzyme inhibition. The precise binding site was determined as Lys100, a conserved active-site residue [2]. Structural features of MJE3 important for PGAM1 labeling have also been reported.

Biological Activity of MJE3
IC_{50} for cell growth: 19 μM [1]
IC_{50} for PGAM1 in situ: 33 μM [1]

References

1. Evans MJ, Saghatelian A, Sorensen EJ, Cravatt BF (2005) Target discovery in small-molecule cell-based screens by *in situ* proteome reactivity profiling. Nat Biotechnol 23:1303–1307
2. Evans MJ, Morris GM, Wu J, Olson AJ, Sorensen EJ, Cravatt BF (2007) Mechanistic and structural requirements for active site labeling of phosphoglycerate mutase by spiroepoxides. Mol BioSyst 3:495–506

MKT-077*

Keywords: [Antitumor] [HSP70 inhibitor] [Telomerase inhibitor] [F-actin binder]

Structure

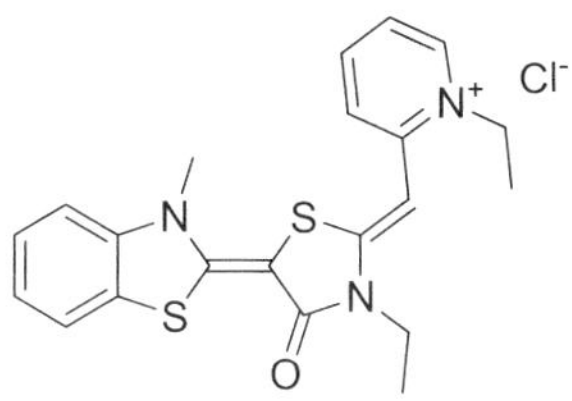

	MKT-077
Molecular formula	$C_{21}H_{22}ClN_3OS_2$
Molecular weight	432.00
Solubility	DMSO, +++; H_2O, +++; MeOH, +++

Discovery/Isolation/Chemical Studies

Rhodacyanine dyes form one group of delocalized lipophilic cations which are accumulated by mitochondria in response to plasma and mitochondrial membrane potentials. MKT-077 was synthesized as a rhodacyanine dye exhibiting marked anticarcinoma activity with relatively low toxicity [1]. This compound is a highly water-soluble (>200 mg/ml) lipophilic compound.

Biological Studies

MKT-077 showed antitumor activity in situ and in vivo, and localization in the mitochondria selectively [2]. This compound disorganized the structure and function of mitochondria in a carcinoma-specific manner [3]. At a pharmacologically toxic dose, MKT-077 significantly but reversibly decreased respiratory rates in rat liver mitochondria in vivo [4].

Two MKT-077 binding proteins were purified by affinity chromatography, actin, and Hsc70 [5, 6]. MKT-077 binds actin directly, bundles actin filaments by cross-linking, and suppresses *ras* transformation by blocking membrane ruffling [6]. MKT-077 also binds to mortalin, an Hsp70 family member, and abrogates the interaction with p53. This inhibition of protein–protein interaction (PPI) results in transcriptional activation of p53 and apoptosis induction in cancer cells [7]. Furthermore, the inhibition of mortalin by MKT-077 induced cell senescence in human tumor cell lines [8]. The MKT-077 binding site has been identified as the nucleotide binding domain of mortalin by a MKT-077 pull-down assay [7], NMR analyses [9], and crystallography [10]. This allosteric binding causes the tertiary structural changes of mortalin and inactivates its chaperone function [8].

In vitro telomerase inhibitory activity of MKT-077 was also reported, but there was no detectable effect on telomerase activity in vivo [11, 12].

Biological Activity of MKT-077
IC_{50} for several cancer cells: 0.15–0.5 μg/ml [2]
Abnormalities in the mitochondria structure in CRK1420 cells: 2 μg/ml [3]

References

1. Kawakami M, Koya K, Ukai T, Tatsuta N, Ikegawa A, Ogawa K, Shishido T, Chen LB (1998) Structure-activity of novel rhodacyanine dyes as antitumor agents. J Med Chem 41:130–142
2. Koya K, Li Y, Wang H, Ukai T, Tatsuta N, Kawakami M, Shishido, Chen LB (1996) MKT-077, a novel rhodacyanine dye in clinical trials, exhibits anticarcinoma activity in preclinical studies based on selective mitochondrial accumulation. Cancer Res 56:538–543
3. Modica-Napolitano JS, Koya K, Weisberg E, Brunelli BT, Li Y, Chen LB (1996) Selective damage to carcinoma mitochondria by the rhodacyanine MKT-077. Cancer Res 56:544–550
4. Weisberg EL, Koya K, Modica-Napolitano J, Li Y, Chen LB (1996) In vivo administration of MKT-077 causes partial yet reversible impairment of mitochondrial function. Cancer Res 56:551–555
5. Maruta H, Tikoo A, Shakri R, Shishido T (1999) The anti-RAS cancer drug MKT-077 is an F-actin cross-linker. Ann N Y Acad Sci 886:283–284
6. Tikoo A, Shakri R, Connolly L, Hirokawa Y, Shishido T, Bowers B, Ye LH, Kohama K, Simpson RJ, Maruta H (2000) Treatment of ras-induced cancers by the F-actin-bundling drug MKT-077. Cancer J 6:162–168
7. Wadhwa R, Sugihara T, Yoshida A, Nomura H, Reddel RR, Simpson R, Maruta H, Kaul SC (2000) Selective toxicity of MKT-077 to cancer cells is mediated by its binding to the hsp70 family protein mot-2 and reactivation of p53 function. Cancer Res 60:6818–6821
8. Deocaris CC, Widodo N, Shrestha BG, Kaur K, Ohtaka M, Yamasaki K, Kaul SC, Wadhwa R (2007) Mortalin sensitizes human cancer cells to MKT-077-induced senescence. Cancer Lett 252:259–269
9. Rousaki A, Miyata Y, Jinwal UK, Dickey CA, Gestwicki JE, Zuiderweg ER (2011) Allosteric drugs: the interaction of antitumor compound MKT-077 with human Hsp70 chaperones. J Mol Biol 411:614–632
10. Amick J, Schlanger SE, Wachnowsky C, Moseng MA, Emerson CC, Dare M, Luo WI, Ithychanda SS, Nix JC, Cowan JA, Page RC, Misra S (2014) Crystal structure of the nucleotide-binding domain of mortalin, the mitochondrial Hsp70 chaperone. Protein Sci 23:833–842
11. Naasani I, Seimiya H, Yamori T, Tsuruo T (1999) FJ5002: a potent telomerase inhibitor identified by exploiting the disease-oriented screening program with COMPARE analysis. Cancer Res 59:4004–4011
12. Wadhwa R, Colgin L, Yaguchi T, Taira K, Reddel RR, Kaul SC (2002) Rhodacyanine dye MKT-077 inhibits in vitro telomerase assay but has no detectable effects on telomerase activity in vivo. Cancer Res 62:4434–4438

Myriocin/ISP-1, FTY720*

Keywords: [Immunosuppressant] [Serine palmitoyltransferase inhibitor] [Sphingolipid metabolism]

Structure

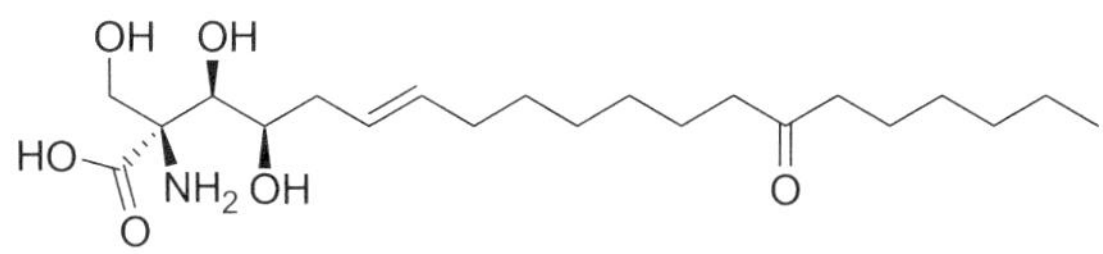

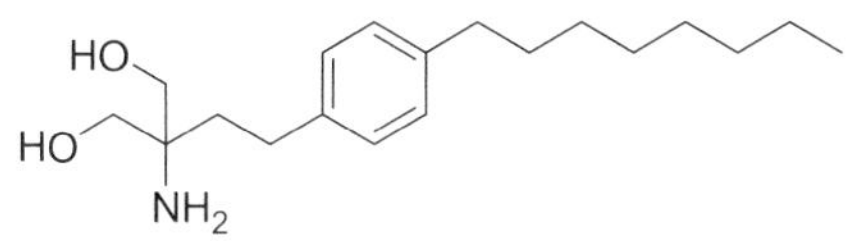

	Myriosin	FTY720
Molecular formula	$C_{21}H_{39}NO_6$	$C_{19}H_{33}NO_2$
Molecular weight	401.54	307.47
Solubility	DMSO, +++; H_2O, −; MeOH, ±	DMSO, ++; H_2O, ? ; MeOH, ?

Discovery/Isolation

Myriocin was originally isolated as an antifungal compound from *Myriococcum albomyces* [1], and reisolated as a potent immunosuppressant [2]. FTY720 is a synthetic immunosuppressant led by modification of ISP-I, and displayed potent immunosuppressive activity both in vitro and in vivo [3].

Biological Studies

Myriocin is structurally related to sphingosine, and inhibited serine palmitoyltransferase (SPT) which catalyzes the first step of sphingolipid biosynthesis [4]. The growth inhibition of CTLL-2 cells induced by myriocin was completely abolished by the addition of sphingosines or sphingosine-1-phosphate, but not by sphingomyelin or glycosphingolipids, suggesting that myriocin suppresses T-cell proliferation by the modulation of sphinogolipid metabolism. The myriocin binding proteins were purified by an affinity purification method and identified to be LCB1 and LCB2, mammalian homologs of two yeast proteins that have been genetically linked to sphingolipid biosynthesis [5]. It has been revealed that SPT consists of three proteins, LCB1/TSC2, LCB2/TSC1, and TSC3 [6–8]; therefore, these results strongly suggest that myriocin is a direct SPT inhibitor. Myriocin has been used not only for the effect of de novo synthesis of ceramide but also for the biological function of several sphingosine-regulated processes.

Compared with myriocin, FTY720 does not inhibit SPT [9]. The structure similarity of FTY720 with sphingosine led to the discovery that FTY720 is phosphorylated by sphingosine kinases and that phosphorylated FTY720 functions as an agonist at sphingosine-1-phosphate (S1P) receptors [10]. Furthermore, phosphorylated FTY720 induced the accumulation of lymphocytes at lymphatic endothelial barriers [11]. Because egress of lymphocytes from lymph nodes requires lymphocytic $S1P_1$ receptors [12], these results suggest that the inhibition of lymphocyte recirculation by activation of S1P receptors results in therapeutically useful immunosuppression.

Biological Activity of Myriocin
MIC for *Candida albicans*: 0.32–25 μg/ml [1]
IC_{50} for mouse allogenic mixed lymphocyte reaction: 10–18 nM [2]
IC_{50} for CTLL-2 cell growth: 15 nM [4]
IC_{50} for SPT: 0.28 nM [4]

Biological Activity of FTY720
IC_{50} for mouse allogenic mixed lymphocyte reaction: 6.1 nM [3]
EC_{50} for agonist activity against $S1P_1$, $S1P_4$, and $S1P_5$ receptors: 0.3–0.6 nM [10]
Lymphopenia in blood and thoracic duct in rats: 2.5 mg/kg, oral administration [11]

References

1. Kluepfel D, Bagli J, Baker H, Charest MP, Kudelski A, Sehgal SN, Vézina C (1972) Myriocin, a new antifungal antibiotic from *Myriococcum albomyces*. J Antibiot (Tokyo) 25:109–115
2. Fujita T, Inoue K, Yamamoto S, Ikumoto T, Sasaki S, Toyama R, Chiba K, Hoshino Y, Okumoto T (1994) Fungal metabolites. Part 11. A potent immunosuppressive activity found in *Isaria sinclairii* metabolite. J Antibiot (Tokyo) 47:208–215
3. Adachi K, Kohara T, Nakao N, Arita M, Chiba K, Mishina T, Sasaki S, Fujita T (1995) Design, synthesis, and structure-activity relationships of 2-substituted-2-amino-1,3-propanediols: discovery of a novel immunosuppressant, FTY720. Bioorg Med Chem Lett 5:853–856
4. Miyake Y, Kozutsumi Y, Nakamura S, Fujita T, Kawasaki T (1995) Serine palmitoyltransferase is the primary target of a sphingosine-like immunosuppressant, ISP-1/myriocin. Biochem Biophys Res Commun 211:396–403
5. Chen JK, Lane WS, Schreiber SL (1999) The identification of myriocin-binding proteins. Chem Biol 6:221–235
6. Buede R, Rinker-Schaffer C, Pinto WJ, Lester RL, Dickson RC (1991) Cloning and characterization of LCB1, a *Saccharomyces* gene required for biosynthesis of the long-chain base component of sphingolipids. J Bacteriol 173:4325–4332

7. Gable K, Slife H, Bacikova D, Monaghan E, Dunn TM (2000) Tsc3p is an 80-amino acid protein associated with serine palmitoyltransferase and required for optimal enzyme activity. J Biol Chem 275:7597–7603
8. Nagiec MM, Baltisberger JA, Wells GB, Lester RL, Dickson RC (1994) The LCB2 gene of *Saccharomyces* and the related LCB1 gene encode subunits of serine palmitoyltransferase, the initial enzyme in sphingolipid synthesis. Proc Natl Acad Sci U S A 91:7899–7902
9. Fujita T, Hirose R, Hamamichi N, Kitao Y, Sasaki S, Yoneta M, Chiba K. (1995) 2-Substituted 2-aminoethanol: minimum essential structure for immunosuppressive activity of ISP-I (Myriocin). Bioorg Med Chem Lett 5:1857–1860
10. Brinkmann V, Davis MD, Heise CE, Albert R, Cottens S, Hof R, Bruns C, Prieschl E, Baumruker T, Hiestand P, Foster CA, Zollinger M, Lynch KR. (2002) The immune modulator FTY720 targets sphingosine 1-phosphate receptors. J Biol Chem 277:21453–21457
11. Mandala S, Hajdu R, Bergstrom J, Quackenbush E, Xie J, Milligan J, Thornton R, Shei GJ, Card D, Keohane C, Rosenbach M, Hale J, Lynch CL, Rupprecht K, Parsons W, Rosen H.(2002) Alteration of lymphocyte trafficking by sphingosine-1-phosphate receptor agonists.Science 296:346–349
12. Matloubian M, Lo CG, Cinamon G, Lesneski MJ, Xu Y, Brinkmann V, Allende ML, Proia RL, Cyster JG. (2004) Lymphocyte egress from thymus and peripheral lymphoid organs is dependent on S1P receptor 1. Nature 427:355–360

N-89*, N-251*

Keywords: [Antimalarial] [Endoplasmic reticulum-resident calcium binding protein (ERC) inhibitor]

Structure

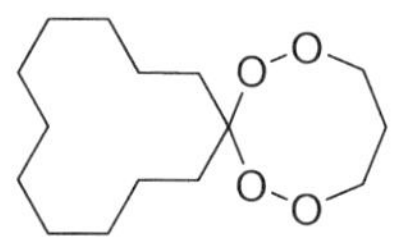

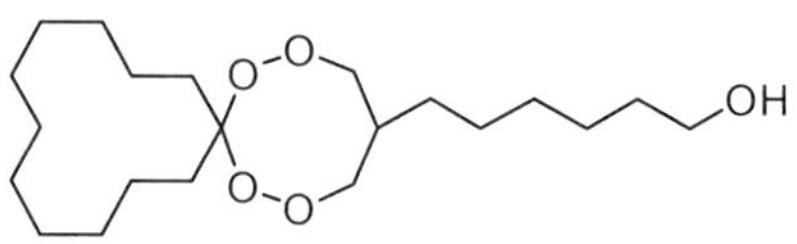

	N-89	N-251
Molecular formula	$C_{15}H_{28}O_4$	$C_{21}H_{40}O_5$
Molecular weight	272.38	372.54
Solubility	DMSO, +; H_2O, ? ; MeOH, ?	DMSO, +; H_2O, ? ; MeOH, ?

Discovery/Isolation/Chemical Studies

N-89 and N-251 are synthetic compounds containing endoperoxide [1, 2].

Biological Studies

Synthetic endoperoxides N-89 and its derivative with hydroxyl group, N-251, had high antimalarial activities both in vivo and in vitro [1, 3]. The target molecules in artemisinin-resitant *Plasmodium falciparum* were surveyed by a chemical proteomics approach using the affinity spin column, and several candidate proteins were determined. One of the candidate proteins, *Pf*ERC, a member of the CREC (Cab45, reticulocalbin, ERC-45, calumenin) protein family, had high affinity to the endoperoxide structure and this interaction was competitively inhibited by N-89 or N-251, suggesting that *Pf*ERC is a target protein of endoperoxide antimalarial compounds.

Biological Activity of N-89 and N-251

EC_{50} values for N-89 against *P. falciparum* and FM3A in vitro: 25 nM and 8.0 μM, respectively [1]

EC_{50} values for N-251 against against *P. falciparum* and FM3A in vitro: 23 nM and 8.0 μM, respectively [3]

K_d values for N-89 and N-251 to recombinant *Pf*ERC in HEPES running buffer: 3.8 and 0.16 mM, respectively [4]

References

1. Kim HS, Nagai Y, Ono K, Begum K, Wataya Y, Hamada Y, Tsuchiya K, Masuyama A, Nojima M, McCullough KJ (2001) Synthesis and antimalarial activity of novel medium-sized 1,2,4,5-tetraoxacycloalkanes. J Med Chem 44:2357–2361
2. Sato A, Kawai S, Hiramoto A, Morita M, Tanigawa N, Nakase Y, Komichi Y, Matsumoto M, Hiraoka O, Hiramoto K, Tokuhara H, Masuyama A, Nojima M, Higaki K, Hayatsu H, Wataya Y, Kim HS (2011) Antimalarial activity of 6-(1,2,6,7-tetraoxaspiro[7.11]nonadec-4-yl)hexan-1-ol (N-251) and its carboxylic acid derivatives. Parasitol Int 60:488–492
3. Sato A, Hiramoto A, Morita M, Matsumoto M, Komich Y, Nakase Y, Tanigawa N, Hiraoka O, Hiramoto K, Hayatsu H, Higaki K, Kawai S, Masuyama A, Nojima M, Wataya Y, Kim HS (2011) Antimalarial activity of endoperoxide compound 6-(1,2,6,7-tetraoxaspiro[7.11]nonadec-4-yl)hexan-1-ol. Parasitol Int 60:270–273
4. Morita M, Sanai H, Hiramoto A, Sato A, Hiraoka O, Sakura T, Kaneko O, Masuyama A, Nojima M, Wataya Y, Kim HS (2012) *Plasmodium falciparum* endoplasmic reticulum-resident calcium binding protein is a possible target of synthetic antimalarial endoperoxides, N-89 and N-251. J Proteome Res 11:5704–5711

Necrostatin-1*

Keywords: [Necroptosis inducer] [RIP1 kinase inhibitor]

Structure

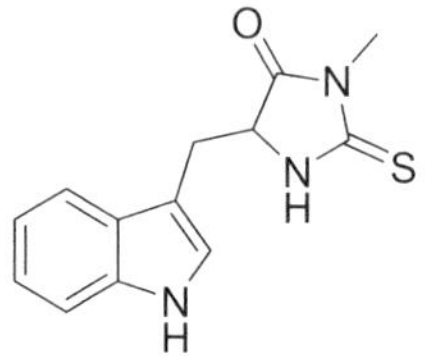

	Necrostatin-1
Molecular formula	$C_{13}H_{13}N_3OS$
Molecular weight	259.33
Solubility	DMSO, +++; H_2O, –; MeOH, –

Discovery/Isolation

Necroptosis is characterized by necrotic cell death morphology and activation of autophagy when undergoing death domain receptor–induced cell death without caspase activation. Necrostatin-1 was identified by screening of inhibitors of the necroptosis of human monocytic U937 cells induced by tumor necrosis factor-α (TNFα) and z-VAD-fmk from a chemical library of ~15,000 chemicals [1].

Biological Studies

Necrostatin-1 inhibited TNFα-stimulated necrosis, but not FasL-cycloheximide-induced apoptosis of FADD-deficinent Jurkat cells [1]. Interestingly, necrostatin-1 failed to inhibit oxidative stress-induced necrosis, suggesting that this compound is a specific inhibitor of necroptosis, and does not act as an antioxidant. An in vitro kinase assay revealed that necrostatin-1 directly inhibits receptor interacting Ser/Thr protein kinase 1 (RIP1) [2]. The crystal structure of necrostatin-1 and RIP1 showed that necrostatin-1 stabilizes RIP1 in an inactive comformation by binding to the activation loop [3].

Necrostatin-1 has been widely used for investigations of necroptosis and RIP kinase–mediated cellular processes [4–7].

Biological Activity of Necrostatin-1

EC_{50} for necroptosis against FADD-deficient Jurkat cells: 494 ± 125 nM [1]
EC_{50} for endogenous RIP1 kinase from Jurkat cells: 182 nM [2]

References

1. Degterev A, Huang Z, Boyce M, Li Y, Jagtap P, Mizushima N, Cuny GD, Mitchison TJ, Moskowitz MA, Yuan J (2005) Chemical inhibitor of nonapoptotic cell death with therapeutic potential for ischemic brain injury. Nat Chem Biol 1:112–119
2. Degterev A, Hitomi J, Germscheid M, Ch'en IL, Korkina O, Teng X, Abbott D, Cuny GD, Yuan C, Wagner G, Hedrick SM, Gerber SA, Lugovskoy A, Yuan J (2008) Identification of RIP1 kinase as a specific cellular target of necrostatins. Nat Chem Biol 4:313–321
3. Xie T, Peng W, Liu Y, Yan C, Maki J, Degterev A, Yuan J, Shi Y (2013) Structural basis of RIP1 inhibition by necrostatins. Structure 21:493–499
4. Trichonas G, Murakami Y, Thanos A, Morizane Y, Kayama M, Debouck CM, Hisatomi T, Miller JW, Vavvas DG (2010) Receptor interacting protein kinases mediate retinal detachment-induced photoreceptor necrosis and compensate for inhibition of apoptosis. Proc Natl Acad Sci U S A 107:21695–21700
5. Motani K, Kushiyama H, Imamura R, Kinoshita T, Nishiuchi T, Suda T (2011) Caspase-1 protein induces apoptosis-associated speck-like protein containing a caspase recruitment domain (ASC)-mediated necrosis independently of its catalytic activity. J Biol Chem 286:33963–33972
6. Linkermann A, Brasen JH, Darding M, Jin MK, Sanz AB, Heller JO, De Zen F, Weinlich R, Ortiz A, Walczak H, Weinberg JM, Green DR, Kunzendorf U, Krautwald S (2013) Two independent pathways of regulated necrosis mediate ischemia-reperfusion injury. Proc Natl Acad Sci U S A 110:12024–12029
7. Nomura M, Ueno A, Saga K, Fukuzawa M, Kaneda Y (2014) Accumulation of cytosolic calcium induces necroptotic cell death in human neuroblastoma. Cancer Res 74:1056–1066

NITD609* (Spiroindolone)

Keywords: [Antimalarial] [PfATP4 inhibitor]

Structure

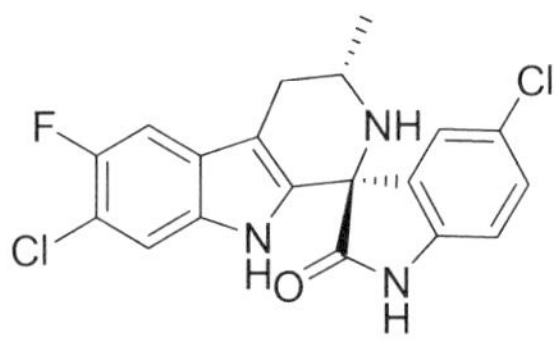

	NITD609
Molecular formula	$C_{19}H_{14}Cl_2FN_3O$
Molecular weight	390.24
Solubility	DMSO, ++; H_2O, ? ; MeOH, ++

Discovery/Isolation/Chemical Study

NITD609 is a synthetic spiroindolone compound optimized from a spiroazepineindole class compound showing antimalarial activity against *Plasmodium falciparum* [1, 2]. NITD609 has good drug-like attributes and physicochemical properties.

Biological Studies

NITD609 showed antimalarial blood-stage activity with no evidence of diminished potency against drug-resistant strains [1]. All stages (rings, trophozoites, and schizonts) were similarly sensitive to NITD609, suggesting that the target molecule is present at all asexual blood stages. Furthermore NITD609 inhibited protein synthesis in *P. falciparum*, suggesting that the mechanism of action is different from those of artemisinin and mefloquine. Mutational analysis, which shows NITD609 resistance, revealed that P-type cation-transporter ATPase4 (PfATP4) mediates the spiroindolone resistance. However, PfATP4 has a high level homology with *Saccharomyces cerevisiae PMR1*, a P-type ATPase required for high-affinity Ca^{2+} and Mn^{2+} transport; this protein of the intraerythrocytic stage actively extrudes Na^{+} against an inward gradient [3]. Therefore, spiroindolones causes a profound disruption of Na^{+} homeostasis in *P. falciparum*.

Biological Activity of NITD609

IC_{50} for antimalarial acitivty against blood-stage *P. falciparum*: 0.5–1.4 nM [1]

References

1. Rottmann M, McNamara C, Yeung BK, Lee MC, Zou B, Russell B, Seitz P, Plouffe DM, Dharia NV, Tan J, Cohen SB, Spencer KR, Gonzalez-Paez GE, Lakshminarayana SB, Goh A, Suwanarusk R, Jegla T, Schmitt EK, Beck HP, Brun R, Nosten F, Renia L, Dartois V, Keller TH, Fidock DA, Winzeler EA, Diagana TT (2010) Spiroindolones, a potent compound class for the treatment of malaria. Science 329:1175–1180
2. Yeung BK, Zou B, Rottmann M, Lakshminarayana SB, Ang SH, Leong SY, Tan J, Wong J, Keller-Maerki S, Fischli C, Goh A, Schmitt EK, Krastel P, Francotte E, Kuhen K, Plouffe D, Henson K, Wagner T, Winzeler EA, Petersen F, Brun R, Dartois V, Diagana TT, Keller TH (2010) Spirotetrahydro β-carbolines (spiroindolones): a new class of potent and orally efficacious compounds for the treatment of malaria. J Med Chem 53:5155–5164
3. Spillman NJ, Allen RJ, McNamara CW, Yeung BK, Winzeler EA, Diagana TT, Kirk K (2013) Na^+ regulation in the malaria parasite *Plasmodium falciparum* involves the cation ATPase PfATP4 and is a target of the spiroindolone antimalarials. Cell Host Microbe 13:227–237

Papuamide A and B

Keywords: [Phosphatidylserine binder] [HIV-1 entry inhibitor]

Structure

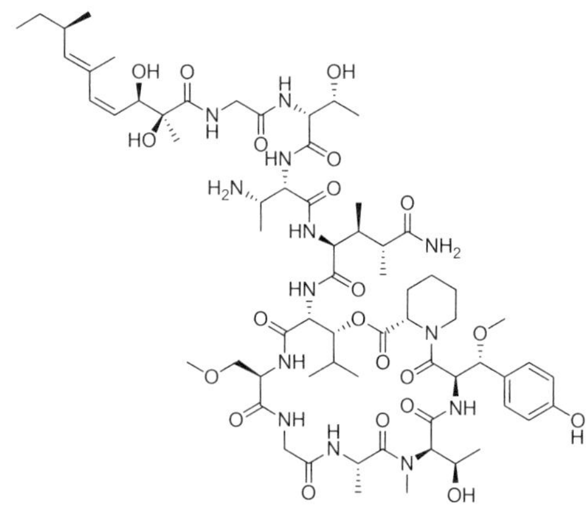

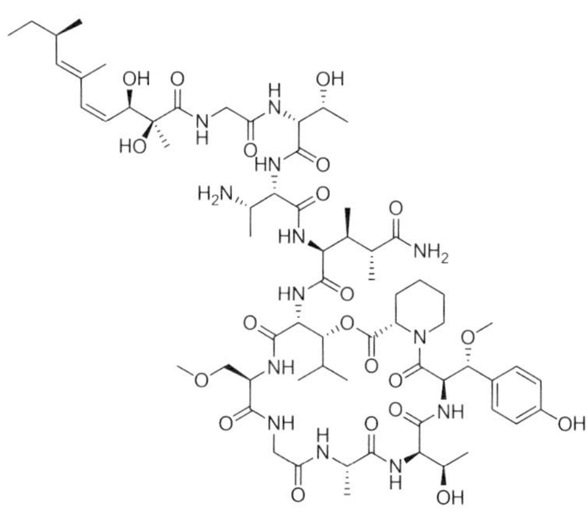

	Papuamide A	Papuamide B
Molecular formula	$C_{66}H_{105}N_{13}O_{21}$	$C_{65}H_{103}N_{13}O_{21}$
Molecular weight	1416.61	1402.59
Solubility	DMSO, ++; H_2O, ? ; MeOH, +++	DMSO, ++; H_2O, ? ; MeOH, +++

Discovery/Isolation/Chemical Studies

Papuamides A and B are cyclic depsipeptides isolated from a Papua New Guinea collection of the sponges *Theonella mirabilis* and *Theonella swinhoei*, as

HIV-inhibitory and cytotoxic compounds [1]. The structures were determined by a combination of spectroscopic analysis, chemical degradation, and derivatization studies.

Biological Studies
Pupuamide A inhibited the infection of human T-lymphoblastoid cells by HIV-1 in vitro [1] by inhibiting HIV-1 entry into host cells [2]. Yeast chemical–genetic profiling, resistant mutant analysis, and an in vitro marker release experiment using liposomes suggested that papuamide B targets phosphatidylserine [3]. Therefore, papuamide B has been used as a probe for investigating the function of phosphatidylserine in a plasma membrane [4, 5].

Biological Activity of Papuamide A
EC_{50} for HIV-1 infection to human T-lymphoblastoid cells: 4 ng/ml [1]
Inhibition of HIV-1 entry (approximately 80% inhibition): 710 nM [2]

References

1. Ford PW, Gustafson KR, McKee TC, Shigematsu N, Maurizi LK, Pannell LK, Williams DE, de Silva ED, Lassota P, Allen TM, Van Soest R, Andersen RJ, Boyd MR (1999) Papuamides A-D, HIV-inhibitory and cytotoxic depsipeptides from the sponges *Theonella mirabilis* and *Theonella swinhoei* collected in Papua New Guinea. J Am Chem Soc 121:5899–5909
2. Andjelic CD, Planelles V, Barrows LR (2008) Characterizing the anti-HIV activity of papuamide A. Mar Drugs 6:528–549
3. Parsons AB, Lopez A, Givoni IE, Williams DE, Gray CA, Porter J, Chua G, Sopko R, Brost RL, Ho CH, Wang JY, Ketela T, Brenner C, Brill JA, Fernandez GE, Lorenz TC, Payne GS, Ishihara S, Ohya Y, Andrews B, Hughes TR, Frey BJ, Graham TR, Andersen RJ, Boone C (2006) Exploring the mode-of-action of bioactive compounds by chemical-genetic profiling in yeast. Cell 126:611–625
4. Chen S, Wang J, Muthusamy BP, Liu K, Zare S, Andersen RJ, Graham TR (2006) Roles for the Drs2p-Cdc50p complex in protein transport and phosphatidylserine asymmetry of the yeast plasma membrane. Traffic 7:1503–151
5. Georgiev AG, Johansen J, Ramanathan VD, Sere YY, Beh CT, Menon AK (2013) Arv1 regulates pm and er membrane structure and homeostasis but is dispensable for intracellular sterol transport. Traffic 14:912–921

Pateamine A

Keywords: [Eukaryotic initiation factor 4A (eIF4A) activator] [Nonsense-mediated mRNA decay (NMD) inhibitior]

Structure

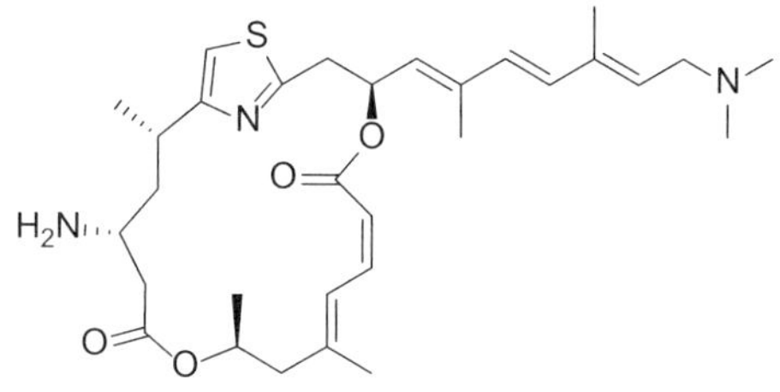

	Pateamine A
Molecular formula	$C_{31}H_{45}N_3O_4S$
Molecular weight	555.78
Solubility	DMSO, ? ; H_2O, ? ; MeOH, ++

Discovery/Isolation/Chemical Studies

Pateamine A, a sulfur-containing heteroaromatic macrolide, was isolated as an antifungal and cytotoxic compound from the sponge *Mycale* sp. [1].

Biological Studies

Pateamine A was reidentified as an inhibitor of eukaryotic translation initiation by high-throughput screening [2]. The pateamine A binding protein was identified to be eIF4AI by an affinity purification method using a matrix containing immobilized pateamine A [2] or biotinylated pateamine A [3]. eIF4A is an RNA- and ATP-dependent helicase that unwinds a local secondary structure in mRNA to allow binding to the 43S ribosomal complex to form the 48S ribosomal complex. Pateamine A bound to and enhanced the ATPase activity of eIF4A [2, 3]. Furthermore, pateamine A inhibited the interaction between eIF4A and eIF4G, then stabilized the ternary complex between eIF4A and eIF4B [4, 3]. These results suggest that pateamine A prevents eIF4A from participating in the recruitment of ribosomes.

Using recombinant protein, it was shown that pateamine A also binds to eIF4AIII which involves a nonsense-mediated mRNA decay (NMD) pathway [2]. Indeed, pateamine A inhibited NMD through direct interaction with eIF4AIII, which is independent of its inhibitory activity of translation initiation [5].

In cells, pateamine A induces stress granules, which are aggregates of small ribosomal subunits, mRNA, and numerous associated RNA binding proteins, by different mechanisms of arsenite [6]. Recently, it has been reported that pateamine

A prevents cachexia-induced muscle wasting in a nontoxic dose in mice [7]. Interestingly, a low dose of pateamine A selectively inhibits the translation of procachectic factors such as iNOS by recruitment of iNOS mRNA to stress granules. These results suggest that pateamine A could be a useful medicine for treatment of cachexia.

Biological Activity of Pateamine A
IC_{50} for P388 murine cells: 0.15 ng/ml [1]
IC_{50} for BSC cells: ~300 ng/ml [1]
Stress granule formation in HeLa cells: 50 nM [6]
Muscle differentiation of C2C12 cells: 12.5 nM [7]
Suppression of tumor-induced muscle wasting in BALB/c mice injected with C26 adenocarcinoma cells: 20 μg/kg [7]

References

1. Northcote PT, Blunt JW, Munro MHG (1991) Pateamine – a potent cytotoxin from the new-zealand marine sponge, *Mycale* sp. Tetrahedron Lett 32:6411–6414
2. Bordeleau ME, Matthews J, Wojnar JM, Lindqvist L, Novac O, Jankowsky E, Sonenberg N, Northcote P, Teesdale-Spittle P, Pelletier J (2005) Stimulation of mammalian translation initiation factor eIF4A activity by a small molecule inhibitor of eukaryotic translation. Proc Natl Acad Sci U S A 102:10460–10465
3. Low WK, Dang Y, Schneider-Poetsch T, Shi Z, Choi NS, Merrick WC, Romo D, Liu JO (2005) Inhibition of eukaryotic translation initiation by the marine natural product pateamine A. Mol Cell 20:709–722
4. Bordeleau ME, Cencic R, Lindqvist L, Oberer M, Northcote P, Wagner G, Pelletier J (2006) RNA-mediated sequestration of the RNA helicase eIF4A by pateamine A inhibits translation initiation. Chem Biol 13:1287–1295
5. Dang Y, Low WK, Xu J, Gehring NH, Dietz HC, Romo D, Liu JO (2009) Inhibition of nonsense-mediated mRNA decay by the natural product pateamine A through eukaryotic initiation factor 4AIII. J Biol Chem 284:23613–23621
6. Dang Y, Kedersha N, Low WK, Romo D, Gorospe M, Kaufman R, Anderson P, Liu JO (2006) Eukaryotic initiation factor 2α-independent pathway of stress granule induction by the natural product pateamine A. J Biol Chem 281:32870–32878
7. Di Marco S, Cammas A, Lian XJ, Kovacs EN, Ma JF, Hall DT, Mazroui R, Richardson J, Pelletier J, Gallouzi IE (2012) The translation inhibitor pateamine A prevents cachexia-induced muscle wasting in mice. Nat Commun 3:896

2-Phenylethynesulfonamide* (PES, Pifithrin-μ)

Keywords: [Heat shock protein 70 (HSP70) inhibitor] [p53-dependent apoptosis inhibitor]

Structure

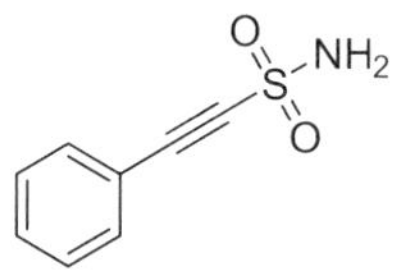

	PES/PFTμ
Molecular formula	$C_8H_7NO_2S$
Molecular weight	181.21
Solubility	DMSO, ++; H_2O, –; MeOH, ++

Discovery/Isolation

2-phenylethynesulfonamide (PES, phenylacetylenylsulfonamide, Pifithrin-μ (PFTμ)) was identified as an inhibitor of apoptosis induced in Saos-2 cells by lentiviral transduction of p53 [1].

Biological Studies

PES inhibited p53-dependent apoptosis by reducing the amount of mitochondria-bound p53 and the binding of p53 to both Bcl-xL and Bcl-2 proteins [1]. Leu et al. identified HPS70 as a PES binding protein using thiol-cleavable biotinylated PES [2]. PES interacted with the carboxy-terminal substrate binding domain, and inhibited the chaperone function of HSP70 by disrupting the association between HSP70 and several of its co-chaperones and substrate proteins. Furthermore, PES treatment induced dysfunction of the autophagy–lysosome system and the proteasome pathway [2, 3]. These results suggest that HSP70 is a crucial factor for the cancer survival pathway and that targeting HSP70 is a potential approach for cancer therapeutics [3, 4].

Prevention of lipopolysaccharide (LPS)–induced liver injury and reduction of proinflammatory factor induction by disruption of the sodium–hydrogen antiporter 1 (NHE1)–HSP70 interaction was also reported [5].

Biological Activity of 2-Phenylethynesulfonamide

Inhibition of p53-dependent cell death: 10 μM [1]

IC_{50} in approximately 50 solid tumor cell lines: 4–10 μM [3]

References

1. Strom E, Sathe S, Komarov PG, Chernova OB, Pavlovska I, Shyshynova I, Bosykh DA, Burdelya LG, Macklis RM, Skaliter R, Komarova EA, Gudkov AV (2006) Small-molecule inhibitor of p53 binding to mitochondria protects mice from gamma radiation. Nat Chem Biol 2:474–479
2. Leu JI, Pimkina J, Frank A, Murphy ME, George DL (2009) A small molecule inhibitor of inducible heat shock protein 70. Mol Cell 36:15–27
3. Leu JI, Pimkina J, Pandey P, Murphy ME, George DL (2011) HSP70 inhibition by the small-molecule 2-phenylethynesulfonamide impairs protein clearance pathways in tumor cells. Mol Cancer Res 9:936–947
4. Balaburski GM, Leu JI, Beeharry N, Hayik S, Andrake MD, Zhang G, Herlyn M, Villanueva J, Dunbrack RL, Jr., Yen T, George DL, Murphy ME (2013) A modified HSP70 inhibitor shows broad activity as an anticancer agent. Mol Cancer Res 11:219–229
5. Huang C, Wang J, Chen Z, Wang Y, Zhang W (2013) 2-phenylethynesulfonamide Prevents Induction of Pro-inflammatory Factors and Attenuates LPS-induced Liver Injury by Targeting NHE1-Hsp70 Complex in Mice. PLoS One 8:e67582

Phoslactomycin A and F

Keywords: [Protein phosphatase type 2A (PP2A) inhibitor]

Structure

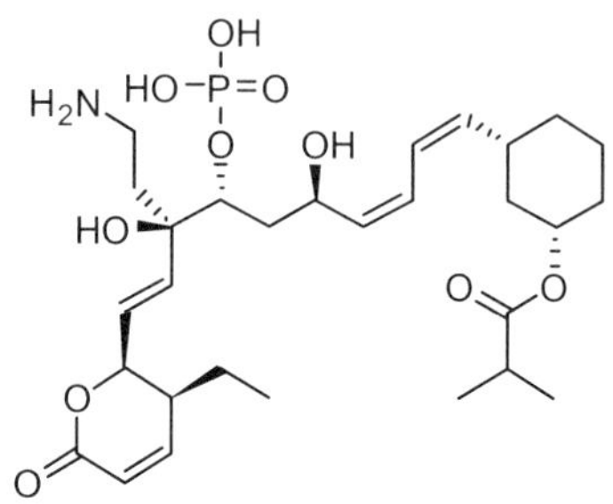

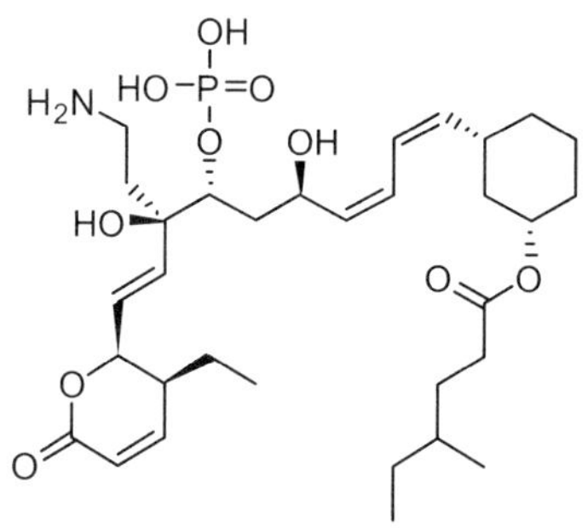

	Phoslactomycin A	Phoslactomycin F
Molecular formula	$C_{29}H_{46}NO_{10}P$	$C_{32}H_{52}NO_{10}P$
Molecular weight	599.65	641.73
Solubility	DMSO, ++; H_2O, +; MeOH, ++	DMSO, ++; H_2O, +; MeOH, ++

Discovery/Isolation/Chemical Studies

Phoslactomycins were originally isolated as antifungal antibiotics from soil *Streptomyces* bacteria species [1, 2]. Several related compounds, phosphazomycins [3, 4], phospholine [5, 6], and leustroducsins [7, 8], were reported as antifungal antibiotics, or as inducers of a colony-stimulating factor in bone marrow stromal cells. These compounds commonly contain not only an α,β-unsaturated δ-lactone, but also an amino group, a phosphate ester, and a cyclohexane ring, and have a different substituents bound to the cyclohexane ring.

Biological Studies

Phoslactomycin F reversibly induces actin filament depolymerization in murine fibroblast NIH/3T3 cells, but did not inhibit actin polymerization in vitro [9]. An in situ and in vitro phosphorylation assay revealed that phoslactomycin F stimulates the phosphorylation of intracellular vimentin by the inhibition of the protein phosphatase type-2A (PP2A). These results suggest that phoslactomycins are PP2A-specific inhibitors and that PP2A is involved in regulation of the organization of the actin cytoskeleton. Because phoslactomycins contain α,β-unsaturated δ-lactone, it has been thought that these compounds covalently bind to their target molecule. Teruya et al. synthesized biotinylated phoslactomycin A and investigated the binding proteins in mouse liver extract [10]. Three proteins to which biotinylated phoslactomycin A covalently binds were identified as PP2A family proteins: PP2Ac (a catalytic subunit of PP2A), PP4, and PP6. Furthermore, the binding site of phoslactomycins on PP2Ac was identified as Cys269 which is located at the entrance of the catalytic site. Interestingly, this cysteine residue is conserved in PP2A family proteins but not even in PP1, the most similar phosphatase. These results strongly suggest that Cys269 is a PP2A-specific gatekeeper residue, and that compounds targeting this residue are expected to be PP2A-specific inhibitors.

Biological Activity of Phoslactomycin F

Disassembly of actin cytoskeleton in NIH/3T3 cells: 10 μM [9]
IC_{50} for PP2A: 4.7 μM [9]

References

1. Fushimi S, Furihata K, Seto H (1989) Studies on new phosphate ester antifungal antibiotics phoslactomycins. II. Structure elucidation of phoslactomycins A to F. J Antibiot (Tokyo) 42:1026–1036
2. Fushimi S, Nishikawa S, Shimazu A, Seto H (1989) Studies on new phosphate ester antifungal antibiotics phoslactomycins. I. Taxonomy, fermentation, purification and biological activities. J Antibiot (Tokyo) 42:1019–1025
3. Tomiya T, Uramoto M, Isono K (1990) Isolation and structure of phosphazomycin C. J Antibiot (Tokyo) 43:118–121
4. Uramoto M, Shen YC, Takizawa N, Kusakabe H, Isono K (1985) A new antifungal antibiotic, phosphazomycin A. J Antibiot (Tokyo) 38:665–668
5. Ozasa T, Tanaka K, Sasamata M, Kaniwa H, Shimizu M, Matsumoto H, Iwanami M (1989) Novel antitumor antibiotic phospholine. 2. Structure determination. J Antibiot (Tokyo) 42:1339–1343
6. Ozasa T, Suzuki K, Sasamata M, Tanaka K, Kobori M, Kadota S, Nagai K, Saito T, Watanabe S, Iwanami M (1989) Novel antitumor antibiotic phospholine. 1. Production, isolation and characterization. J Antibiot (Tokyo) 42:1331–1338

7. Kohama T, Nakamura T, Kinoshita T, Kaneko I, Shiraishi A (1993) Novel microbial metabolites of the phoslactomycins family induce production of colony-stimulating factors by bone marrow stromal cells. II. Isolation, physico-chemical properties and structure determination. J Antibiot (Tokyo) 46:1512–1519
8. Kohama T, Enokita R, Okazaki T, Miyaoka H, Torikata A, Inukai M, Kaneko I, Kagasaki T, Sakaida Y, Satoh A, Shiraishi A (1993) Novel microbial metabolites of the phoslactomycins family induce production of colony-stimulating factors by bone marrow stromal cells. I. Taxonomy, fermentation and biological properties. J Antibiot (Tokyo) 46:1503–1511
9. Usui T, Marriott G, Inagaki M, Swarup G, Osada H (1999) Protein phosphatase 2A inhibitors, phoslactomycins. Effects on the cytoskeleton in NIH/3T3 cells. J Biochem 125:960–965
10. Teruya T, Simizu S, Kanoh N, Osada H (2005) Phoslactomycin targets cysteine-269 of the protein phosphatase 2A catalytic subunit in cells. FEBS Lett 579:2463–2468

Piperlongumine

Keywords: [Antitumor] [Glutathione-*S*-transferase P1 (GSTP1) inhibitor]

Structure

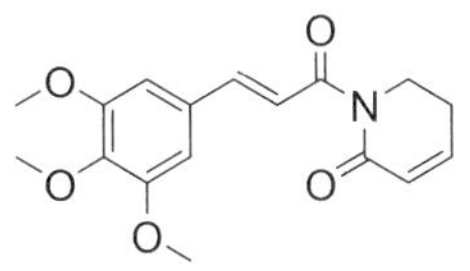

	Piperlongumine
Molecular formula	$C_{17}H_{19}NO_5$
Molecular weight	317.34
Solubility	DMSO, +++; H_2O, −; MeOH, ++

Discovery/Isolation/Chemical Studies

Piperlongumine was originally isolated from the root of *Piper longum* Linn [1]. This compound was screened by a cell-based and quantitative proteomics approach and showed a cancer-cell-selective killing property [2]. SAR analysis suggested that the electrophilicity of the C2-C3 olefin of piperlogumine is critical for selective toxicity against cancer cells [3].

Biological Studies

Piperlongumine shows several biological activities including insecticidal activity [4], platelet aggregation inhibition [5], and antitumor activity [6]. Piperlongumine increases the level of reactive oxygen species and apoptotic cell death in cells with a cancer genotype, but have little effect on primary normal cells. Furthermore, a combination method of affinity enrichment with stable-isotope labeling and quantitative proteomics revealed that GSTP1 is the target molecule of piperlongumine, suggesting that piperlongumine induces apoptosis by interfering with redox and GSTP1 reactive oxygen species (ROS) homeostatic regulators. Further detailed analysis using synthetic analogs suggested that piperlongumine possesses two sites of reactivity, one with great and the other with lesser electrophilicity [3]. Great electrophilicity is required for covalent binding with glutathione, and less is probably required for cellular cross-linking with proteins, which is associated with toxicity. Therefore, both electrophilic and oxidative stress phenotypes can contribute to cancer-selective toxicity of piperlongumine.

Inhibitory activity of piperlongumine on atherosclerotic plaque formation and vascular smooth muscle cell proliferation has also been reported [7].

Biological Activity of Piperlongumine

IC_{50} for cell proliferation of HL-60, HCT-8, SF295, and MDA-MB-435: >25 μg/ml [6]

Cellular ROS production in EJ, MDA-MB-231, U2OS, and MDA-MB-435: 10 μM [2]
EC_{50} values for cellular ATP level in H1703 and HeLa cells: 2.8 and 7.1 μM, respectively [3]

References

1. Chatterjee A, Dutta CP (1967) Alkaloids of *Piper longum* Linn–I: Structure and synthesis of piperlongumine and piperlonguminine. Tetrahedron 23:1769–1781
2. Raj L, Ide T, Gurkar AU, Foley M, Schenone M, Li X, Tolliday NJ, Golub TR, Carr SA, Shamji AF, Stern AM, Mandinova A, Schreiber SL, Lee SW (2011) Selective killing of cancer cells by a small molecule targeting the stress response to ROS. Nature 475:231–234
3. Adams DJ, Dai M, Pellegrino G, Wagner BK, Stern AM, Shamji AF, Schreiber SL (2012) Synthesis, cellular evaluation, and mechanism of action of piperlongumine analogs. Proc Natl Acad Sci U S A 109:15115–15120
4. Bernard CB, Krishanmurty HG, Chauret D, Durst T, Philogene BJR, Sanchez-Vindas P, Hasbun C, Poveda L, San Roman L, Arnason JT (1995) Insecticidal defenses of Piperaceae from the neotropics. J Chem Ecol 21:801–814
5. Iwashita M, Oka N, Ohkubo S, Saito M, Nakahata N (2007) Piperlongumine, a constituent of *Piper longum* L., inhibits rabbit platelet aggregation as a thromboxane A2 receptor antagonist. Eur J Pharmacol 570:38–42
6. Bezerra DP, Pessoa C, Moraes MO, Alencar NM, Mesquita RO, Lima MW, Alves AP, Pessoa OD, Chaves JH, Silveira ER, Costa-Lotufo LV (2008) In vivo growth inhibition of sarcoma 180 by piperlonguminine, an alkaloid amide from the *Piper* species. J Appl Toxicol 28:599–607
7. Son DJ, Kim SY, Han SS, Kim CW, Kumar S, Park BS, Lee SE, Yun YP, Jo H, Park YH (2012) Piperlongumine inhibits atherosclerotic plaque formation and vascular smooth muscle cell proliferation by suppressing PDGF receptor signaling. Biochem Biophys Res Commun 427:349–354

Pironetin

Keywords: [Tubulin polymerization inhibitor]

Structure

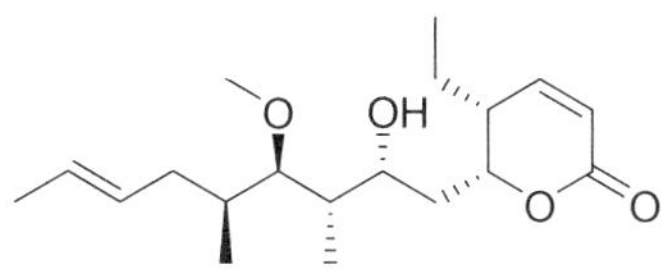

	Pironetin
Molecular formula	$C_{19}H_{32}O_4$
Molecular weight	324.46
Solubility	DMSO, +++; H_2O, –; MeOH, ++

Discovery/Isolation

Pironetin was isolated as a plant growth regulator which induced shortening of the plant height of rice [1, 2], and as an immunosuppressor which inhibited the blastogenesis of both T and B cells.

Biological Studies

Pironetin showed antitumor activity against a murine tumor cell line, P388 leukemia, transplanted in mice [3], and induced apoptosis via disruption of the cellular microtubule network in situ [4]. The compound inhibited the polymerization of pure tubulin, showing that pironetin is a tubulin inhibitor which inhibits tubulin polymerization by binding to tubulin directly. Binding competition experiments with [^{3}H]colchicine and [^{3}H]vinblastine strongly suggested that pironetin binds on the vinblastine binding site on β-tubulin.

According to structure–activity relationship analyses, it was speculated that pironetin covalently binds to tubulin by Michael addition via its α,β-unsaturated δ-lactone [5]. The biotinylated pironetin (**2**), which inhibited tubulin assembly both in vitro and in situ, covalently bound with tubulin, and its binding was inhibited by natural pironetin [6]. The finding that inactive pironetin derivatives failed to bind to tubulin in situ suggested that covalent binding is important to inhibit tubulin polymerization. Partial proteolytic analyses of biotinylated pironetin–bound peptides, followed by systematic alanine scanning of both cysteine and lysine residues, strongly suggested that Lys352 of α-tubulin is the pironetin binding site. This is surprising, because the reactivity of the ε-amino group of lysine is quite low compared with that of the sulfhydryl group of cysteine. The in silico analysis of the pironetin–tubulin complex suggested that α helixes 8/10 and β sheet 8 form a cavity that fits pironetin well, and that pironetin binding with Lys352 and Asn258 disrupts the hydrogen network among Glu254, Lys352, and Asn258. Discrepancy with [^{3}H]-vinblastine competition experiments were explained by the structure of vinblastine bound to the tubulin–colchicine:RB3-SLD complex [7, 8]. Gigant et al. revealed that vinblastine binds not only to β-tubulin but also on a hydrophobic groove on the α-tubulin surface that is located at an intermolecular contact in microtubules, and pironetin binding residue, Lys352, is part of the vinblastine binding site. These results strongly suggest that the binding sites of pironetin and vinblastine are overlapped, and that these drugs turn out each other.

Biological Activity of Pironetin

23% inhibition of growth of rice plants without any loss of crop yield 9 days before heading: 10 g/a [1]

Complete inhibition of cell proliferation of 3Y1 cells: 10–20 ng/ml [3]

Antiproliferative effects in several tumor cell lines: 5–25 ng/ml [3]

References

1. Kobayashi S, Tsuchiya K, Harada T, Nishide M, Kurokawa T, Nakagawa T, Shimada N, Kobayashi K (1994) Pironetin, a novel plant growth regulator produced by *Streptomyces* sp. NK10958. I. Taxonomy, production, isolation and preliminary characterization. J Antibiot (Tokyo) 47:697–702
2. Kobayashi S, Tsuchiya K, Kurokawa T, Nakagawa T, Shimada N, Iitaka Y (1994) Pironetin, a novel plant growth regulator produced by *Streptomyces* sp. NK10958. II. Structural elucidation. J Antibiot (Tokyo) 47:703–707
3. Kondoh M, Usui T, Kobayashi S, Tsuchiya K, Nishikawa K, Nishikiori T, Mayumi T, Osada H (1998) Cell cycle arrest and antitumor activity of pironetin and its derivatives. Cancer Lett 126:29–32
4. Kondoh M, Usui T, Nishikiori T, Mayumi T, Osada H (1999) Apoptosis induction via microtubule disassembly by an antitumour compound, pironetin. Biochem J 340:411–416
5. Watanabe H, Watanabe H, Usui T, Kondoh M, Osada H, Kitahara T (2000) Synthesis of pironetin and related analogs: studies on structure-activity relationships as tubulin assembly inhibitors. J Antibiot (Tokyo) 53:540–545
6. Usui T, Watanabe H, Nakayama H, Tada Y, Kanoh N, Kondoh M, Asao T, Takio K, Watanabe H, Nishikawa K, Kitahara T, Osada H (2004) The anticancer natural product pironetin selectively targets Lys352 of α-tubulin. Chem Biol 11:799–806
7. Gigant B, Cormier A, Dorleans A, Ravelli RBG, Knossow M (2009) Microtubule-destabilizing agents: structural and mechanistic insights from the interaction of colchicine and vinblastine with tubulin. Top Curr Chem 286:259–278
8. Gigant B, Wang C, Ravelli RBG, Roussi F, Steinmetz MO, Curmi PA, Sobel A, Knossow M (2005) Structural basis for the regulation of tubulin by vinblastine. Nature 435:519–522

Pladienolide B

Keywords: [Antitumor] [Splicing inhibitor] [SF3b]

Structure

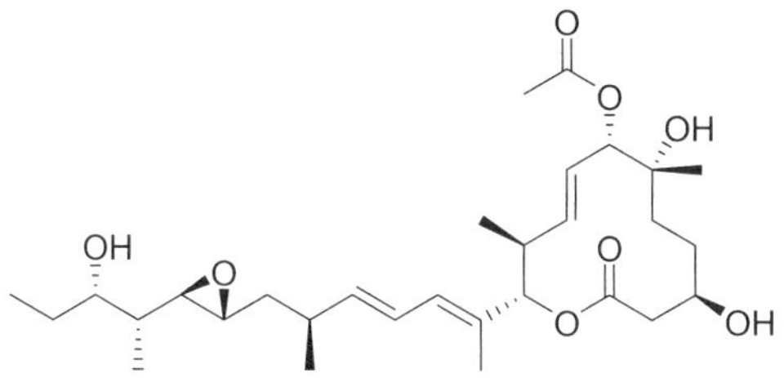

	Pladienolide B
Molecular formula	$C_{30}H_{48}O_8$
Molecular weight	536.70
Solubility	DMSO, +++; H_2O, ±; MeOH, +++

Discovery/Isolation/Chemical Studies
Pladienolide B is a 12-member macrolide purified from *Streptomyces platensis* Mer-11107, possessing inhibitory activity against vascular endothelial growth factor (VEGF) expression and cancer cell proliferation [1]. Absolute configurations at 10 chiral centers were determined on the basis of spectral data of pladienolide B and its chemical transformation products [2]. Total synthesis has done by Kanada et al. [3].

Biological Studies
Pladienolide B potently inhibited hypoxia-induced VEGF expression and proliferation of the U251 cancer cell line with strong antitumor activities in vitro and in vivo [4]. Using ^{3}H-labeled, fluorescence-tagged and photoaffinity/biotin (PB)–tagged "chemical probes," the target protein of pladienolide was determined as a 140 kDa protein in splicing factor SF3b [5]. Pladienolide B impaired in vivo splicing in a dose-dependent manner. The pladienolide-resistant clones have a mutation at Arg1074 in the gene for SF3B1, which encodes a subunit of the SF3b complex and impaired the binding of pladienolide [6]. These results clearly demonstrate that pladienolide exerts its potent activity by targeting SF3b and also suggest that inhibition of SF3b is a promising drug target for anticancer therapy.

Biological Activity of Pladienolide B
IC_{50} values for anti-VEGF-PLAP and antiproliferative activities in U251 cells: 1.8 and 3.5 nM, respectively [7]
Inhibition of splicing of *DNAJB1* and *RIOK3* in HeLa cells for 4 hours: 10 nM [5]

References

1. Sakai T, Asai N, Okuda A, Kawamura N, Mizui Y (2004) Pladienolides, new substances from culture of *Streptomyces platensis* Mer-11107. II. Physico-chemical properties and structure elucidation. J Antibiot (Tokyo) 57:180–187
2. Asai N, Kotake Y, Niijima J, Fukuda Y, Uehara T, Sakai T (2007) Stereochemistry of pladienolide B. J Antibiot (Tokyo) 60:364–369
3. Kanada RM, Itoh D, Nagai M, Niijima J, Asai N, Mizui Y, Abe S, Kotake Y (2007) Total synthesis of the potent antitumor macrolides pladienolide B and D. Angew Chem Int Ed Eng 46:4350–4355
4. Mizui Y, Sakai T, Iwata M, Uenaka T, Okamoto K, Shimizu H, Yamori T, Yoshimatsu K, Asada M (2004) Pladienolides, new substances from culture of *Streptomyces platensis* Mer-11107 III. In vitro and in vivo antitumor activities . J Antibiot (Tokyo) 57:188–196
5. Kotake Y, Sagane K, Owa T, Mimori-Kiyosue Y, Shimizu H, Uesugi M, Ishihama Y, Iwata M, Mizui Y (2007) Splicing factor SF3b as a target of the antitumor natural product pladienolide. Nat Chem Biol 3:570–575
6. Yokoi A, Kotake Y, Takahashi K, Kadowaki T, Matsumoto Y, Minoshima Y, Sugi NH, Sagane K, Hamaguchi M, Iwata M, Mizui Y (2011) Biological validation that SF3b is a target of the antitumor macrolide pladienolide. FEBS J 278:4870–4880
7. Sakai T, Sameshima T, Matsufuji M, Kawamura N, Dobashi K, Mizui Y (2004) Pladienolides, new substances from culture of *Streptomyces platensis* Mer-11107. I. Taxonomy, fermentation, isolation and screening. J Antibiot (Tokyo) 57:173–179

Prodigiosin 25-C

Keywords: [Antitumor] [Immunosuppressant]

Structure

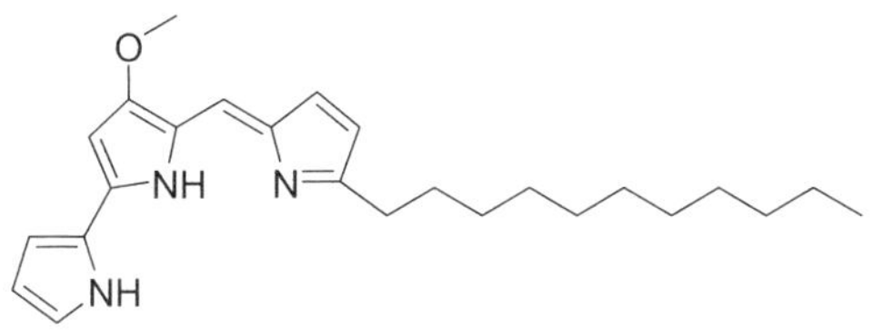

	Prodigiosin 25-C
Molecular formula	$C_{25}H_{35}N_3O$
Molecular weight	393.57
Solubility	DMSO, +++; H_2O, −; MeOH, ++

Discovery/Isolation

Prodigiosins were originally isolated as pigments from *Serratia marcescens* and *Streptomyces* sp. Prodigiosin 25-C was reisolated as an immunomodulating substance showing inhibition of cytotoxic T-cell induction in a mixed lymphocyte reaction [1].

Biological Studies

Prodigiosin 25-C completely inhibited induction of cytotoxic T cells in vitro and in vivo [2], and moderately prolonged survival of MHC-mismatched skin grafts [3]. This compound uncoupled vacuolar H^+-ATPase (V-ATPase), inhibited vacuolar acidification, and affected glycoprotein processing [4]. Further analyses suggested that prodigiosins promote H^+/Cl^- symport (or OH^-/Cl^- exchange, in its equivalence) across vesicular membranes [5]. Because prodigiosin facilitates chloride/bicarbonate exchange (antiport) across liposomal membranes [6], the disturbance of ions endows these immunosuppressive activities.

Suppressive activity on osteoclast bone resorption [7], and antitumor activity by mTOR inhibition [8] and by up-regulating p73 in p53-deficient cells [9] have been reported.

Biological Activity of Prodigiosins:
Prodigiosin 25-C

IC_{50} for proton pump activity in rat liver lysosomes: 30 nM [4]
Inhibition of ATPase activity in rat liver lysosomes: >1 μM [4]

Prodigiosin

IC_{50} for intralysosomal pH through inhibition of lysosomal acidification driven by V-ATPase: 30–120 pmol/mg of protein [10]

Activation of both autophagic and apoptotic mechanisms in SK-MEL-28 and SK-MEL-5 cells: 4.51 ± 0.47 μM and 1.02 ± 0.15 μM, respectively [8]
Induction of p53 target proteins and PARP cleavage in various cancer cells: 1 μM [9]

References

1. Nakamura A, Nagai K, Ando K, Tamura G (1986) Selective suppression by prodigiosin of the mitogenic response of murine splenocytes. J Antibiot (Tokyo) 39:1155–1159
2. Nakamura A, Magae J, Tsuji RF, Yamasaki M, Nagai K (1989) Suppression of cytotoxic T cell induction in vivo by prodigiosin 25-C. Transplantation 47:1013–1016
3. Tsuji RF, Magae J, Yamashita M, Nagai K, Yamasaki M (1992) Immunomodulating properties of prodigiosin 25-C, an antibiotic which preferentially suppresses induction of cytotoxic T cells. J Antibiot (Tokyo) 45:1295–1302
4. Kataoka T, Muroi M, Ohkuma S, Waritani T, Magae J, Takatsuki A, Kondo S, Yamasaki M, Nagai K (1995) Prodigiosin 25-C uncouples vacuolar type H(+)-ATPase, inhibits vacuolar acidification and affects glycoprotein processing. FEBS Lett 359:53–59
5. Sato T, Konno H, Tanaka Y, Kataoka T, Nagai K, Wasserman HH, Ohkuma S (1998) Prodigiosins as a new group of H^+/Cl^- symporters that uncouple proton translocators. J Biol Chem 273:21455–21462
6. Davis JT, Gale PA, Okunola OA, Prados P, Iglesias-Sanchez JC, Torroba T, Quesada R (2009) Using small molecules to facilitate exchange of bicarbonate and chloride anions across liposomal membranes. Nat Chem 1:138–144
7. Woo JT, Ohba Y, Tagami K, Sumitani K, Kataoka T, Nagai K (1997) Prodigiosin 25-C and metacycloprodigiosin suppress the bone resorption by osteoclasts. Biosci Biotechnol Biochem 61:400–402
8. Espona-Fiedler M, Soto-Cerrato V, Hosseini A, Lizcano JM, Guallar V, Quesada R, Gao T, Perez-Tomas R (2012) Identification of dual mTORC1 and mTORC2 inhibitors in melanoma cells: prodigiosin vs. obatoclax. Biochem Pharmacol 83:489–496
9. Hong B, Prabhu VV, Zhang S, van den Heuvel AP, Dicker DT, Kopelovich L, El-Deiry WS (2014) Prodigiosin rescues deficient p53 signaling and antitumor effects via upregulating p73 and disrupting its interaction with mutant p53. Cancer Res 74:1153–1165
10. Ohkuma S, Sato T, Okamoto M, Matsuya H, Arai K, Kataoka T, Nagai K, Wasserman HH (1998) Prodigiosins uncouple lysosomal vacuolar-type ATPase through promotion of H^+/Cl^- symport. Biochem J 334:731–741

Pseudolaric Acid B

Keywords: [Antifungal] [Tubulin inhibitor]

Structure

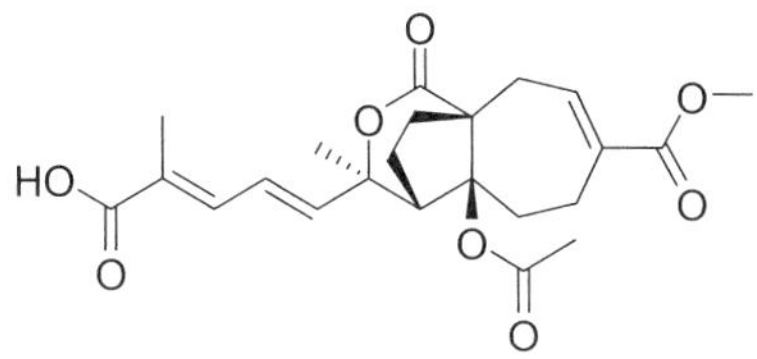

	Pseudolaric acid B
Molecular formula	$C_{23}H_{28}O_8$
Molecular weight	432.47
Solubility	DMSO, ++; H_2O, ±; MeOH, ++

Discovery/Isolation/Chemical Study

Pseudolaric acid B is a diterpene acid isolated from Tu-Jin-Pi, *Pseudolarix kaempferi* Golden (Pinaceae), which has been used in traditional Chinese medicine for skin diseases caused by fungi. The chemical structure and total synthesis have been reported [1, 2].

Biological Studies

Several biological activities of pseudolaric acid B including cytotoxicity against cancer cells [3], antifungal activities [4], antiangiogenesis activity [5], apoptosis induction [6], and anti-inflammatory activity [7] have been reported.

Wong et al. investigated the molecular target of pseudolaric acid B in detail [8]. They found that pseudolaric acid B induced cell cycle arrest at G2/M transition, leading to apoptosis. Furthermore, this compound dose-dependently inhibited not only cellular microtubule networks but also tubulin polymerization in vitro. The binding site of pseudolaric acid B is thought to be the distinct binding site on tubulin compared with those of colchicine and vinblastine [9].

Biological Activity of Pseudolaric Acid B

IC_{50} for VEGF-stimulated migration of HUVECs: 2.92 μM [5]
IC_{50} for FBS-stimulated tube formation of HUVECs: 2.60 μM [5]
IC_{50} for cytotoxicities against several cancer cell lines: ~3 μM [8]
IC_{50} for in vitro microtubule polymerization: 10.9 ± 1.8 μM [8]
Depolymerization of micortubule network in HMEC cells: 1 μM [9]

References

1. Li Z-L, Pan D-J, Chang-Qi H, Qin-Li W, Song-Song Y, Guang-Yi X (1982) Studies on the novel diterpenic constituents of Tu-Jin-Pi. I. Strucutre of pseudolaric acid A and pseudolaric acid B. Acta Chim Sin 05
2. Trost BM, Waser J, Meyer A (2008) Total synthesis of (−)-pseudolaric acid B. J Am Chem Soc 130:16424–16434
3. Pan DJ, Li ZL, Hu CQ, Chen K, Chang JJ, Lee KH (1990) The cytotoxic principles of *Pseudolarix kaempferi*: pseudolaric acid-A and -B and related derivatives. Planta Med 56:383–385
4. Li E, Clark AM, Hufford CD (1995) Antifungal evaluation of pseudolaric acid B, a major constituent of *Pseudolarix kaempferi*. J Nat Prod 58:57–67
5. Li MH, Miao ZH, Tan WF, Yue JM, Zhang C, Lin LP, Zhang XW, Ding J (2004) Pseudolaric acid B inhibits angiogenesis and reduces hypoxia-inducible factor 1α by promoting proteasome-mediated degradation. Clin Cancer Res 10:8266–8274
6. Gong X, Wang M, Wu Z, Tashiro S, Onodera S, Ikejima T (2004) Pseudolaric acid B induces apoptosis via activation of c-Jun N-terminal kinase and caspase-3 in HeLa cells. Exp Mol Med 36:551–556
7. Li T, Wong VK, Yi XQ, Wong YF, Zhou H, Liu L (2009) Pseudolaric acid B suppresses T lymphocyte activation through inhibition of NF-κB signaling pathway and p38 phosphorylation. J Cell Biochem 108:87–95
8. Wong VK, Chiu P, Chung SS, Chow LM, Zhao YZ, Yang BB, Ko BC (2005) Pseudolaric acid B, a novel microtubule-destabilizing agent that circumvents multidrug resistance phenotype and exhibits antitumor activity in vivo. Clin Cancer Res 11:6002–6011
9. Tong YG, Zhang XW, Geng MY, Yue JM, Xin XL, Tian F, Shen X, Tong LJ, Li MH, Zhang C, Li WH, Lin LP, Ding J (2006) Pseudolarix acid B, a new tubulin-binding agent, inhibits angiogenesis by interacting with a novel binding site on tubulin. Mol Pharmacol 69:1226–1233

Pyrrolizilactone

Keywords: [Proteasome inhibitor]

Structure

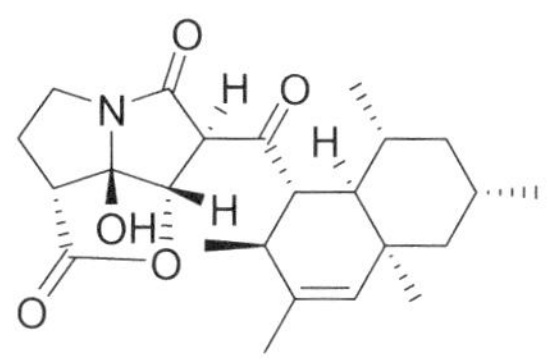

	Pyrrolizilactone
Molecular formula	$C_{24}H_{33}NO_5$
Molecular weight	415.53
Solubility	DMSO, ++; H_2O, –; MeOH, +++

Discovery/Isolation/Chemical Studies

Pyrrolizilactone was isolated from uncharacterized fungus in the course of screening program to find structurally unique metabolites from microorganisms on the basis of spectral data collected through LC/MS analysis [1].

Biological Studies

A molecular target of pyrrolizilactone was identified as a proteasome inhibitor by using two phenotypic profiling systems, cell morphology–based profiling analysis and proteome-based profiling analysis. Pyrrolizilactone is a novel type of proteasome inhibitor inhibiting the trypsin-like activity of the proteasome [2].

Biological Activity of Pyrrolizilactone

IC_{50} values for HeLa and HL60: 9.5 and 2.5 μM, respectively [2]

IC_{50} values for trypsin-, chymotrypsin-, and caspase-like protease activity of proteasome: 1.6, 29, and 84 μM, respectively [2]

Inhibition of cell cycle progression of HeLa in G2/M phase: 10 μM [2]

References

1. Nogawa T, Kawatani M, Uramoto M, Okano A, Aono H, Futamura Y, Koshino H, Takahashi S, Osada H (2013) Pyrrolizilactone, a new pyrrolizidinone metabolite produced by a fungus. J Antibiot (Tokyo) 66:621–623
2. Futamura Y, Kawatani M, Muroi M, Aono H, Nogawa T, Osada H (2013) Identification of a molecular target of a novel fungal metabolite, pyrrolizilactone, by phenotypic profiling systems. ChemBioChem 14:2456–2463

QS11*

Keywords: [Wnt/β-catenin signaling activator] [GTPase activating protein of ADP-ribosylation factor 1 (ARFGAP1) inhibitor]

Structure

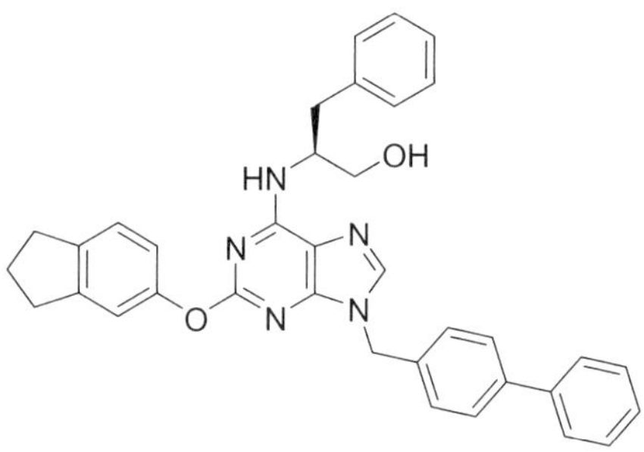

	QS11
Molecular formula	$C_{36}H_{33}N_5O_2$
Molecular weight	567.68
Solubility	DMSO, +++; H_2O, −; MeOH, +++

Discovery/Isolation

QS11 is a purine derivative identified as a Wnt-3a synergist by a cell-based assay of Wnt/β-catenin signal transduction [1].

Biological Studies

QS11 specifically activated the reporter activity of Wnt/β-catenin signal transduction in the presence, but not in the absence, of Wnt-3a [1]. GTPase activating protein of ADP-ribosylation factor 1 (ARFGAP1) was identified as the QS11-target protein by an affinity-probe approach. Therefore, it is thought that QS11 activates ARF activity by inhibition of ARFGAP1, and modulates the localization of β-catenin in the cells. Further analysis suggested that the activation of ARF is necessary for both the production of phosphatidylinositol-4,5-bisphosphate and LRP6 phosphorylation, which is a critical initiation step for the transmission of Wnt/β-catenin signaling [2].

QS11 has also been used for investigation of other ARF-mediated processes including intercellular trafficking [3], interleukin signaling pathway [4], and endothelin-1-induced lipolysis [5].

Biological Activity of QS11

EC_{50} for activation of the Wnt signal reporter in the presence of Wnt-3a: 0.5 μM [1]
K_d for ARFGAP1: 620 nM [1]
80% inhibition of MDA-MB-231 migration: 2.5 μM [1]

References

1. Zhang Q, Major MB, Takanashi S, Camp ND, Nishiya N, Peters EC, Ginsberg MH, Jian X, Randazzo PA, Schultz PG, Moon RT, Ding S (2007) Small-molecule synergist of the Wnt/β-catenin signaling pathway. Proc Natl Acad Sci U S A 104:7444–7448
2. Kim W, Kim SY, Kim T, Kim M, Bae DJ, Choi HI, Kim IS, Jho E (2013) ADP-ribosylation factors 1 and 6 regulate Wnt/β-catenin signaling via control of LRP6 phosphorylation. Oncogene 32:3390–3396
3. Kanamarlapudi V, Thompson A, Kelly E, Lopez Bernal A (2012) ARF6 activated by the LHCG receptor through the cytohesin family of guanine nucleotide exchange factors mediates the receptor internalization and signaling. J Biol Chem 287:20443–20455
4. Zhu W, London NR, Gibson CC, Davis CT, Tong Z, Sorensen LK, Shi DS, Guo J, Smith MC, Grossmann AH, Thomas KR, Li DY (2012) Interleukin receptor activates a MYD88-ARNO-ARF6 cascade to disrupt vascular stability. Nature 492:252–255
5. Davies JC, Bain SC, Kanamarlapudi V (2014) ADP-ribosylation factor 6 regulates endothelin-1-induced lipolysis in adipocytes. Biochem Pharmacol 90:406–413

Radicicol

Keywords: [HSP90 inhibitor] [ATP citrate lyase inhibitor (ACL)]

Structure

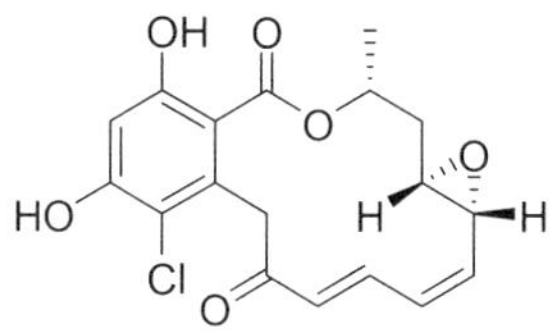

	Radicicol
Molecular formula	$C_{18}H_{17}O_6Cl$
Molecular weight	364.78
Solubility	DMSO, ++; H_2O, +; MeOH, +++

Discovery/Isolation/Chemical Studies

Radicicol was isolated as an antifungal substance from *Monosporuim bonorden* [1]. The compound was assigned the molecular formula $C_{17}H_{16}O_7$ on the basis of preliminary analytical data, but McCapra et al. revised the formula to $C_{18}H_{17}O_6Cl$ [2].

Biological Studies

Radicicol was reisolated as the compound that induces the reversal of transformed phenotypes of v-*src*-transformed fibroblasts [3]. This compound showed various effects on mammalian cells, such as antiangiogenesis activity [4], suppression of transformation by RAS and MOS [5], differentiation of HL60 cells [6], and selective depletion of Raf-1 kinase [7].

The first report of the molecular target of radicicol was made by Sharma et al. [8]. They synthesized a biotinylated radicicol, and identified HSP90 as the molecular target of radicicol. Radicicol binds on the *N*-terminal ATP/ADP binding domain of HSP90 as a nucleotide mimic and inhibits inherent ATPase activity [9, 10]. Radicicol destabilizes Hsp90 client proteins including telomerase, mutated p53, Bcr-Abl, Raf-1, Akt, HER2/Neu, mutated B-Raf, mutated EGF receptor, and HIF-1α, and shows promising antitumor activity in various preclinical tumor models [11].

The second report was made by Ki et al. [12]. They also synthesized two functional biotinylated derivatives, and found that one derivative binds HSP90, but the other derivative binds another protein, ATP citrate lyase (ACL), a potential target for hypolipidaemic intervention. Because ACL was inhibited noncompetitively in vitro, these results imply that a different part in radicicol is required for each specific binding and that unmodified radicicol can bind both the proteins. It is likely that the introduction of the bulky biotin probe prevents these biotinylated compounds from access to one of these target molecules. It was also reported that radiciol binds on an ATP binding motif belonging to the GHKL superfamily, for example in archeal topoisomerase VI; however, radicicol inhibits the ATPase activity of yeast HSP90 at least 1000-fold more effectively than is seen for archaeal TopoVI [13].

Biological Activity of Radicicol

Morphological reversion and inhibition of *src* kinase of *src*-transformed 3Y1 cells: 0.1–1 μg/ml [3]

ID_{50} for inhibition embryonic angiogenesis: 200 ng/egg [4]

IC_{50} for Raf-1 depletion in KNRK5.2 cells: 3.1 μM [7]

K_d for HSP90: 19 nM [9]

K_i values for citrate and ATP against ACL: 13 and 7 μM, respectively [12]

References

1. Delmotte P, Delmotte-Plaque J (1953) A new antifungal substance of fungal origin. Nature 171:344
2. McCapra F, Scott AI (1964) The constitution of monorden, an antibiotic with tranquilising action. Tetrahedron Lett 15:869–875
3. Kwon HJ, Yoshida M, Abe K, Horinouchi S, Beppu T (1992) Radicicol, an agent inducing the reversal of transformed phenotypes of *src*-transformed fibroblasts. Biosci Biotechnol Biochem 56:538–539
4. Oikawa T, Ito H, Ashino H, Toi M, Tominaga T, Morita I, Murota S (1993) Radicicol, a microbial cell differentiation modulator, inhibits in vivo angiogenesis. Eur J Pharmacol 241:221–227
5. Zhao JF, Nakano H, Sharma S (1995) Suppression of RAS and MOS transformation by radicicol. Oncogene 11:161–173
6. Shimada Y, Ogawa T, Sato A, Kaneko I, Tsujita Y (1995) Induction of differentiation of HL-60 cells by the anti-fungal antibiotic, radicicol. J Antibiot (Tokyo) 48:824–830
7. Soga S, Kozawa T, Narumi H, Akinaga S, Irie K, Matsumoto K, Sharma SV, Nakano H, Mizukami T, Hara M (1998) Radicicol leads to selective depletion of Raf kinase and disrupts K-Ras-activated aberrant signaling pathway. J Biol Chem 273:822–828
8. Sharma SV, Agatsuma T, Nakano H (1998) Targeting of the protein chaperone, HSP90, by the transformation suppressing agent, radicicol. Oncogene 16:2639–2645
9. Roe SM, Prodromou C, O″Brien R, Ladbury JE, Piper PW, Pearl LH (1999) Structural basis for inhibition of the Hsp90 molecular chaperone by the antitumor antibiotics radicicol and geldanamycin. J Med Chem 42:260–266
10. Schulte TW, Akinaga S, Soga S, Sullivan W, Stensgard B, Toft D, Neckers LM (1998) Antibiotic radicicol binds to the N-terminal domain of Hsp90 and shares important biologic activities with geldanamycin. Cell Stress Chaperones 3:100–108
11. Neckers L (2006) Using natural product inhibitors to validate Hsp90 as a molecular target in cancer. Curr Top Med Chem 6:1163–1171
12. Ki SW, Ishigami K, Kitahara T, Kasahara K, Yoshida M, Horinouchi S (2000) Radicicol binds and inhibits mammalian ATP citrate lyase. J Biol Chem 275:39231–39236
13. Gadelle D, Bocs C, Graille M, Forterre P (2005) Inhibition of archaeal growth and DNA topoisomerase VI activities by the Hsp90 inhibitor radicicol. Nucleic Acids Res 33:2310–2317

Rapamycin

Keywords: [Antifungal] [Immunosuppressant] [mTOR inhibitor]

Structure

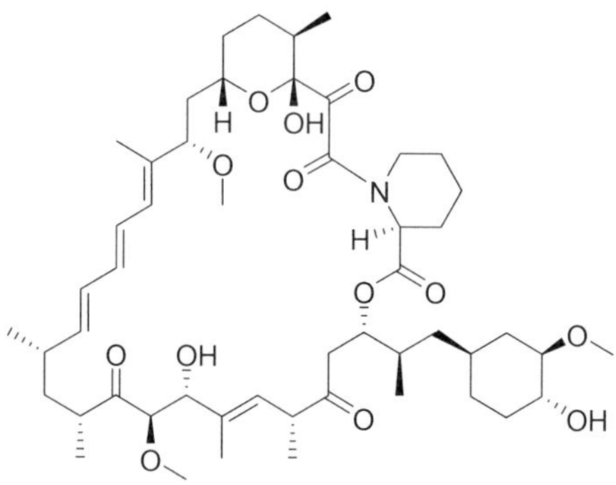

	Rapamycin
Molecular formula	$C_{51}H_{79}NO_{13}$
Molecular weight	914.19
Solubility	DMSO, ++; H_2O, +; MeOH, ++

Discovery/Isolation

Rapamycin was isolated as an antifungal antibiotic produced by *Streptomyces hygroscopicus* [1, 2].

Biological Studies

Rapamycin inhibited the immune response in rats and the immunosuppressant activity of rapamycin appears to be related to inhibition of the lymphatic system [3]. Interestingly, FK506, which is structurally related to rapamycin, showed immunosuppressive properties by a different mechanism. Remarkably, these two drugs inhibit each other's actions, raising the possibility that both act by means of a common immunophilin, FKBP [4]. Heitman et al. found that *Saccharomyces cerevisiae* treated with rapamycin was irreversibly arrested in the G1 phase and isolated TOR1 and TOR2 genes, which participate in rapamycin toxicity [5]. Brown et al. and Sabatini et al. identified a mammalian protein targeted by the G1-arresting rapamycin–receptor complex, FRAP, that is highly related to TOR1/2 [6, 7]. The rapamycin/FKBP complex selectively inhibits the $p70^{S6k}$ activation cascade; however, the kinase activity of FRAP alone is not sufficient for control of $p70^{S6k}$ [8–10]. This inhibition cascade results in the decrease of cap-dependent, but not cap-independent, translation, which is related to dephosphorylation and consequent activation of 4E–BP1/PHAS-I, a protein identified as a repressor of the cap binding protein, eIF-4E [11, 12].

Now it is known that mTOR is an atypical serine/threonine kinase that nucleates two distinct multiprotein complexes commonly known as mTORC1 that is generally associated with cell growth, and mTORC2 that promotes cell proliferation and survival [13]. Rapamycin selectively inhibits mTORC1, but not mTORC2.

Biological Activity of Rapamycin

K_d for rapamycin to FKBP: 0.2 nM [5]

Inhibition of IL-2-induced S-phase entry of T cells: 0.05–0.2 nM [8]

References

1. Sehgal SN, Baker H, Vezina C (1975) Rapamycin (AY-22,989), a new antifungal antibiotic. II. Fermentation, isolation and characterization. J Antibiot (Tokyo) 28:727–732
2. Vezina C, Kudelski A, Sehgal SN (1975) Rapamycin (AY-22,989), a new antifungal antibiotic. I. Taxonomy of the producing streptomycete and isolation of the active principle. J Antibiot (Tokyo) 28:721–726
3. Martel RR, Klicius J, Galet S (1977) Inhibition of the immune response by rapamycin, a new antifungal antibiotic. Can J Physiol Pharmacol 55:48–51
4. Bierer BE, Mattila PS, Standaert RF, Herzenberg LA, Burakoff SJ, Crabtree G, Schreiber SL (1990) Two distinct signal transmission pathways in T lymphocytes are inhibited by complexes formed between an immunophilin and either FK506 or rapamycin. Proc Natl Acad Sci U S A 87:9231–9235
5. Heitman J, Movva NR, Hall MN (1991) Targets for cell cycle arrest by the immunosuppressant rapamycin in yeast. Science 253:905–909
6. Brown EJ, Albers MW, Shin TB, Ichikawa K, Keith CT, Lane WS, Schreiber SL (1994) A mammalian protein targeted by G1-arresting rapamycin-receptor complex. Nature 369:756–758
7. Sabatini DM, Erdjument-Bromage H, Lui M, Tempst P, Snyder SH (1994) RAFT1: a mammalian protein that binds to FKBP12 in a rapamycin-dependent fashion and is homologous to yeast TORs. Cell 78:35–43
8. Brown EJ, Beal PA, Keith CT, Chen J, Shin TB, Schreiber SL (1995) Control of p70 S6 kinase by kinase activity of FRAP in vivo. Nature 377:441–446
9. Chung J, Kuo CJ, Crabtree GR, Blenis J (1992) Rapamycin-FKBP specifically blocks growth-dependent activation of and signaling by the 70 kd S6 protein kinases. Cell 69:1227–1236
10. Kuo CJ, Chung J, Fiorentino DF, Flanagan WM, Blenis J, Crabtree GR (1992) Rapamycin selectively inhibits interleukin-2 activation of p70 S6 kinase. Nature 358 (6381):70–73
11. Beretta L, Gingras AC, Svitkin YV, Hall MN, Sonenberg N (1996) Rapamycin blocks the phosphorylation of 4E–BP1 and inhibits cap-dependent initiation of translation. EMBO J 15:658–664
12. Brunn GJ, Hudson CC, Sekulic A, Williams JM, Hosoi H, Houghton PJ, Lawrence JC, Jr., Abraham RT (1997) Phosphorylation of the translational repressor PHAS-I by the mammalian target of rapamycin. Science 277:99–101
13. Laplante M, Sabatini DM (2012) mTOR signaling in growth control and disease. Cell 149:274–293

Rebeccamycin

Keywords: [Topoisomerase I inhibitor] [Chk1 inhibitor]

Structure

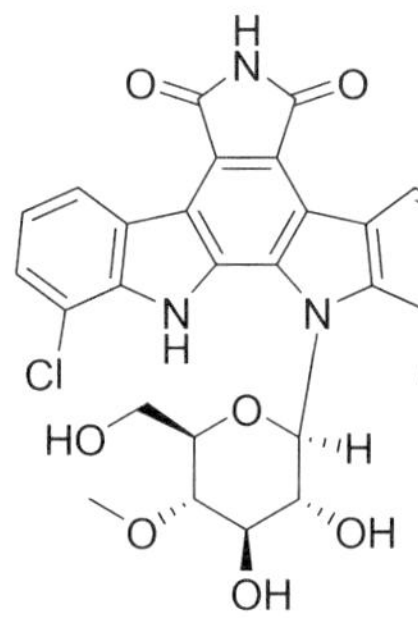

	Rebeccamycin
Molecular formula	$C_{27}H_{21}Cl_2N_3O_7$
Molecular weight	570.38
Solubility	DMSO, ++; H_2O, ? ; MeOH, ?

Discovery/Isolation

Rebeccamycin is an indolocarbazole which was isolated from the mycelium of an actinomycete, strain C-38,383 as a novel antitumor agents [1].

Biological Studies

Rebeccamycin showed antitumor activity against P388 leukemia, L1210 leukemia, and B16 melanoma implanted in mice [1]. However rebeccamycin is structurally related to staurosporine, this compound has no inhibitory activity on protein kinases [2]. Instead, rebeccamycin induced topoisomerase I–mediated DNA cleavage [2]. Structure–activity relationship analyses using rebeccamycin analogs suggested that the stereochemistry of the sugar residue is the most important factor for interaction with DNA [3, 4].

Biological Activity of Rebeccamycin

DNA break in A549 cells: 0.15–5.0 μg/mL [1]
IC_{50} for protein kinase A and protein kinase C in vitro: >100 μM [2]
Topoisomerase I–mediated DNA cleavage: 5 μM [3]

References

1. Bush JA, Long BH, Catino JJ, Bradner WT, Tomita K. (1987) Production and biological activity of rebeccamycin, a novel antitumor agent. J Antibiot (Tokyo) 40:668–678
2. Sancelme M, Fabre S, Prudhomme M. (1994) Antimicrobial activities of indolocarbazole and bis-indole protein kinase C inhibitors. J Antibiot (Tokyo) 47:792–798
3. Yamashita Y, Fujii N, Murakata C, Ashizawa T, Okabe M, Nakano H. (1992) Induction of mammalian DNA topoisomerase I mediated DNA cleavage by antitumor indolocarbazole derivatives. Biochemistry 31:12069–12075
4. Bailly C, Qu X, Graves DE, Prudhomme M, Chaires JB. (1999) Calories from carbohydrates: energetic contribution of the carbohydrate moiety of rebeccamycin to DNA binding and the effect of its orientation on topoisomerase I inhibition. Chem Biol 6:277–286

Reveromycin A

Keywords: [Antiosteoporotic] [Detransforming agent] [Protein translation inhibitor]

Structure

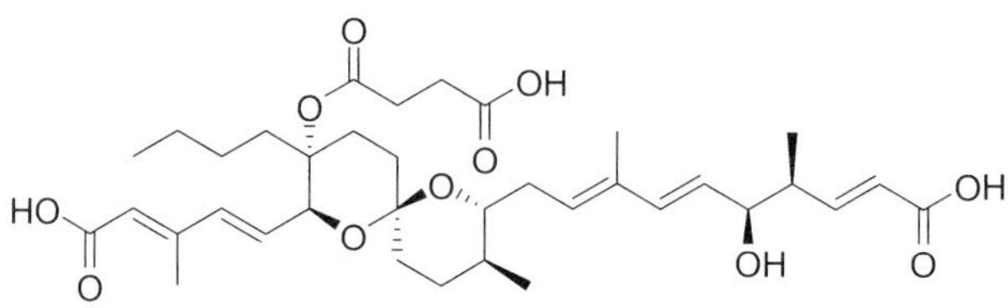

	Reveromycin A
Molecular formula	$C_{36}H_{52}O_{11}$
Molecular weight	660.80
Solubility	DMSO, ++; H_2O, +; MeOH, ++

Discovery/Isolation

Reveromycin A was isolated as an inhibitor of mitogenic activity induced by EGF in a mouse epidermal keratinocyte [1, 2].

Biological Studies

Reveromycin A showed inhibitory activity against EGF-stimulated mitogen response in Balb/MK cells and exhibited morphological reversion of *src*^{ts}-NRK cells, antiproliferative activity against human tumor cell lines, and antifungal activity. In vitro studies revealed that reveromycin A is a selective inhibitor of protein synthesis in eukaryotic cells [3], and showed little antitumor effect against three murine tumors tested, but showed a strong antitumor effect against a human ovarian carcinoma BG-1 [4].

The molecular target of reveromycin A was determined by yeast genetics [5]. A *Saccharomyces cerevisiae* dominant mutant specifically resistant to reveromycin A was isolated from a strain in which the genes implicated in nonspecific multidrug resistance had been deleted. The mutant gene responsible for the resistance was identified as an allele of the ILS1 gene encoding tRNA^{Ile} synthetase (IleRS). Reveromycin A inhibited the activity of IleRS specifically and the mutation IleRS(N660D), near the KMSKS consensus sequence commonly found in the class I aminoacyl transferases, was found to be responsible for reveromycin A resistance. These results indicated that IleRS is a target of reveromycin A in vivo.

Because reveromycin A has three carboxyl groups in its structure, this compound accumulates in an acidic microenvironment, a prominent characteristic of mature bone-resorbing osteoclasts. In fact, reveromycin A induced apoptosis specifically in mature bone-resorbing osteoclasts, but not in osteoclast progenitors, nonfunctional osteoclasts, or osteoblasts [6]. Furthermore, reveromycin A inhibited the osteolytic bone metastasis of human small lung cancer cells in vivo [7]. These findings suggest that reveromycin A may inhibit osteolytic bone metastasis through suppression of osteoclast activity by directly inducing apoptosis and indirectly inhibiting tumor cell–derived parathyroid hormone–related peptide production.

Biological Activity of Reveromycin A
IC_{50} for inhibition of protein synthesis in *src*^{ts}-NRK cells: 1 μM [3]
IC_{50} for inhibition of protein synthesis in rabbit reticulocyte lysate: 40 nM [3]
Inhibition of cell proliferation induced by TGF-α in BG-1 cells: 30–300 nM [4]
Growth inhibition of *Saccharomyces cerevisiae* on YPD plate adjusted at pH 4.5: 1 μg/ml [8]

References

1. Osada H, Koshino H, Isono K, Takahashi H, Kawanishi G (1991) Reveromycin A, a new antibiotic which inhibits the mitogenic activity of epidermal growth factor. J Antibiot (Tokyo) 44:259–261
2. Takahashi H, Osada H, Koshino H, Kudo T, Amano S, Shimizu S, Yoshihama M, Isono K (1992) Reveromycins, new inhibitors of eukaryotic cell growth. I. Producing organism, fermentation, isolation and physico-chemical properties. J Antibiot (Tokyo) 45:1409–1413
3. Takahashi H, Osada H, Koshino H, Sasaki M, Onose R, Nakakoshi M, Yoshihama M, Isono K (1992) Reveromycins, new inhibitors of eukaryotic cell growth. II Biological activities. J Antibiot (Tokyo) 45:1414–1419
4. Takahashi H, Yamashita Y, Takaoka H, Nakamura J, Yoshihama M, Osada H (1997) Inhibitory action of reveromycin A on TGF-α-dependent growth of ovarian carcinoma BG-1 in vitro and in vivo. Oncol Res 9:7–11
5. Miyamoto Y, Machida K, Mizunuma M, Emoto Y, Sato N, Miyahara K, Hirata D, Usui T, Takahashi H, Osada H, Miyakawa T (2002) Identification of *Saccharomyces cerevisiae* isoleucyl-tRNA synthetase as a target of the G1-specific inhibitor Reveromycin A. J Biol Chem 277:28810–28814
6. Woo JT, Kawatani M, Kato M, Shinki T, Yonezawa T, Kanoh N, Nakagawa H, Takami M, Lee KH, Stern PH, Nagai K, Osada H (2006) Reveromycin A, an agent for osteoporosis, inhibits bone resorption by inducing apoptosis specifically in osteoclasts. Proc Natl Acad Sci U S A 103:4729–4734
7. Muguruma H, Yano S, Kakiuchi S, Uehara H, Kawatani M, Osada H, Sone S (2005) Reveromycin A inhibits osteolytic bone metastasis of small-cell lung cancer cells, SBC-5, through an antiosteoclastic activity. Clin Cancer Res 11:8822–8828
8. Cui Z, Hirata D, Tsuchiya E, Osada H, Miyakawa T (1996) The multidrug resistance-associated protein (MRP) subfamily (Yrs1/Yor1) of *Saccharomyces cerevisiae* is important for the tolerance to a broad range of organic anions. J Biol Chem 271:14712–14716

R-Roscovitine*

Keywords: [Cyclin-dependent kinase (CDK) inhibitor]

Structure

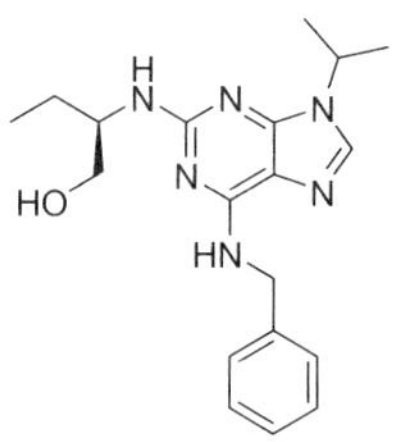

	R-Roscovitine
Molecular formula	$C_{19}H_{26}N_6O$
Molecular weight	354.46
Solubility	DMSO, +++; H_2O, +; MeOH, +++

Discovery

R-roscovitine (named after Roscoff—a French town, where the lab that discovered the compound was located; seliciclib, CYC202) is the second CDK inhibitor to enter clinical trials. Roscovitine belongs to the family of purines and acts through competion with ATP for binding at the ATP binding site of CDKs. Consistent with these structural motifs, R-roscovitine inhibits the activity of CDK2 and CDK1, 5, 7, 8, and 9. R-roscovitine exposure to human diffuse large B-cell lymphoma (DLBCL) cells results in G1- and G2/M-phase arrest and induction of apoptosis independently of any underlying chromosomal translocations [1, 2].

Biological Studies

Roscovitine is a selective inhibitor of cyclin-dependent kinase (CDK), which has a key role in cell cycle regulation. Since CDKs are also known to be involved in the disregulation of cell growth and tumorgenesis, roscovitine is used for cancer therapy. In a study of the effects on the cell cycle in Lovo colorectal carcinoma cells, the major effect was the induction of cell death at all phases of the cell cycle—not merely in specific phases. Roscovitine has also been found to inhibit transition from prophase to metaphase at a range of micromolar concentrations. It has also been shown that it inhibits progesterone-induced oocyte maturation of *Xenopus* oocytes [3].

Biological Activity of R-Roscovitine

IC_{50} values for cdc2 (cdc2/cyclin B), cdk2 (cdk2/cyclin A and cdk2/cyclin E), and cdk5 (cdk5/p35): 0.65 μM, 0.7 μM, 0.7 μM, and 0.16 μM respectively [3]
IC_{50} values for Erk1 and Erk2: 34 μM and 14 μM respectively [3]

Average IC_{50} for CDK2/cyclin E kinase activity in a panel of 19 human tumor cell lines: 15.2 μM [4]

References

1. Lacrima K, Rinaldi A, Vignati S, Martin V, Tibiletti MG, Gaidano G, Catapano CV, Bertoni F (2007) Cyclin-dependent kinase inhibitor seliciclib shows in vitro activity in diffuse large B-cell lymphomas. Leuk Lymphoma 48:158–167
2. Cicenas J, Kalyan K, Sorokinas A, Stankunas E, Levy J, Meskinyte I, Stankevicius V, Kaupinis A, Valius M (2015) Roscovitine in cancer and other diseases. Ann Transl Med 3:135
3. Meijer L, Borgne A, Mulner O, Chong JP, Blow JJ, Inagaki N, Inagaki M, Delcros JG, Moulinoux JP (1997) Biochemical and cellular effects of roscovitine, a potent and selective inhibitor of the cyclin-dependent kinases cdc2, cdk2 and cdk5. Eur J Biochem 243:527–536
4. McClue SJ, Blake D, Clarke R, Cowan A, Cummings L, Fischer PM, MacKenzie M, Melville J, Stewart K, Wang S, Zhelev N, Zheleva D, Lane DP (2002) In vitro and in vivo antitumor properties of the cyclin dependent kinase inhibitor CYC202 (R-roscovitine). Int J Cancer 102: 463–468

Rottlerin

Keywords: [Autophagy inducer] [Kinase inhibitor]

Structure

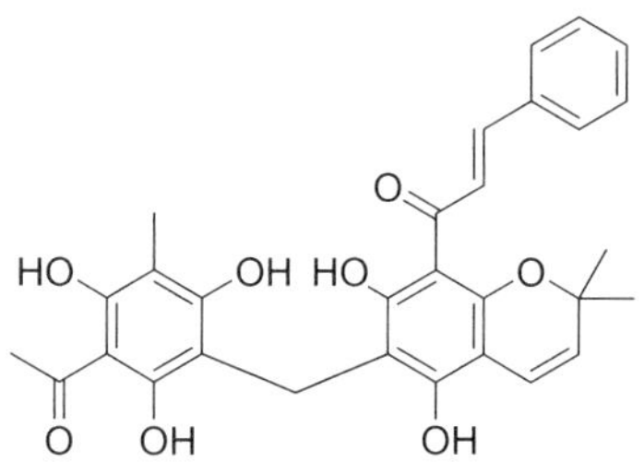

	Rottlerin
Molecular formula	$C_{30}H_{28}O_8$
Molecular weight	516.55
Solubility	DMSO, +++; H_2O, –; MeOH, +

Discovery/Isolation

The first description of rottlerin dates from 1855. In this paper, Anderson isolated a crystalline compound named rottlerin from the Indian coloring and anthelmintic drug "kamala," which is obtained from the fruit glands of *Mallotus philippensis* (Muell) [1].

Biological Studies

Rottlerin has been reported as a protein kinase C (PKC) δ inhibitor and many researchers have used rottlerin as a specific inhibitor of PKCδ [2]. However, a lot of reports strongly suggest that rottlerin is not a PKCδ-specific inhibitor [3]. Rottlerin inhibits mitochondrial function independently of PKCδ, and induces apoptosis [4,5].

Recently, rottlerin was screened as a modulator of autophagy [6]. Rottlerin inhibited mTORC1 signaling and this inhibition was dependent on the presence of TSC2, a negative regulator of mTORC1.

Inhibitory activity of rottlerin on amyloid polymerization was also reported [7].

Biological Activity of Rottlerin

Rapid reduction of cellular ATP in parotid acinar cells: 10 μM [4]
Inhibition of amyloid polymerization: 30 μM [7]
Autophagy induction in MCF-7 cells: 3 μM [6]

References

1. Anderson (1855) The Edinburgh new philosophical journal 1:300
2. Gschwendt M, Muller HJ, Kielbassa K, Zang R, Kittstein W, Rincke G, Marks F (1994) Rottlerin, a novel protein kinase inhibitor. Biochem Biophys Res Commun 199:93–98
3. Soltoff SP (2007) Rottlerin: an inappropriate and ineffective inhibitor of PKCδ. Trends Pharmacol Sci 28:453–458
4. Soltoff SP (2001) Rottlerin is a mitochondrial uncoupler that decreases cellular ATP levels and indirectly blocks protein kinase Cδ tyrosine phosphorylation. J Biol Chem 276:37986–37992
5. Tillman DM, Izeradjene K, Szucs KS, Douglas L, Houghton JA (2003) Rottlerin sensitizes colon carcinoma cells to tumor necrosis factor-related apoptosis-inducing ligand-induced apoptosis via uncoupling of the mitochondria independent of protein kinase C. Cancer Res 63:5118–5125
6. Balgi AD, Fonseca BD, Donohue E, Tsang TC, Lajoie P, Proud CG, Nabi IR, Roberge M (2009) Screen for chemical modulators of autophagy reveals novel therapeutic inhibitors of mTORC1 signaling. PLoS One 4:e7124
7. Feng BY, Toyama BH, Wille H, Colby DW, Collins SR, May BC, Prusiner SB, Weissman J, Shoichet BK (2008) Small-molecule aggregates inhibit amyloid polymerization. Nat Chem Biol 4:197–199

S1201*

Keywords: [Lysine-specific demethylase I inhibitor]

Structure

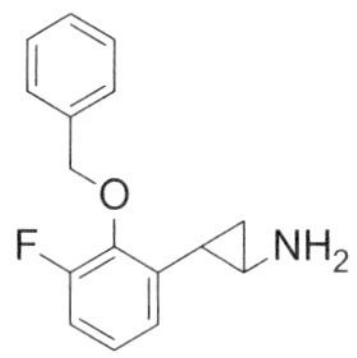

	S1201
Molecular formula	$C_{16}H_{16}FNO$
Molecular weight	257.30
Solubility	DMSO, +++; H_2O, +; MeOH, +

Discovery

S1201 was synthesized as a compound to inhibit lysine-specific demethylase-1 (LSD1), based on the structures of LSD1 and monoamine oxidase B (MAO-B), in complex with an antidepressant tranylcypromine derivative [1].

Biological Studies

LSD1, a unique nuclear protein that utilizes FAD as a cofactor, regulates eukaryotic gene expression through chromatin structure alteration by demethylates histone H3 and DNA methyltransferase 1 (Dnmt1). S1201 strongly and specifically inhibited LSD1 [1]. The S1201-bound LSD1 structure revealed the enhanced stability of the S1201-FAD adducts by their interactions with the surrounding LSD1 residues. S1201 treatments resulted in the activation of mitochondrial respiration through the induction of a number of regulators of energy expenditure and mitochondrial methabolism in adipocytes [2]. Therefore, LSD1 is thought to regulate cellular energy balance through coupling with cellular FAD biosynthesis.

Biological Activity of S1201

k_{inact}/K_i for demethylation activity of LSD1: 4560 M^{-1} s^{-1} [1]

References

1. Mimasu S, Umezawa N, Sato S, Higuchi T, Umehara T, Yokoyama S (2010) Structurally designed *trans*-2-phenylcyclopropylamine derivatives potently inhibit histone demethylase LSD1/KDM1. Biochemistry 49:6494–6503
2. Hino S, Sakamoto A, Nagaoka K, Anan K, Wang Y, Mimasu S, Umehara T, Yokoyama S, Kosai K, Nakao M (2012) FAD-dependent lysine-specific demethylase-1 regulates cellular energy expenditure. Nat Commun 3:758

SC1* (Pluripotin)

Keywords: [Self-renewal] [Dual inhibitor of RasGAP and ERK1]

Structure

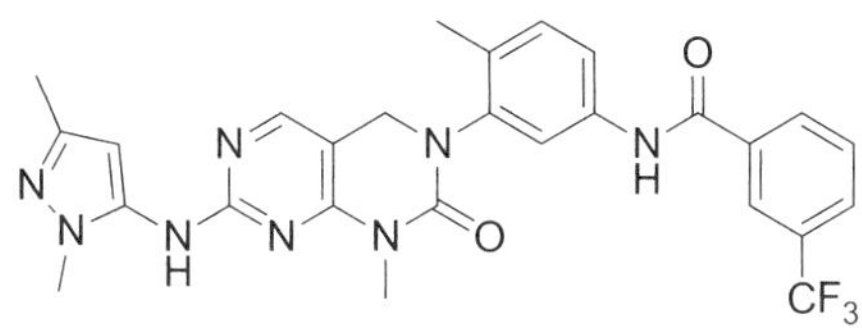

	SC1 (pluripotin)
Molecular formula	$C_{27}H_{25}F_3N_8O_2$
Molecular weight	550.54
Solubility	DMSO, +++; H_2O, ±; MeOH, +

Discovery/Isolation

SC1/pluripotin was discovered by a cell-based screen of chemical libraries to identify small molecules that control the self-renewal of ES cells [1].

Biological Studies

SC1 propagated murine ES cells (mES) in an undifferentiated, pluripotent state for >10 passages under chemically defined conditions in the absence of feeder cells, serum, and LIF. SC1 inhibited differentiation of mES induced by either 20% FBS or 1 μM retinoic acid. After removal of SC1, SC1-expanded mES cells (>10 passages) were differentiated into neural/neuronal, cardiac muscle, and endodermal cells and could generate healthy chimeric mice in vivo.

Using SC1-affinity beads, ERK1 and RasGAP were identified as the SC1 binding proteins. Biochemical and cellular experiments suggest that SC1 acts through two targets: by inhibiting RasGAP, SC1 blocks RasGAPs differentiation-inducing activity and activates self-renewal pathways, while at the same time blocking ERK-dependent differentiation.

SC1 could be used for the efficient establishment of ES cells from immunodeficient mice on refractory backgrounds [2] and a tractable cancer stem cell model [3].

Biological Activity of SC1 (Pluripotin)

Self-renewal of mES for >10 passages in an undifferentiated/pluripotent state: 1 μM [1]

K_d values for ERK1 and RasGAP: 98 and 212 nM, respectively [1]

References

1. Chen S, Do JT, Zhang Q, Yao S, Yan F, Peters EC, Scholer HR, Schultz PG, Ding S (2006) Self-renewal of embryonic stem cells by a small molecule. Proc Natl Acad Sci U S A 103:17266–17271
2. Yang W, Wei W, Shi C, Zhu J, Ying W, Shen Y, Ye X, Fang L, Duo S, Che J, Shen H, Ding S, Deng H (2009) Pluripotin combined with leukemia inhibitory factor greatly promotes the derivation of embryonic stem cell lines from refractory strains. Stem Cells 27:383–389
3. Mertins SD, Scudiero DA, Hollingshead MG, Divelbiss RD, Jr., Alley MC, Monks A, Covell DG, Hite KM, Salomon DS, Niederhuber JE (2013) A small molecule (pluripotin) as a tool for studying cancer stem cell biology: proof of concept. PLoS One 8:e57099

Sirtinol*

Keywords: [NAD-dependent lysine deacetylase inhibitor]

Structure

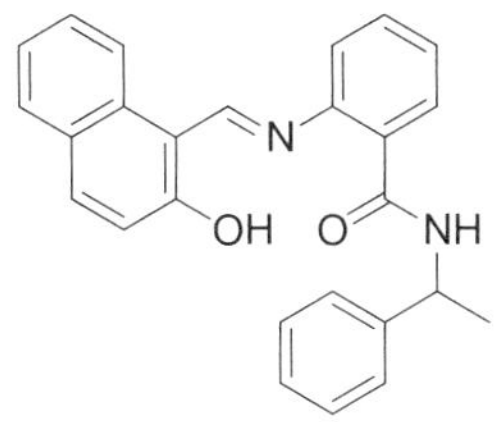

	Sirtinol
Molecular formula	$C_{26}H_{22}N_2O_2$
Molecular weight	394.47
Solubility	DMSO, +++; H_2O, ? ; MeOH, ?

Discovery/Isolation

Sirtinol is an NAD-dependent lysine deacetylase inhibitor with a structurally related α-substituted β-naphthol analog, which was identified by high-throughput cell-based phenotypic screening using yeast [1].

Biological Studies

Sirtinol inhibits Sir2p-mediated transcriptional silencing at three loci in yeast [2]. A chemical–genetic screen for *Arabidopsis* mutants resistant to sirtinol identified SIR1 as an upstream regulator in the auxin signaling pathway [2]. Sirtinol activates auxin-signal transduction and inhibits root elongation in *Arabidopsis* [1, 2]. In mammals, sirtinol has been shown to induce senescence-like growth arrest by impairing activation of MAPK pathways in human breast and lung cancer cells [3] and to induce apoptosis and autophagic cell death in human breast cancer cells [4]. Anti-inflammatory activity of sirtinol in human dermal microvascular endothelial cells was also reported [5].

Biological Activity of Sirtinol

In vitro IC_{50} for yeast Sir2p and human SIRT2: 68 μM and 38 μM [1]

IC_{50} for proliferation of human breast cancer MCF-7 cells: 48.6 μM (24 hours) and 43.5 μM (48 hours) [4]

References

1. Grozinger CM, Chao ED, Blackwell HE, Moazed D, Schreiber SL (2001) Identification of a class of small molecule inhibitors of the sirtuin family of NAD-dependent deacetylases by phenotypic screening. J Biol Chem 276:38837–38843
2. Zhao Y, Dai X, Blackwell HE, Schreiber SL, Chory J (2003) SIR1, an upstream component in auxin signaling identified by chemical genetics. Science 301:1107–1110
3. Ota H, Tokunaga E, Chang K, Hikasa M, Iijima K, Eto M, Kozaki K, Akishita M, Ouchi Y, Kaneki M (2006) Sirt1 inhibitor, Sirtinol, induces senescence-like growth arrest with attenuated Ras-MAPK signaling in human cancer cells. Oncogene 25:176–85
4. Wang J, Kim TH, Ahn MY, Lee J, Jung JH, Choi WS, Lee BM, Yoon KS, Yoon S, Kim HS (2011) Sirtinol, a class III HDAC inhibitor, induces apoptotic and autophagic cell death in MCF-7 human breast cancer cells. Int J Oncol 41:1101–1109
5. Orecchia A, Scarponi C, Di Felice F, Cesarini E, Avitabile S, Mai A, Mauro ML, Sirri V, Zambruno G, Albanesi C, Camilloni G, Failla CM (2011) Sirtinol treatment reduces inflammation in human dermal microvascular endothelial cells. PLoS One 6:e24307

Spliceostatin A*

Keywords: [Antitumor] [Splicing inhibitor] [SF3b]

Structure

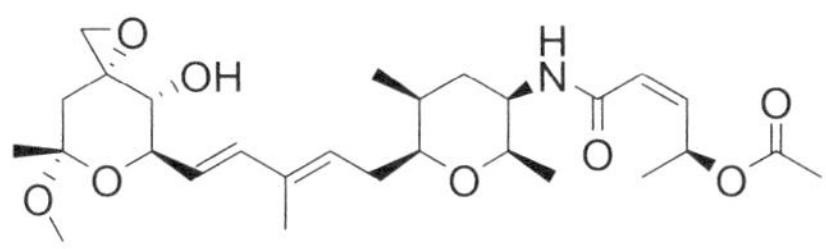

	Spliceostatin A
Molecular formula	$C_{28}H_{43}NO_8$
Molecular weight	521.64
Solubility	DMSO, +++; H_2O, ±; MeOH, +

Discovery/Isolation/Chemical Study
Spliceostatin A is a methylated derivative of FR901464, a natural product isolated from the culture broth of a bacterium of *Pseudomonas* sp. No.2663 as an antitumor compound [1–4].

Biological Studies
FR901464 inhibited cell cycle progression at G1 and G2 phases by producing a C-terminally truncated p27, named p27* [5]. Spliceostatin A inhibited in vitro splicing and promoted pre-mRNA accumulation by binding to SF3b, a subcomplex of the U2 small nuclear ribonucleoprotein in the spliceosome. Detailed investigation of molecular mechanism revealed that spliceostatin A prevents interaction of an SF3b 155 kDa subunit with the pre-mRNA, concomitant with nonproductive recruitment of U2 snRNP to sequences 5′ of the branch point [6], and abolishes 3′ end pausing of polymerases [7]. Furthermore, this inhibition induces leakage of unspliced transcripts to the nucleoplasm [7] and to the cytoplasm [5]. These results clearly indicate that spliceostatin A targets SF3b and inhibits splicing of pre-mRNA, and that the inhibition of pre-mRNA splicing at least during early steps involving SF3b allows unspliced mRNA leakage and translation.

Biological Activity of Spliceostatin A
Inhibition of in vitro splicing: >60 nM [5]
Inhibition of in vivo splicing and pre-mRNA translation: 100 ng/ml [5]
Inhibition of in vivo tumor angiogenesis in CAM assay: 2 nM [8]

References

1. Motoyoshi H, Horigome M, Ishigami K, Yoshida T, Horinouchi S, Yoshida M, Watanabe H, Kitahara T (2004) Structure-activity relationship for FR901464: a versatile method for the conversion and preparation of biologically active biotinylated probes. Biosci Biotechnol Biochem 68:2178–2182
2. Nakajima H, Hori Y, Terano H, Okuhara M, Manda T, Matsumoto S, Shimomura K (1996) New antitumor substances, FR901463, FR901464 and FR901465. II. Activities against experimental tumors in mice and mechanism of action. J Antibiot (Tokyo) 49:1204–1211
3. Nakajima H, Sato B, Fujita T, Takase S, Terano H, Okuhara M (1996) New antitumor substances, FR901463, FR901464 and FR901465. I. Taxonomy, fermentation, isolation, physico-chemical properties and biological activities. J Antibiot (Tokyo) 49:1196–1203
4. Nakajima H, Takase S, Terano H, Tanaka H (1997) New antitumor substances, FR901463, FR901464 and FR901465. III. Structures of FR901463, FR901464 and FR901465. J Antibiot (Tokyo) 50:96–99
5. Kaida D, Motoyoshi H, Tashiro E, Nojima T, Hagiwara M, Ishigami K, Watanabe H, Kitahara T, Yoshida T, Nakajima H, Tani T, Horinouchi S, Yoshida M (2007) Spliceostatin A targets SF3b and inhibits both splicing and nuclear retention of pre-mRNA. Nat Chem Biol 3:576–583
6. Corrionero A, Minana B, Valcarcel J (2011) Reduced fidelity of branch point recognition and alternative splicing induced by the anti-tumor drug spliceostatin A. Genes Dev 25:445–459
7. Martins SB, Rino J, Carvalho T, Carvalho C, Yoshida M, Klose JM, de Almeida SF, Carmo-Fonseca M (2011) Spliceosome assembly is coupled to RNA polymerase II dynamics at the 3′ end of human genes. Nat Struct Mol Biol 18:1115–1123
8. Furumai R, Uchida K, Komi Y, Yoneyama M, Ishigami K, Watanabe H, Kojima S, Yoshida M (2010) Spliceostatin A blocks angiogenesis by inhibiting global gene expression including VEGF. Cancer Sci 101:2483–2489

Staurosporine

Keywords: [Cell cycle inhibitor] [Kinase inhibitor]

Structure

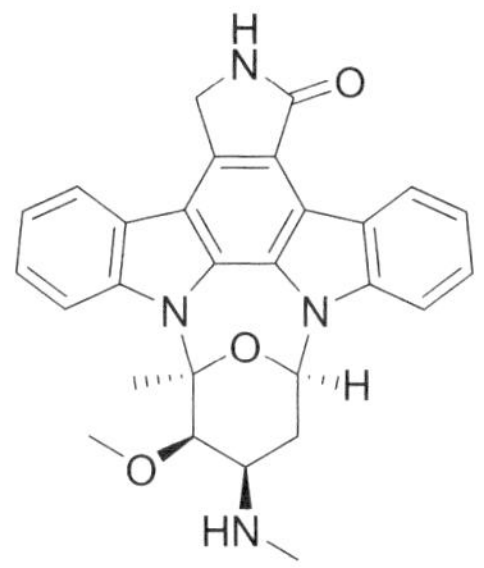

	Staurosporine
Molecular formula	$C_{28}H_{26}N_4O_3$
Molecular weight	466.53
Solubility	DMSO, ++; H_2O, –; MeOH, +

Discovery/Isolation
Staurosporine is the first natural indolocarbazole compound [1]. Thereafter, a lot of indolocarbazoles have been isolated from a variety of organisms [2] or synthesized by chemical synthesis and by engineered biosynthetic pathways.

Biological Studies
This compound was originally discovered as an antifungal antibiotic [1]. However, it was shown that staurosporine possesses various biological activities, including induction of differentiation [2], apoptosis induction [3], cell cycle inhibition [4], and so on. It is important to use this compound as a potent but nonspecific protein kinase inhibitor [5–7], and 100 kinases were identified as staurosporine binding kinases in HepG2 cells [8].

Recently, there has been application of analog-sensitive (AS) kinase technology of staurosporine to investigation of phospho-signaling pathways in diverse organisms and physiological processes [9].

There are good reviews about the characteristics, biological activities, and usage of staurosporine [10].

Biological Activity of Staurosporine
IC_{50} for protein kinase C from rat brain: 2.7 nM [5]
Reversible cell cycle arrest of rat 3Y1 fibroblasts at the early G1 phase: 1–10 ng/ml [4]
Reversible cell cycle arrest of rat 3Y1 fibroblasts at G2: 100 ng/ml [4]
Apoptosis induction in MOLT-4 cells: 10–200 nM [3]

References

1. Omura S, Iwai Y, Hirano A, Nakagawa A, Awaya J, Tsuchya H, Takahashi Y, Masuma R (1977) A new alkaloid AM-2282 of *Streptomyces* origin. Taxonomy, fermentation, isolation and preliminary characterization. J Antibiot (Tokyo) 30:275–282
2. Morioka H, Ishihara M, Shibai H, Suzuki T (1985) Staurosporine-induced differentiation in a human neuroblastoma cell line, NB-1. Agric Biol Chem 49:1959–1963
3. Falcieri E, Martelli AM, Bareggi R, Cataldi A, Cocco L (1993) The protein kinase inhibitor staurosporine induces morphological changes typical of apoptosis in MOLT-4 cells without concomitant DNA fragmentation. Biochem Biophys Res Commun 193:19–25
4. Abe K, Yoshida M, Usui T, Horinouchi S, Beppu T (1991) Highly synchronous culture of fibroblasts from G2 block caused by staurosporine, a potent inhibitor of protein kinases. Exp Cell Res 192:122–127
5. Tamaoki T, Nomoto H, Takahashi I, Kato Y, Morimoto M, Tomita F (1986) Staurosporine, a potent inhibitor of phospholipid/Ca^{2+} dependent protein kinase. Biochem Biophys Res Commun 135:397–402
6. Nakano H, Kobayashi E, Takahashi I, Tamaoki T, Kuzuu Y, Iba H (1987) Staurosporine inhibits tyrosine-specific protein kinase activity of Rous sarcoma virus transforming protein p60. J Antibiot (Tokyo) 40:706–708
7. Fujita-Yamaguchi Y, Kathuria S (1988) Characterization of receptor tyrosine-specific protein kinases by the use of inhibitors. Staurosporine is a 100-times more potent inhibitor of insulin receptor than IGF-I receptor. Biochem Biophys Res Commun 157:955–962
8. Fischer JJ, Graebner OY, Dalhoff C, Michaelis S, Schrey AK, Ungewiss J, Andrich K, Jeske D, Kroll F, Glinski M, Sefkow M, Dreger M, Koester H (2010) Comprehensive identification of staurosporine-binding kinases in the hepatocyte cell line HepG2 using Capture Compound Mass Spectrometry (CCMS). J Proteome Res 9:806–817
9. Lopez MS, Choy JW, Peters U, Sos ML, Morgan DO, Shokat KM (2013) Staurosporine-derived inhibitors broaden the scope of analog-sensitive kinase technology. J Am Chem Soc 135:18153–18159
10. Nakano H, Omura S (2009) Chemical biology of natural indolocarbazole products: 30 years since the discovery of staurosporine. J Antibiot (Tokyo) 62:17–26

TC11*

Keywords: [Antitumor] [Nucleophosmin 1 inhibitor]

Structure

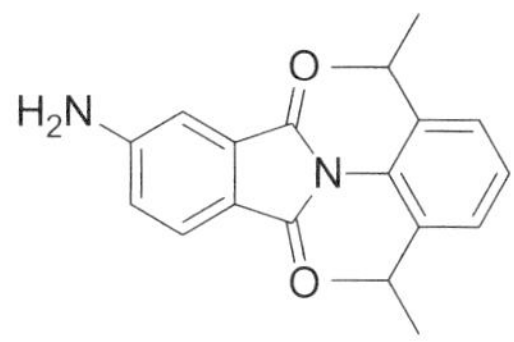

	TC11
Molecular formula	$C_{20}H_{22}N_2O_2$
Molecular weight	322.41
Solubility	DMSO, +++; H_2O, −; MeOH, ±

Discovery/Isolation

TC11 is a phthalimide derivative identified as antitumor compound against several multiple myeloma cell line [1].

Biological Studies

TC11 is a potent inhibitor of tumor cell proliferation and showed antitumor activity against multiple myeloma cell line KMS34 tumor xenografts in ICR/SCID mice [1]. The target protein was identified to be nucleophosmin 1 by mRNA display selection on a microfluidic chip. TC11 treatment and knockdown of nucleophosmin 1 inhibited centrosomal clustering, and induced multipolar cells. These results suggest that nucleophosmin 1 is a novel antitumor target protein. Further analyses revealed that TC11 also significantly inhibited osteoclasts and induced apoptosis of myeloma [2].

Biological Activity of TC11

IC_{50} for several multiple myeloma cell lines: 4–8 μM [1]
K_d values for oligomeric and monomeric fractions of nucleophosmin 1: 1.3×10^{-4} and 6.6×10^{-8} M, respectively [1]
Formation of multipolar spindle in HeLa cells: 5 μM [1]

References

1. Shiheido H, Terada F, Tabata N, Hayakawa I, Matsumura N, Takashima H, Ogawa Y, Du W, Yamada T, Shoji M, Sugai T, Doi N, Iijima S, Hattori Y, Yanagawa H (2012) A phthalimide derivative that inhibits centrosomal clustering is effective on multiple myeloma. PLoS One 7:e38878
2. Matsushita M, Ozaki Y, Hasegawa Y, Terada F, Tabata N, Shiheido H, Yanagawa H, Oikawa T, Matsuo K, Du W, Yamada T, Hozumi M, Ichikawa D, Hattori Y. (2015) A novel phthalimide derivative, TC11, has preclinical effects on high-risk myeloma cells and osteoclasts. PLoS One 10: e0116135

Terpendole E

Keywords: [Kinesin spindle protein (KSP) inhibitor]

Structure

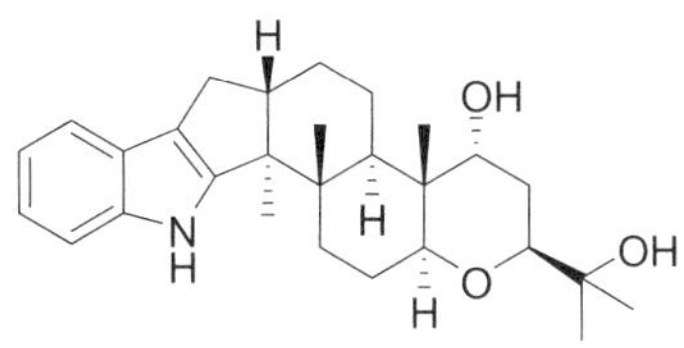

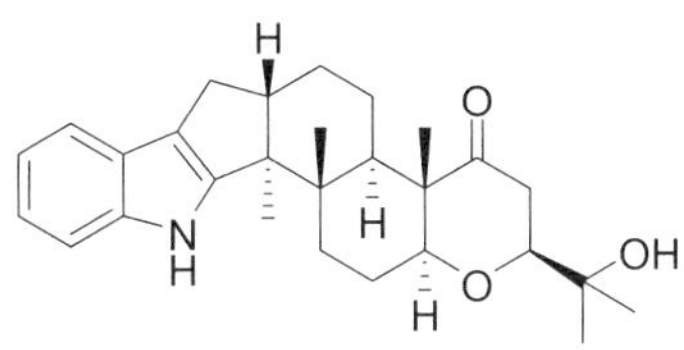

	Terpendole E	11-Ketopaspaline
Molecular formula	$C_{28}H_{39}NO_3$	$C_{28}H_{37}NO_3$
Molecular weight	437.62	435.61
Solubility	DMSO, +++; H_2O, ±; MeOH, +	DMSO, +++; H_2O, ±; MeOH, +

Discovery/Isolation/Chemical Studies

Terpendoles were isolated as acyl-CoA:cholesterol acyltransferase (ACAT) inhibitors from *Albophoma yamanashiensis* [1–3]. 11-Ketopaspaline was isolated from the strain whose biosynthetic pathway of terpendole E was genetically modified [4].

Biological Studies

Terpendole E is a minor derivative and possesses only weak ACAT inhibitory activity [3]. However, this compound was rediscovered as an inhibititor of cell cycle progression in M phase from the Ascomycota fungus *Tolypocladium album* (syn. *Chaunopycnis alba*) [5]. Terpendole E accumulated M-phase cells containing monoastar spindle by inhibiting the bipolar spindle formation. Detailed in vitro analyses revealed that terpendole E is a natural Eg5/KSP-specific inhibitor. Terpendole E and its derivative 11-ketopaspaline inhibit the ATPase activity not only of wild-type but also *S*-trityl-L-cysteine- or GSK1-resistant Eg5, suggesting that these compounds have a different binding site and/or inhibitory mechanism from *S*-trityl-L-cysteine or GSK1 [4].

Biological Activity of Terpendole E

IC_{50} for in vitro ACAT activity: 228 μM [3]

Inhibition of cell cycle progression in M phase in 3Y1, A549, and HeLa cells: 50 μM [5]

IC_{50} for microtubule-stimulated ATPase activity of recombinant Eg5: 23 μM [5]

References

1. Huang XH, Tomoda H, Nishida H, Masuma R, Omura S (1995) Terpendoles, novel ACAT inhibitors produced by *Albophoma yamanashiensis*. I. Production, isolation and biological properties. J Antibiot (Tokyo) 48: 1–4
2. Huang XH, Nishida H, Tomoda H, Tabata N, Shiomi K, Yang DJ, Takayanagi H, Omura S (1995) Terpendoles, novel ACAT inhibitors produced by *Albophoma yamanashiensis*. II. Structure elucidation of terpendoles A, B, C and D. J Antibiot (Tokyo) 48: 5–11
3. Tomoda H, Tabata N, Yang DJ, Takayanagi H, Omura S (1995) Terpendoles, novel ACAT inhibitors produced by *Albophoma yamanashiensis*. III. Production, isolation and structure elucidation of new components. J Antibiot (Tokyo) 48: 793–804
4. Tarui Y, Chinen T, Nagumo Y, Motoyama T, Hayashi T, Hirota H, Muroi M, Ishii Y, Kondo H, Osada H, Usui T (2014) Terpendole E and its derivative inhibit STLC- and GSK-1-resistant Eg5. ChemBioChem 15: 934–938
5. Nakazawa J, Yajima J, Usui T, Ueki M, Takatsuki A, Imoto M, Toyoshima YY, Osada H (2003) A novel action of terpendole E on the motor activity of mitotic Kinesin Eg5. Chem Biol 10: 131–137

Terpestacin

Keywords: [Antiagiogenesis] [HIV-1 inhibitor] [Mitochondria complex III inhibitor]

Structure

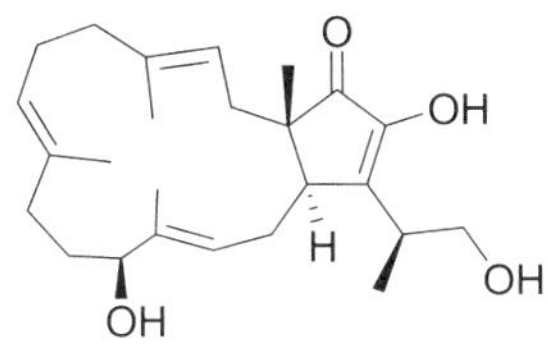

	Terpestacin
Molecular formula	$C_{25}H_{38}O_4$
Molecular weight	402.57
Solubility	DMSO, ++; H_2O, ±; MeOH, +++

Discovery/Isolation/Chemical Study

Terpestacin was isolated from *Arthrinium* sp. FA1744 (ATCC 74132) as an inhibitor of syncytium formation of HIV-infected T4 cells [1]. Total synthesis of natural terpestacin [2] and enantioselective synthesis [3] have been reported.

Biological Studies

Terpestacin potently inhibited the syncytium formation of BSC-1 cells infected with recombinant vaccinia virus expressing HIV-1 gp120 and gp41 and HeLa-T4 cells expressing CD4 antigen [1]. This compound also showed antiangiogenic activity [4]. A 13.4 kDa subunit of mitochondrial complex III (ubiquinol–cytochrome *c* reductase binding protein, UQCRB) was identified as a molecular target protein of terpestacin by phage-display affinity selection using biotinylated terpestacin [5]. The UQCRB binding of terpestacin resulted in the inhibition of hypoxia-induced ROS generation, followed by blocking of HIF-1α activation and tumor angiogenesis in vivo. Further analyses revealed that UQCRB plays a crucial role in the oxygen-sensing mechanism that regulates responses to hypoxia including VEGF expression [5, 6].

Biological Activity of Terpestacin

ID_{50} for syncytium formation: 0.46 μg/ml [1]
Inhibition of BAECs tube formation stimulated by 30 ng/ml bFGF: 10 μg/ml [4]
Instability of HIF-1α under hypoxia conditions: 50 μM [5]

References

1. Oka M, Iimura S, Tenmyo O, Sawada Y, Sugawara M, Ohkusa N, Yamamoto H, Kawano K, Hu SL, Fukagawa Y, et al. (1993) Terpestacin, a new syncytium formation inhibitor from *Arthrinium* sp. J Antibiot (Tokyo) 46:367–373
2. Tatsuta K, Masuda N (1998) The first total synthesis of natural (+)-terpestacin, syncytium formation inhibitor. J Antibiot (Tokyo) 51:602–606
3. Myers AG, Siu M, Ren F (2002) Enantioselective synthesis of (−)-terpestacin and (−)-fusaproliferin: clarification of optical rotational measurements and absolute configurational assignments establishes a homochiral structural series. J Am Chem Soc 124:4230–4232
4. Jung HJ, Lee HB, Kim CJ, Rho JR, Shin J, Kwon HJ (2003) Anti-angiogenic activity of terpestacin, a bicyclo sesterterpene from *Embellisia chlamydospora*. J Antibiot (Tokyo) 56:492–496
5. Jung HJ, Shim JS, Lee J, Song YM, Park KC, Choi SH, Kim ND, Yoon JH, Mungai PT, Schumacker PT, Kwon HJ (2010) Terpestacin inhibits tumor angiogenesis by targeting UQCRB of mitochondrial complex III and suppressing hypoxia-induced reactive oxygen species production and cellular oxygen sensing. J Biol Chem 285:11584–11595
6. Cho YS, Jung HJ, Seok SH, Payumo AY, Chen JK, Kwon HJ (2013) Functional inhibition of UQCRB suppresses angiogenesis in zebrafish. Biochem Biophys Res Commun 433:396–400

Thalidomide*, Lenalidomide*

Keywords: [Cereblon inhibitor] [E3 ubiquitin ligase inhibitor]

Structure

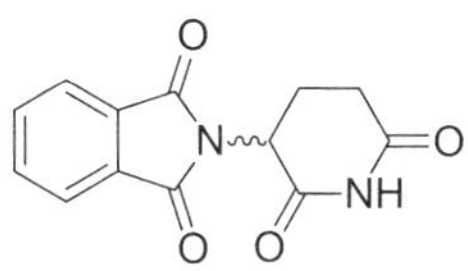

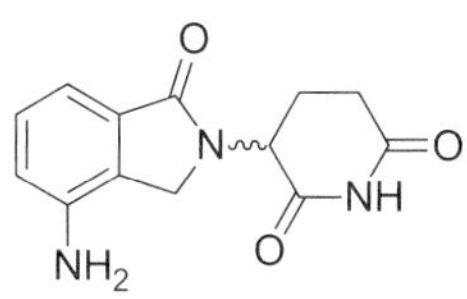

	Thalidomide	Lenalidomide
Molecular formula	$C_{13}H_{10}N_2O_4$	$C_{13}H_{13}N_3O_3$
Molecular weight	258.23	259.26
Solubility	DMSO, +++; H_2O, –; MeOH, +	DMSO, +++; H_2O, –; MeOH, +

Discovery

Thalidomide is a synthetic drug marketed in 1957 by Chemie Grunenthal as a sedative, hypnotic, nausea, and morning sickness drug for pregnant women [1], but the use of thalidomide was prohibited in many countries because of its teratogenic properties [2].

Biological Studies

Thalidomide caused many infants to be born with phocomelia and amelia because of its teratogenic effects. However, the discovery of benefical effects in erythema nodosum leprosum remarked the potential of thalidomide as a medicine again [3]. Furthermore, its antiangiogenesis activity has suggested that thalidomide is a promising medicine for cancer therapy [4]. Today, thalidomide and its derivatives, for example lenalidomide [5], have been used as a treatment for a complication of leprosy and multiple myeloma.

Using high-performance affinity beads, cereblon (CRBN), a component of E3 ubiquitin ligase complex, was identified as the binding protein of thalidomide [6]. This E3 complex has been shown to be important for limb outgrowth and expression of FGF8 in zebrafish and chicks. Thalidomide initiates its teratogenic effects by binding to CRBN and inhibiting the associated ubiquitin ligase activity.

Thalidomide and its derivatives promote the degradation of Ikaros proteins, transcription factors IKZF1 and IKZF3, in a CRBN-dependent fashion by altering the specificity of the E3 ubiquitin ligase complex [7, 8]. The crystal structure of the human CRBN-DDB1-thalidomide derivative complex was reported [9]. Based on the complex structure, it has revealed that the key drug binding residues in the thalidomide binding domain are critical for antiproliferative effects. These studies reveals a basis for thalidomide teratogenicity and may contribute to the development of new thalidomide derivatives without teratogenic activity.

Biological Activity of Thalidomide and Lenalidomide:

Thalidomide
Thalidomide-induced teratogenicity in zebrafish: 400 μM [6]

Lenalidomide
Down-regulation of Ikaros proteins by lenalidomide: 2 μM [8]

References

1. Franks ME, Macpherson GR, Figg WD. (2004) Thalidomide. Lancet 363:1802–1811
2. Speirs AL (1962) Thalidomide and congenital abnormalities. Lancet 279:303–305
3. Sheskin L. Thalidomide in the treatment of lepra reaction. (1965) Clin Pharmacol Ther 6303–306
4. D'Amato RJ, Loughnan MS, Flynn E, Folkman J.(1994) Thalidomide is an inhibitor of angiogenesis. Proc Natl Acad Sci U S A 91:4082–4085
5. Tsenova L, Mangaliso B, Muller G, Chen Y, Freedman VH, Stirling D, Kaplan G. (2002) Use of IMiD3, a thalidomide analog, as an adjunct to therapy for experimental tuberculous meningitis. Antimicrob Agents Chemother 46:1887–1895
6. Ito T, Ando H, Suzuki T, Ogura T, Hotta K, Imamura Y, Yamaguchi Y, Handa H (2010) Identification of a primary target of thalidomide teratogenicity. Science 327:1345–1350
7. Licht JD, Shortt J, Johnstone R (2014) From anecdote to targeted therapy: the curious case of thalidomide in multiple myeloma. Cancer Cell 25:9–11
8. Lu G, Middleton RE, Sun H, Naniong M, Ott CJ, Mitsiades CS, Wong KK, Bradner JE, Kaelin WG, Jr. (2014) The myeloma drug lenalidomide promotes the cereblon-dependent destruction of Ikaros proteins. Science 343:305–309
9. Chamberlain PP, Lopez-Girona A, Miller K, Carmel G, Pagarigan B, Chie-Leon B, Rychak E, Corral LG, Ren YJ, Wang M, Riley M, Delker SL, Ito T, Ando H, Mori T, Hirano Y, Handa H, Hakoshima T, Daniel TO, Cathers BE (2014) Structure of the human Cereblon-DDB1-lenalidomide complex reveals basis for responsiveness to thalidomide analogs. Nat Struct Mol Biol 21:803–809

Theonellamide F

Keywords: [Antifungal] [3β-hydroxysterol]

Structure

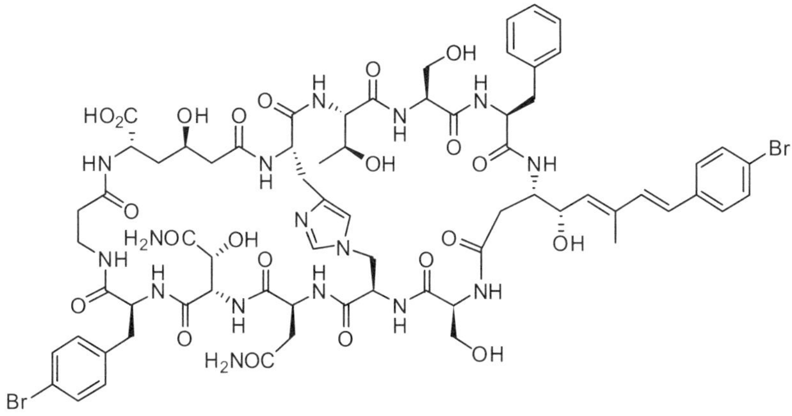

	Theonellamide F
Molecular formula	$C_{69}H_{86}Br_2N_{16}O_{22}$
Molecular weight	1651.35
Solubility	DMSO, +++; H_2O, ++; MeOH, +++

Discovery/Isolation/Chemical Study

Theonellamide F was isolated as an antifungal dodecapeptide from a marine sponge, genus *Theonella*. Its bicyclic structure, which contains an unprecedented histidinoalanine, was unequivocally determined [1].

Biological Studies

Theonellamide F induced the formation of many abnormally large vacuoles around the nucleus in the cells [2]. This effect was more drastic in an amino acid–deficient medium, suggesting that theonellamide F affects cellular autophagy by inhibiting the degradation of the organelles and turnover of proteins. Acidification by vacuolar ATPase (V-ATPase) was required to maintain the induced vacuoles [3].

Using affinity bead covalently bound theonellamide A, 17β-hydroxysteroid dehydrogenase IV and glutamate dehydrogenase were identified as the binding proteins [4]. However, the in vitro inhibitory activity could not account for the effects of theollenamide F. A chemical–genomic profile of theonellamide F used a collection of fission yeast strains in which each open reading frame is expressed under the control of an inducible promoter. Clustering analysis of the Gene Ontology terms associated with the genes that alter drug sensitivity suggested a mechanistic link between theonellamide and 1,3-β-D-glucan synthesis. Furthermore analysis revealed that theonellamides specifically bind to 3β-hydroxysterols, including ergosterol, and cause membrane damage. These results suggest that theonellamides are a new class of sterol binding molecules that induce membrane damage and activate Rho1-mediated 1,3-β-D-glucan synthesis [5].

Biological Activity of Theonellamide F

Vacuolation in exponentially growing 3Y1 cells: 6 μM [2]

Elevation of 1,3-β-D-glucan synthesis in *Schizosaccharomyces pombe*: 5 μg/ml [5]

References

1. Matsunaga S, Fusetani N, Hashimoto K, Walchli M (1989) Theonellamide F. A novel antifungal bicyclic peptide from a marine sponge *Theonella* sp. J Am Chem Soc 111:2582–2588
2. Wada S, Matsunaga S, Fusetani N, Watabe S (1999) Theonellamide F, a bicyclic peptide marine toxin, induces formation of vacuoles in 3Y1 rat embryonic fibroblast. Mar Biotechnol (NY) 1:337–341
3. Wada S, Kantha S, Yamashita T, Matsunaga S, Fusetani N, Watabe S (2002) Accumulation of H^+ in vacuoles induced by a marine peptide toxin, theonellamide F, in rat embryonic 3Y1 fibroblasts. Mar Biotechnol (NY) 4:571–582
4. Wada S, Matsunaga S, Fusetani N, Watabe S (2000) Interaction of cytotoxic bicyclic peptides, theonellamides A and F, with glutamate dehydrogenase and 17β-hydroxysteroid dehydrogenase IV. Mar Biotechnol (NY) 2:285–292
5. Nishimura S, Arita Y, Honda M, Iwamoto K, Matsuyama A, Shirai A, Kawasaki H, Kakeya H, Kobayashi T, Matsunaga S, Yoshida M (2010) Marine antifungal theonellamides target 3β-hydroxysterol to activate Rho1 signaling. Nat Chem Biol 6:519–526

Trapoxin B

Keywords: [Cell cycle inhibitor] [Histone deacetylase (HDAC) inhibitor]

Structure

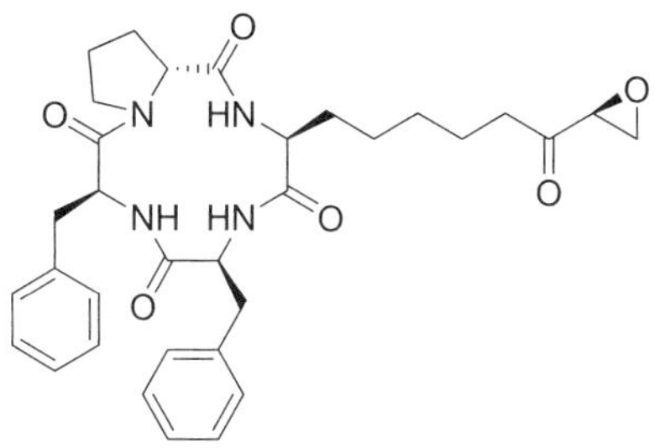

	Trapoxin B
Molecular formula	$C_{33}H_{40}N_4O_6$
Molecular weight	588.71
Solubility	DMSO, +++; H_2O, ±; MeOH, ++

Discovery/Isolation

Trapoxin B, a cyclotetrapeptide, was isolated from the culture broth of *Helicoma ambiens* RF-1023 as antitumor compound [1].

Biological Studies

Trapoxin B exhibits detransformation activities against v-*sis* oncogene-transformed NIH/3T3 cells as antitumor agents [1]. Kijima et al. found that trapoxin B accumulated highly acetylated core histones in a variety of mammalian cell lines [2]. In vitro experiments using partially purified mouse histone deacetylase (HDAC) showed that trapoxin B irreversibly inhibited deacetylation of acetylated histone molecules probably by covalent binding to histone deacetylase via the epoxide. This covalent binding ability made it possible to isolate histone deacetylase HDAC1 by using a synthesized trapoxin affinity matrix [3]. Isolated histone deacetylase protein is very similar to the yeast transcriptional regulator Rpd3p; these results support a role for histone deacetylase as a key regulator of eukaryotic transcription, and expression of several proteins has been shown, including transcription factor [4], cell cycle inhibitor [5, 6], and so on.

It was also revealed that HDAC proteins form complexes with protein phosphatases [7] or hypoxia-inducible factor 1α [8], suggesting that HDACs regulate not only protein expression but also phosphorylation/stabilization of protein complexes.

Another HDAC inhibitor, trichostatin A, inhibits all histone deacetylases (HDACs) of both class I and II, whereas trapoxin B cannot inhibit HDAC6, a cytoplasmic member of class II HDACs. Using this differential sensitivity to HDAC6, it was revealed that the acetylation and deacetylation are coupled to the microtubule turnover and that HDAC6 plays a key regulatory role in the stability of the dynamic microtubules [9].

Biological Activity of Trapoxin B

ID_{50} for growth of *sis*-transformed NIH3T3 cells: 200 ng/ml [1]
Histone hyperacetylation in various cell lines: 100 nM [2]
MIC values for FM3A and TR303, trichostatin-resistant mutant of FM3A: 2 and 50 ng/ml, respectively [2]

References

1. Itazaki H, Nagashima K, Sugita K, Yoshida H, Kawamura Y, Yasuda Y, Matsumoto K, Ishii K, Uotani N, Nakai H, et al. (1990) Isolation and structural elucidation of new cyclotetrapeptides, trapoxins A and B, having detransformation activities as antitumor agents. J Antibiot (Tokyo) 43:1524–1532
2. Kijima M, Yoshida M, Sugita K, Horinouchi S, Beppu T (1993) Trapoxin, an antitumor cyclic tetrapeptide, is an irreversible inhibitor of mammalian histone deacetylase. J Biol Chem 268:22429–22435
3. Taunton J, Hassig CA, Schreiber SL (1996) A mammalian histone deacetylase related to the yeast transcriptional regulator Rpd3p. Science 272:408–411
4. Hassig CA, Fleischer TC, Billin AN, Schreiber SL, Ayer DE (1997) Histone deacetylase activity is required for full transcriptional repression by mSin3A. Cell 89:341–347
5. Kim YB, Ki SW, Yoshida M, Horinouchi S (2000) Mechanism of cell cycle arrest caused by histone deacetylase inhibitors in human carcinoma cells. J Antibiot (Tokyo) 53:1191–1200
6. Sambucetti LC, Fischer DD, Zabludoff S, Kwon PO, Chamberlin H, Trogani N, Xu H, Cohen D (1999) Histone deacetylase inhibition selectively alters the activity and expression of cell cycle proteins leading to specific chromatin acetylation and antiproliferative effects. J Biol Chem 274:34940–34947
7. Brush MH, Guardiola A, Connor JH, Yao TP, Shenolikar S (2004) Deacetylase inhibitors disrupt cellular complexes containing protein phosphatases and deacetylases. J Biol Chem 279:7685–7691
8. Qian DZ, Kachhap SK, Collis SJ, Verheul HM, Carducci MA, Atadja P, Pili R (2006) Class II histone deacetylases are associated with VHL-independent regulation of hypoxia-inducible factor 1α. Cancer Res 66:8814–8821
9. Matsuyama A, Shimazu T, Sumida Y, Saito A, Yoshimatsu Y, Seigneurin-Berny D, Osada H, Komatsu Y, Nishino N, Khochbin S, Horinouchi S, Yoshida M (2002) In vivo destabilization of dynamic microtubules by HDAC6-mediated deacetylation. EMBO J 21:6820–6831

Trichostatin A

Keywords: [Cell cycle inhibitor] [Histone deacetylase (HDAC) inhibitor]

Structure

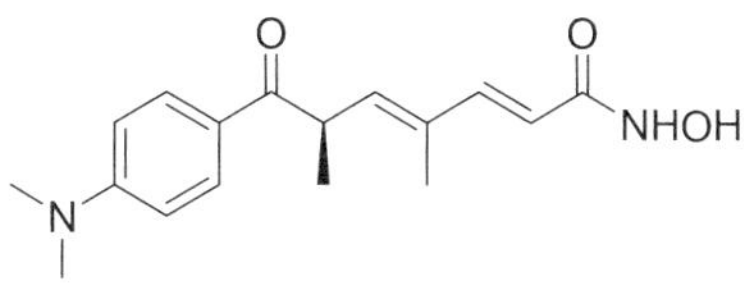

	Trichostatin A
Molecular formula	$C_{17}H_{22}N_2O_3$
Molecular weight	302.37
Solubility	DMSO, ++; H_2O, –; MeOH, ++

Discovery/Isolation

Trichostatin A was isolated from the metabolites of *Streptomyces hygroscopicus* as active compounds against trichophytons and some fungi [1], and reisolated as a potent inducer of differentiation in murine erythroleukemia cells [2].

Biological Studies

Trichostatin A blocked cell proliferation by inhibiting cell cycle progression specifically at the G1 and G2 phases [3]. Interestingly, when trichostatin A–arrested cells at the G2 phase were released from the inhibition, cells with 4C DNA entered a new S phase without passage through the M phase, resulting in the formation of proliferative tetraploid cells. Trichostatin A induces not only the morphological reversion of *sis*- or *ras*-transformed NIH/3T3, T24, and HeLa [4, 5], but also apoptosis [6, 7].

The molecular target of trichostatin A was biochemically determined as histone deacetylase (HDAC) [8]. Now it has been revealed that trichostatin A is a potent and specific inhibitor of the class I and II HDACs, and induces histone hyperacetylation. Because transcriptionally active chromatin is associated with hyperacetylated histones in general, the treatment of trichostatin A drastically changes the gene expression profiles [9, 10], and in vitro effects of trichostatin A on cell proliferation and differentiation can be attributed to the inhibition of HDACs [8].

Trichostatin A is a good bioprobe for investigating the functions of acetylation of histone on gene expression [9, 11], the relationships of histone acetylation and methylation [12, 13], and so on. Furthermore, it is known that HDAC deacetylates not only acetylated histones, but also the other proteins including p53 [14, 15], NF-κB [16], sp3 [17], Bcl-6 [18], and α-tubulin [19]. The functions of these proteins are also posttranslationally regulated by reversible acetylation; the regulatory mechanisms has been investigated using trichostatin A.

Biological Activity of Trichostatin A
Differentiation in murine erythroleukemia cells: 15 nM [2]
K_i for partially purified histone deacetylase from FM3A cells, noncompetitive: 3.4 nM [8]
Induction of normal and flat phenotype in *sis*-transformed NIH3T3 cells: 1 ng/ml [4]

References

1. Tsuji N, Kobayashi M, Nagashima K, Wakisaka Y, Koizumi K (1976) A new antifungal antibiotic, trichostatin. J Antibiot (Tokyo) 29:1–6
2. Yoshida M, Nomura S, Beppu T (1987) Effects of trichostatins on differentiation of murine erythroleukemia cells. Cancer Res 47:3688–3691
3. Yoshida M, Beppu T (1988) Reversible arrest of proliferation of rat 3Y1 fibroblasts in both the G1 and G2 phases by trichostatin A. Exp Cell Res 177:122–131
4. Sugita K, Koizumi K, Yoshida H (1992) Morphological reversion of sis-transformed NIH3T3 cells by trichostatin A. Cancer Res 52:168–172
5. Hoshikawa Y, Kwon HJ, Yoshida M, Horinouchi S, Beppu T (1994) Trichostatin A induces morphological changes and gelsolin expression by inhibiting histone deacetylase in human carcinoma cell lines. Exp Cell Res 214:189–197
6. Lee E, Furukubo T, Miyabe T, Yamauchi A, Kariya K (1996) Involvement of histone hyperacetylation in triggering DNA fragmentation of rat thymocytes undergoing apoptosis. FEBS Lett 395:183–187
7. McBain JA, Eastman A, Nobel CS, Mueller GC (1997) Apoptotic death in adenocarcinoma cell lines induced by butyrate and other histone deacetylase inhibitors. Biochem Pharmacol 53:1357–1368
8. Yoshida M, Kijima M, Akita M, Beppu T (1990) Potent and specific inhibition of mammalian histone deacetylase both in vivo and in vitro by trichostatin A. J Biol Chem 265:17174–17179
9. Bernstein BE, Tong JK, Schreiber SL (2000) Genomewide studies of histone deacetylase function in yeast. Proc Natl Acad Sci U S A 97:13708–13713
10. Schlake T, Klehr-Wirth D, Yoshida M, Beppu T, Bode J (1994) Gene expression within a chromatin domain: the role of core histone hyperacetylation. Biochemistry 33:4197–4206
11. Sheridan PL, Mayall TP, Verdin E, Jones KA (1997) Histone acetyltransferases regulate HIV-1 enhancer activity in vitro. Genes Dev 11:3327–3340
12. Cameron EE, Bachman KE, Myohanen S, Herman JG, Baylin SB (1999) Synergy of demethylation and histone deacetylase inhibition in the re-expression of genes silenced in cancer. Nat Genet 21:103–107

13. Nan X, Ng HH, Johnson CA, Laherty CD, Turner BM, Eisenman RN, Bird A (1998) Transcriptional repression by the methyl-CpG-binding protein MeCP2 involves a histone deacetylase complex. Nature 393:386–389
14. Roy S, Tenniswood M (2007) Site-specific acetylation of p53 directs selective transcription complex assembly. J Biol Chem 282:4765–4771
15. Zhao Y, Lu S, Wu L, Chai G, Wang H, Chen Y, Sun J, Yu Y, Zhou W, Zheng Q, Wu M, Otterson GA, Zhu WG (2006) Acetylation of p53 at lysine 373/382 by the histone deacetylase inhibitor depsipeptide induces expression of p21(Waf1/Cip1). Mol Cell Biol 26:2782–2790
16. Chen L, Fischle W, Verdin E, Greene WC (2001) Duration of nuclear NF-κB action regulated by reversible acetylation. Science 293:1653–1657
17. Ammanamanchi S, Freeman JW, Brattain MG (2003) Acetylated Sp3 is a transcriptional activator. J Biol Chem 278:35775–35780
18. Bereshchenko OR, Gu W, Dalla-Favera R (2002) Acetylation inactivates the transcriptional repressor BCL6. Nat Genet 32:606–613
19. Matsuyama A, Shimazu T, Sumida Y, Saito A, Yoshimatsu Y, Seigneurin-Berny D, Osada H, Komatsu Y, Nishino N, Khochbin S, Horinouchi S, Yoshida M (2002) In vivo destabilization of dynamic microtubules by HDAC6-mediated deacetylation. EMBO J 21:6820–6831

Tunicamycin

Keywords: [*N*-linked glycosylation inhibitor] [Secretion inhibitor]

Structure

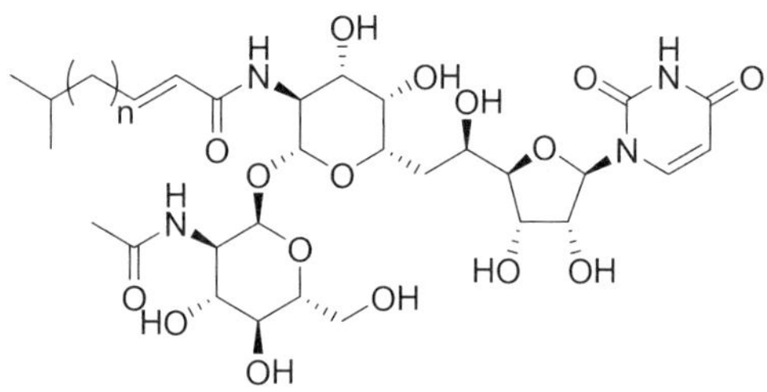

	Tunicamycin			
	n = 8	n = 9	n = 10	n = 11
Molecular formula	$C_{37}H_{60}N_4O_{16}$	$C_{38}H_{62}N_4O_{16}$	$C_{39}H_{64}N_4O_{16}$	$C_{40}H_{66}N_4O_{16}$
Molecular weight	816.90	830.93	844.95	858.99
Solubility	DMSO, +++; H_2O, ±; MeOH, +			

Discovery/Isolation/Chemical Studies

Tunicamycin was isolated as an antiviral antibiotic produced by *Streptomyces lysosuperificus*, nov. sp. [1]. Analytical studies on fatty acids which were obtained by hydrolysis of tunicamycin suggested that tunicamycin is a mixture of at least 10 homologous antibiotics [2].

Biological Studies

Tunicamycin is active against not only Gram-positive bacteria, yeast, and fungi but also animal and plant viruses [1]. Interestingly, this antiviral activity was partially reversed with some amino sugars and their derivatives [3], and the tunicamycin inhibition point was identified as the transfer of *N*-acetylglucosamine from UDP-GlcNAc to a lipid intermediate [4]. An in vitro glycolipid synthesis system revealed that tunicamycin inhibits UDP-*N*-acetylglucosamine:dolichol phosphate GlcNAc-1-P transferase (GPT) that catalyzes the first step in protein *N*-glycosylation [5, 6].

Because tunicamycin blocks the transfer of the common pentasaccharide core of *N*-glycan on asparagin residues of newly synthesized proteins, tunicamycin has been used as a bioprobe for investigating the numerous aspects of *N*-glycan function, including protein folding [7] and trafficking [8]. Furthermore, tunicamycin has been used as the reagent to induce the unfolded protein response (UPR)/ER stress [9, 10] and cellular response against UPR [11–13].

Biological Activity of Tunicamycin

MIC for Newcastle disease virus production by the agar-diffusion plaque inhibition method: 0.5 mg/ml [1]

Apoptosis induction in HL-60 cells: 0.05 μg/ml [14]

Inhibition of VEGF glycosylation in RPE-J cells: 10 μg/ml [8]

Induction of GRP78 mRNA in HeLa cells: 10 μg/ml [10]
Induction of autophagy in yeast: 2 μg/ml [12]

References

1. Takatsuki A, Arima K, Tamura G (1971) Tunicamycin, a new antibiotic. I. Isolation and characterization of tunicamycin. J Antibiot (Tokyo) 24:215–223
2. Takatsuki A, Kawamura K, Okina M, Kodama Y, Ito T, Tamura G (1977) The structure of tunicamycin. Agric Biol Chem 41:2307–2309
3. Takatsuki A, Tamura G (1971) Tunicamycin, a new antibiotic. 3. Reversal of the antiviral activity of tunicamycin by aminosugars and their derivatives. J Antibiot (Tokyo) 24:232–238
4. Takatsuki A, Kohno K, Tamura G (1975) Inhibition of biosynthesis of polyisoprenol sugars in chick embryo microsomes by tunicamycin. Arg BiolChem 39:2089–2091
5. Datema R, Schwarz RT (1978) Formation of 2-deoxyglucose-containing lipid-linked oligosaccharides. Interference with glycosylation of glycoproteins. Eur J Biochem 90:505–516
6. Xu L, Appell M, Kennedy S, Momany FA, Price NP (2004) Conformational analysis of chirally deuterated tunicamycin as an active site probe of UDP-*N*-acetylhexosamine: polyprenol-P *N*-acetylhexosamine-1-P translocases. Biochemistry 43:13248–13255
7. McCormick LM, Urade R, Arakaki Y, Schwartz AL, Bu G (2005) Independent and cooperative roles of *N*-glycans and molecular chaperones in the folding and disulfide bond formation of the low-density lipoprotein (LDL) receptor-related protein. Biochemistry 44:5794–5803
8. Marmorstein AD, Csaky KG, Baffi J, Lam L, Rahaal F, Rodriguez-Boulan E (2000) Saturation of, and competition for entry into, the apical secretory pathway. Proc Natl Acad Sci U S A 97:3248–3253
9. Brewer JW, Cleveland JL, Hendershot LM (1997) A pathway distinct from the mammalian unfolded protein response regulates expression of endoplasmic reticulum chaperones in non-stressed cells. EMBO J 16:7207–7216
10. Shinjo S, Mizotani Y, Tashiro E, Imoto M (2013) Comparative analysis of the expression patterns of UPR-target genes caused by UPR-inducing compounds. Biosci Biotechnol Biochem 77:729–735
11. Brewer JW, Hendershot LM, Sherr CJ, Diehl JA (1999) Mammalian unfolded protein response inhibits cyclin D1 translation and cell-cycle progression. Proc Natl Acad Sci U S A 96:8505–8510
12. Yorimitsu T, Nair U, Yang Z, Klionsky DJ (2006) Endoplasmic reticulum stress triggers autophagy. J Biol Chem 281:30299–30304
13. Kim H, Tu HC, Ren D, Takeuchi O, Jeffers JR, Zambetti GP, Hsieh JJ, Cheng EH (2009) Stepwise activation of BAX and BAK by tBID, BIM, and PUMA initiates mitochondrial apoptosis. Mol Cell 36:487–499
14. Perez-Sala D, Mollinedo F (1995) Inhibition of N-linked glycosylation induces early apoptosis in human promyelocytic HL-60 cells. J Cell Physiol 163:523–531

TWS119*

Keywords: [Glycogen synthase kinase-3β (GSK-3β) inhibitor] [Neurogenesis inducer]

Structure

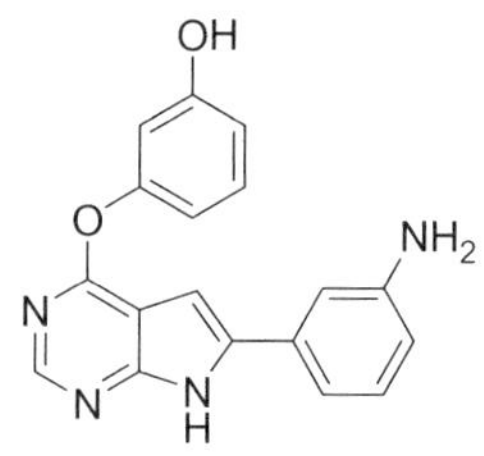

	TWS119
Molecular formula	$C_{18}H_{14}N_4O_2$
Molecular weight	318.33
Solubility	DMSO, +++; H_2O, –; MeOH, –

Discovery/Isolation/Chemical Studies

TWS119 is a 4,6-disubstituted pyrrolopyrimidine identified as a neurogenesis inducer of ES cells by high-throughput phenotypic cell-based screening of kinase-directed combinatorial libraries [1].

Biological Studies

Affinity-based and biochemical methods revealed that the target molecule of TWS119 is glycogen synthase kinase-3β (GSK-3β) [1]. GSK-3β stmulates the degradation of β-catenin in the Wnt signaling pathway by phosphorylation. TWS119 treatment resulted in increases in the amount and nuclear translocation of β-catenin, resulting in the activation of β-catenin-induced TCF/LEF reporter activity in a dose-dependent manner.

GSK-3β is a multifunctional Ser/Thr kinase involved in pattern formation during embryonic development, cell fate determination, transcriptional control, metabolism, oncogenesis, and neurological diseases. Therefore, TWS119 has been used for the investigation of biological events involving GSK-3β kinase [2–4].

Biological Activity of TWS119

Neuronal differentiation of P19 EC cells and primary mouse ES cells: 0.4–1 μM [1]
K_d for GSK-3β: 126 nM [1]
IC_{50} for kinase activity of GSK-3β: 30 nM [1]

References

1. Ding S, Wu TY, Brinker A, Peters EC, Hur W, Gray NS, Schultz PG (2003) Synthetic small molecules that control stem cell fate. Proc Natl Acad Sci U S A 100:7632–7637
2. Morioka N, Abe H, Araki R, Matsumoto N, Zhang FF, Nakamura Y, Hisaoka-Nakashima K, Nakata Y (2014) A β1/2 adrenergic receptor-sensitive intracellular signaling pathway modulates CCL2 production in cultured spinal astrocytes. J Cell Physiol 229:323–332
3. Muralidharan S, Hanley PJ, Liu E, Chakraborty R, Bollard C, Shpall E, Rooney C, Savoldo B, Rodgers J, Dotti G (2011) Activation of Wnt signaling arrests effector differentiation in human peripheral and cord blood-derived T lymphocytes. J Immunol 187:5221–5232
4. Zeng FY, Dong H, Cui J, Liu L, Chen T (2010) Glycogen synthase kinase 3 regulates PAX3-FKHR-mediated cell proliferation in human alveolar rhabdomyosarcoma cells. Biochem Biophys Res Commun 391:1049–1055

UCN-01

Keywords: [G2 checkpoint inhibitor] [Chk1 inhibitor] [α1-acid glycoprotein binder]

Structure

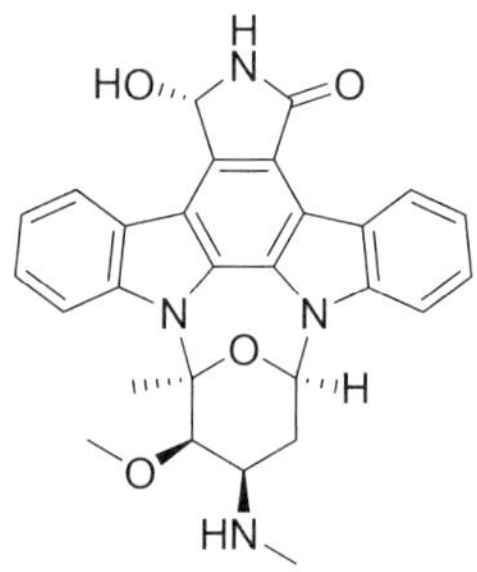

	UCN-01
Molecular formula	$C_{28}H_{26}N_4O_4$
Molecular weight	482.53
Solubility	DMSO, +++; H_2O, –; MeOH, +

Discovery/Isolation/Chemical Studies

UCN-01 was found in a culture broth of *Streptomyces* sp., which produces staurosporine. The structure of UCN-01 differs from staurosporine in that the C-7 carbon bears a hydroxyl group [1].

Biological Studies

UCN-01 was originally found as a protein kinase C–specific inhibitor compared with staurosporine [1] and inhibits cPKC [2]. UCN-01 shows strong cytotoxic effects on the growth of mammalian cells and shows antitumor activity in vivo against three human tumor xenografts [3]. UCN-01 retards but does not prevent cell cycle progression through phase, but cells are clearly blocked from exit of G1 and entry into S phase [4]. In clinically, UCN-01 sensitizes cells to ionizing radiation and chemotherapeutic agents. The important determinants of the sensitivity of cancer cells to UCN-01 are apoptosis induction and cell cycle arrest in G1 [5]. Synergic effects with histone deacetylase inhibitors were also reported [6].

UCN-01 abrogates G2 arrest following DNA damage through inhibition of a kinase residing upstream of Cdc2, Wee1Hu, and Cdc25C [7]. Because DNA damage induces G2 checkpoint activation by preventing the activation of Cdc25C and arrests cell cycle progression in G2 phase to provide time for repair, it has been thought that UCN-01 inhibits an upstream kinase of Cdc25C. Indeed, UCN-01 potently inhibited the ability of Chk1 to phosphorylate Cdc25C in vitro and caused loss of both serine 216 phosphorylation and 14-3-3 binding to Cdc25C in

DNA-damaged cells, strongly suggesting the Chk1 kinase and the Cdc25C pathway as potential targets of G2 checkpoint abrogation by UCN-01 [8]. UCN-01 also potently inhibits p38MAPK/MK2, a protein kinase involved in the cell cycle checkpoint pathway, suggesting that UCN-01 simultaneously disrupts two critical checkpoint pathways [9]. Furthermore, high concentrations of UCN-01 caused rapid premature mitosis by inhibiting C-TAK1, a constitutive kinase phosphorylating Cdc25C in S phase [10]. These results suggest that UCN-01 shows multiple phenotypes in a dose-dependent manner. The crystal structure of the Chk1 kinase domain in the complex revealed that UCN-01 binds the ATP binding pocket of Chk1. The potency and selectivity of UCN-01 to Chk1 is accounted for by high structural complementarity of both hydrophobic and hydrogen-bonding interactions [11]. UCN-01 also potently inhibits 3-phosphoinositide-dependent protein kinase-1 (PDK1). The crystal structure of UCN-01 in complex with the kinase domain of PDK1 has been reported [12].

UCN-01 shows unusual pharmacokinetics with extremely low distribution volumes, low systemic clearance, and long elimination half-lives in humans. Human α1-acid glycoprotein, but not human serum albumin, completely inhibited the initial uptake of UCN-01 into isolated rat hepatocytes. Therefore, the high-affinity binding of UCN-01 to α1-acid glycoprotein at least partly results in high plasma concentrations in humans [13, 14]. The binding site was thought to be around Trp160 by photoaffinity labeling and site-directed mutation analysis [15].

Biological Activity of UCN-01

IC_{50} values for protein kinase C, protein kinase A, and $pp60^{src}$ kinase: 4.1, 42, and 45 nM, respectively [1]

IC_{50} values for cdk2, 4, and 6: 42, 32, and 58 nM, respectively [16]

IC_{50} for Chk1 autophosphorylation and Cdc25C-serine 216 phosphorylation: ~25 nM [8]

References

1. Takahashi I, Kobayashi E, Asano K, Yoshida M, Nakano H. (1987) UCN-01, a selective inhibitor of protein kinase C from *Streptomyces*. J Antibiot (Tokyo) 40:1782–1784
2. Mizuno K, Noda K, Ueda Y, Hanaki H, Saido TC, Ikuta T, Kuroki T, Tamaoki T, Hirai S, Osada S, Ohno S (1995) UCN-01, an anti-tumor drug, is a selective inhibitor of the conventional PKC subfamily. FEBS Lett 359:259–261
3. Akinaga S, Gomi K, Morimoto M, Tamaoki T, Okabe M. (1991) Antitumor activity of UCN-01, a selective inhibitor of protein kinase C, in murine and human tumor models. Cancer Res 51:4888–4892
4. Seynaeve CM, Stetler-Stevenson M, Sebers S, Kaur G, Sausville EA, Worland PJ. (1993) Cell cycle arrest and growth inhibition by the protein kinase antagonist UCN-01 in human breast carcinoma cells. Cancer Res 53:2081–2086
5. Sugiyama K, Akiyama T, Shimizu M, Tamaoki T, Courage C, Gescher A, Akinaga S. (1999) Decrease in susceptibility toward induction of apoptosis and

alteration in G1 checkpoint function as determinants of resistance of human lung cancer cells against the antisignaling drug UCN-01 (7-Hydroxystaurosporine). Cancer Res 59:4406–4412
6. Lee JH, Choy ML, Ngo L, Venta-Perez G, Marks PA. (2011) Role of checkpoint kinase 1 (Chk1) in the mechanisms of resistance to histone deacetylase inhibitors. Proc Natl Acad Sci U S A 108:19629–19634
7. Shao RG, Cao CX, Zhang H, Kohn KW, Wold MS, Pommier Y. (1999) Replication-mediated DNA damage by camptothecin induces phosphorylation of RPA by DNA-dependent protein kinase and dissociates RPA:DNA-PK complexes. EMBO J. 18:1397–1406. EMBO J 1999 Mar 1;18(5):1397–406
8. Graves PR, Yu L, Schwarz JK, Gales J, Sausville EA, O'Connor PM, Piwnica-Worms H. (2000) The Chk1 protein kinase and the Cdc25C regulatory pathways are targets of the anticancer agent UCN-01. J Biol Chem 275:5600–5605
9. Reinhardt HC, Aslanian AS, Lees JA, Yaffe MB. (2007) p53-deficient cells rely on ATM- and ATR-mediated checkpoint signaling through the p38MAPK/MK2 pathway for survival after DNA damage. Cancer Cell 11:175–189
10. Kohn EA, Ruth ND, Brown MK, Livingstone M, Eastman A. (2002) Abrogation of the S phase DNA damage checkpoint results in S phase progression or premature mitosis depending on the concentration of 7-hydroxystaurosporine and the kinetics of Cdc25C activation. J Biol Chem 277:26553–26564
11. Zhao B, Bower MJ, McDevitt PJ, Zhao H, Davis ST, Johanson KO, Green SM, Concha NO, Zhou BB. (2002) Structural basis for Chk1 inhibition by UCN-01. J Biol Chem 277:46609–46615
12. Komander D, Kular GS, Bain J, Elliott M, Alessi DR, Van Aalten DM. (2003) Structural basis for UCN-01 (7-hydroxystaurosporine) specificity and PDK1 (3-phosphoinositide-dependent protein kinase-1) inhibition. Biochem J 375:255–262
13. Fuse E, Tanii H, Kurata N, Kobayashi H, Shimada Y, Tamura T, Sasaki Y, Tanigawara Y, Lush RD, Headlee D, Figg WD, Arbuck SG, Senderowicz AM, Sausville EA, Akinaga S, Kuwabara T, Kobayashi S. (1998) Unpredicted clinical pharmacology of UCN-01 caused by specific binding to human α1-acid glycoprotein. Cancer Res 58:3248–3253
14. Fuse E, Tanii H, Takai K, Asanome K, Kurata N, Kobayashi H, Kuwabara T, Kobayashi S, Sugiyama Y. (1999) Altered pharmacokinetics of a novel anticancer drug, UCN-01, caused by specific high affinity binding to α1-acid glycoprotein in humans. Cancer Res 59:1054–1060
15. Katsuki M, Chuang VT, Nishi K, Kawahara K, Nakayama H, Yamaotsu N, Hirono S, Otagiri M. (2005) Use of photoaffinity labeling and site-directed mutagenesis for identification of the key residue responsible for extraordinarily high affinity binding of UCN-01 in human α1-acid glycoprotein. J Biol Chem 280:1384–1391
16. Kawakami K, Futami H, Takahara J, Yamaguchi K. (1996) UCN-01, 7-hydroxyl-staurosporine, inhibits kinase activity of cyclin-dependent kinases and reduces the phosphorylation of the retinoblastoma susceptibility gene product in A549 human lung cancer cell line. Biochem Biophys Res Commun 219:778–783

UNC0638*

Keywords: [Protein lysine methyltransferases G9a and GLP inhibitor]

Structure

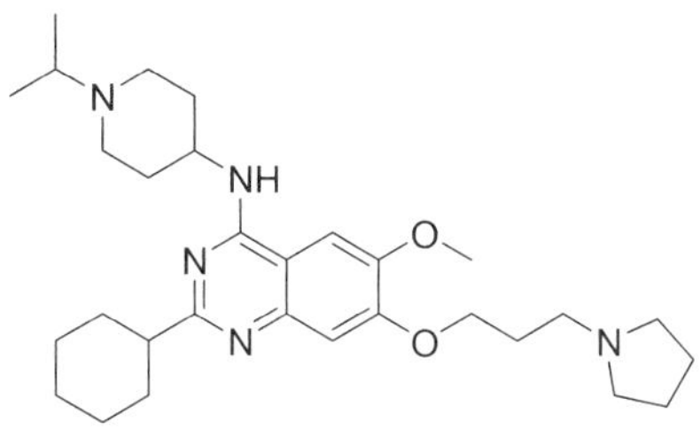

	UNC0638
Molecular formula	$C_{30}H_{47}N_5O_2$
Molecular weight	509.73
Solubility	DMSO, +++; H_2O, −; MeOH, ±

Discovery/Isolation/Chemical Studies

UNC0638 is a synthetic compound from the structure–activity relationships study of UNC0321, a potent and selective G9a and GLP inhibitor [1] which was synthesized based on the X-ray structures of complexes of GLP-BIX01294 and G9a-UNC0224 [2]. UNC0638 has balanced in vitro potency and physicochemical properties aiding cell penetration.

Biological Studies

UNC0638 potently and selectively inhibits G9a and GLP activities in a substrate-competitive manner [2]. The crystal structure of G9a-UNC0638-SAH (*S*-adenosyl-L-homocysteine) revealed that UCN0638 occupies the substrate binding groove without interaction with the SAM (*S*-adenosyl-L-methionine) binding pocket. Because UCN0638 possesses the cell penetration potency and is stable under cellular assay conditions, this compound potently inhibits cellular G9a and GLP without cytotoxicity and reactivated the silenced gene expression in mES cell by decreasing H3K9me2 (Lys9-dimethylated histone H3) levels.

It is thought that histone modification by protein methyltransfereases is associated with transcriptional repression during development and this reinforces lineage choice by affecting high-order chromatin structure. UNC0638-treated hematopoietic stem and progenitor cells retained stem cell–like phenotypes and function during in vitro expansion, suggesting that this compound is useful for stem cell therapy [3].

Chromatin modification is also seen in human cancers and is thought to be an attractive drug target for cancer therapy. The report that methyltransferase activity is connected with the activity of the leukomogenic transcription factor HoxA9 suggests a clinical potential of G9A inhibition as a means to counteract the proliferation and self-renewal of AML cells by attenuating HoxA9-dependent transcription [4].

Biological Activity of UNC0638
IC_{50} values for G9a and GLP: <15 and 19 ± 1 nM, respectively [2]
Stem cell–like phenotypes maintenance in hematopoietic stem and progenitor cells: 1 μM [3]
Growth inhibition and induction of myeloid differentiation of AML cells: 1 μM [4]

References

1. Liu F, Chen X, Allali-Hassani A, Quinn AM, Wigle TJ, Wasney GA, Dong A, Senisterra G, Chau I, Siarheyeva A, Norris JL, Kireev DB, Jadhav A, Herold JM, Janzen WP, Arrowsmith CH, Frye SV, Brown PJ, Simeonov A, Vedadi M, Jin J (2010) Protein lysine methyltransferase G9a inhibitors: design, synthesis, and structure activity relationships of 2,4-diamino-7-aminoalkoxy-quinazolines. J Med Chem 53:5844–5857
2. Vedadi M, Barsyte-Lovejoy D, Liu F, Rival-Gervier S, Allali-Hassani A, Labrie V, Wigle TJ, Dimaggio PA, Wasney GA, Siarheyeva A, Dong A, Tempel W, Wang SC, Chen X, Chau I, Mangano TJ, Huang XP, Simpson CD, Pattenden SG, Norris JL, Kireev DB, Tripathy A, Edwards A, Roth BL, Janzen WP, Garcia BA, Petronis A, Ellis J, Brown PJ, Frye SV, Arrowsmith CH, Jin J (2011) A chemical probe selectively inhibits G9a and GLP methyltransferase activity in cells. Nat Chem Biol 7:566–574
3. Chen X, Skutt-Kakaria K, Davison J, Ou YL, Choi E, Malik P, Loeb K, Wood B, Georges G, Torok-Storb B, Paddison PJ (2012) G9a/GLP-dependent histone H3K9me2 patterning during human hematopoietic stem cell lineage commitment. Genes Dev 26:2499–2511
4. Lehnertz B, Pabst C, Su L, Miller M, Liu F, Yi L, Zhang R, Krosl J, Yung E, Kirschner J, Rosten P, Underhill TM, Jin J, Hebert J, Sauvageau G, Humphries RK, Rossi FM (2014) The methyltransferase G9a regulates HoxA9-dependent transcription in AML. Genes Dev 28:317–327

UTKO1*

Keywords: [Migration inhibitor] [14-3-3ζ] [Protein-Protein interaction inhibitor]

Structure

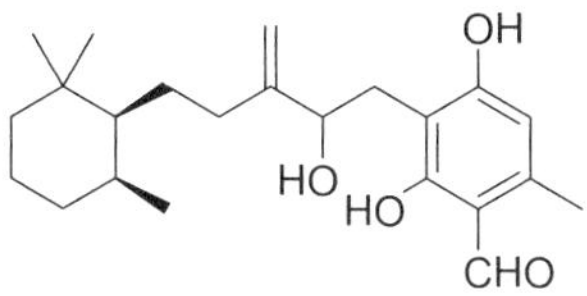

	UTKO1
Molecular formula	$C_{23}H_{34}O_4$
Molecular weight	374.52
Solubility	DMSO, +++; H_2O, ±; MeOH, +

Discovery/Isolation

UTKO1 is a synthetic analog of moverastins which were isolated from *Aspergillus* sp. F7720 as inhibitors of cancer cell migration [1, 2].

Biological Studies

UTKO1 was found to have the most potent inhibitory activity against the migration of human esophageal tumor EC17 cells in a chemotaxis cell chamber assay [1]. Interestingly, although moverastins are considered to inhibit tumor cell migration by inhibiting farnesyltransferase (FTase), UTKO1 did not inhibit FTase.

During the investigation of EGF-stimulating migration, Kobayashi et al. found that EGF induces biphasic Rac1 activation in the process of cell migration, and UTKO1 inhibits the second EGF-induced wave of Rac1 activation but not the first wave [3]. They focused on 14-3-3ζ, one of the UTKO1 binding proteins which were identified by affinity purification using a biotinylated UTKO1 probe, because it has been previously reported to relate to the formation of lamellipodia. Further investigation revealed that UTKO1 abrogated the binding of 14-3-3ζ to Tiam1 that was responsible for the second wave of Rac1 activation, suggesting that the interaction of 14-3-3ζ with Tiam1 is involved in this event.

Biological Activity of UTKO1

IC_{50} values for cell migration and cell viability against EC17 cells: 1.98 and 46 μM, respectively [1]

IC_{50} for second EGF-induced wave of lamellipodia formation in A431 cells: 0.78 μM [3]

IC_{50} for cell migration induced by EGF against A431 cells: 0.67 μM [3]

References

1. Sawada M, Kubo S, Matsumura K, Takemoto Y, Kobayashi H, Tashiro E, Kitahara T, Watanabe H, Imoto M (2011) Synthesis and anti-migrative evaluation of moverastin derivatives. Bioorg Med Chem Lett 21:1385–1389
2. Takemoto Y, Watanabe H, Uchida K, Matsumura K, Nakae K, Tashiro E, Shindo K, Kitahara T, Imoto M (2005) Chemistry and biology of moverastins, inhibitors of cancer cell migration, produced by *Aspergillus*. Chem Biol 12:1337–1347
3. Kobayashi H, Ogura Y, Sawada M, Nakayama R, Takano K, Minato Y, Takemoto Y, Tashiro E, Watanabe H, Imoto M (2011) Involvement of 14-3-3 proteins in the second epidermal growth factor-induced wave of Rac1 activation in the process of cell migration. J Biol Chem 286:39259–39268

Withaferin A

Keywords: [Antiangiogenesis] [Anti-inflammation] [Antitumor]

Structure

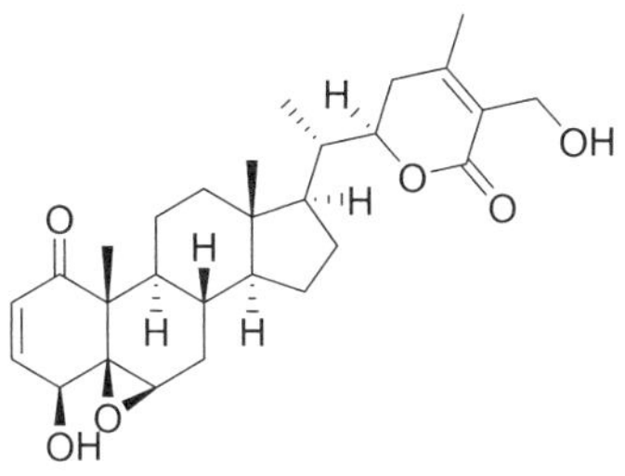

	Withaferin A
Molecular formula	$C_{28}H_{38}O_6$
Molecular weight	470.61
Solubility	DMSO, ++; H_2O, ±; MeOH, ++

Discovery/Isolation/Chemical Studies

Withaferin A was isolated from *Withania somnifera* Dun (*Solanaceae*) in 1962 [1], and its structure was reported in 1965 [2].

Biological Studies

There are many reports about biological activities of withaferin A: antitumor [3], antiangiogenesis [4], anti-inflammation, and so on. Several proteins including annexin II [5], protein kinase C [6], tumor proteasome [7], IKKβ [8], vimentin [9], HSP90 [10], glial fibrillary acidic protein (GFAP) [11], and β-tubulin [12] were suggested as target molecules of withaferin A. Because withaferin A possesses the α,β-unsaturated ketone in its structure, it has been thought that some of these proteins are inhibited by covalent binding of withaferin A. The binding sites of annexin II (Cys133), GFAP (Cys294), β-tubulin (Cys303), and IKKβ (Cys179) have also been reported [12, 11, 5, 13].

Biological Activity of Withaferin A

Actin filament disruption in WI-38 and HepG2 cells: 4 μM [5]
IC_{50} for protein kinase C activity of *Leishmania*: ~5 μM [6]
IC_{50} for chymotrypsin-like activity of a purified rabbit 20S proteasome: 4.5 μM [7]
IC_{50} for pancreatic cancer cells: 1.24–2.93 μM [10]

References

1. Yarden A, Lavie D (1962) Withaferin – a new constituent of *Withania somnifera*. B Res Counc Israel A 11:35
2. Lavie D, Glotter E, Shvo Y (1965) Constituents of *Withania somnifera* Dun .4. Structure of withaferin A. J Chem Soc (Dec):7517–7531
3. Shohat B, Gitter S, Abraham A, Lavie D (1967) Antitumor activity of withaferin A (NSC-101088). Cancer Chemother Rep 51:271–276
4. Mohan R, Hammers HJ, Bargagna-Mohan P, Zhan XH, Herbstritt CJ, Ruiz A, Zhang L, Hanson AD, Conner BP, Rougas J, Pribluda VS (2004) Withaferin A is a potent inhibitor of angiogenesis. Angiogenesis 7:115–122
5. Falsey RR, Marron MT, Gunaherath GM, Shirahatti N, Mahadevan D, Gunatilaka AA, Whitesell L (2006) Actin microfilament aggregation induced by withaferin A is mediated by annexin II. Nat Chem Biol 2:33–38
6. Sen N, Banerjee B, Das BB, Ganguly A, Sen T, Pramanik S, Mukhopadhyay S, Majumder HK (2007) Apoptosis is induced in leishmanial cells by a novel protein kinase inhibitor withaferin A and is facilitated by apoptotic topoisomerase I-DNA complex. Cell Death Differ 14:358–367
7. Yang H, Shi G, Dou QP (2007) The tumor proteasome is a primary target for the natural anticancer compound Withaferin A isolated from "Indian winter cherry". Mol Pharmacol 71:426–437
8. Kaileh M, Vanden Berghe W, Heyerick A, Horion J, Piette J, Libert C, De Keukeleire D, Essawi T, Haegeman G (2007) Withaferin A strongly elicits IκB kinase β hyperphosphorylation concomitant with potent inhibition of its kinase activity. J Biol Chem 282:4253–4264
9. Bargagna-Mohan P, Hamza A, Kim YE, Khuan Abby Ho Y, Mor-Vaknin N, Wendschlag N, Liu J, Evans RM, Markovitz DM, Zhan CG, Kim KB, Mohan R (2007) The tumor inhibitor and antiangiogenic agent withaferin A targets the intermediate filament protein vimentin. Chem Biol 14:623–634
10. Yu Y, Hamza A, Zhang T, Gu M, Zou P, Newman B, Li Y, Gunatilaka AA, Zhan CG, Sun D (2010) Withaferin A targets heat shock protein 90 in pancreatic cancer cells. Biochem Pharmacol 79:542–551
11. Bargagna-Mohan P, Paranthan RR, Hamza A, Dimova N, Trucchi B, Srinivasan C, Elliott GI, Zhan CG, Lau DL, Zhu H, Kasahara K, Inagaki M, Cambi F, Mohan R (2010) Withaferin A targets intermediate filaments glial fibrillary acidic protein and vimentin in a model of retinal gliosis. J Biol Chem 285:7657–7669
12. Antony ML, Lee J, Hahm ER, Kim SH, Marcus AI, Kumari V, Ji X, Yang Z, Vowell CL, Wipf P, Uechi GT, Yates NA, Romero G, Sarkar SN, Singh SV (2014) Growth arrest by the antitumor steroidal lactone withaferin A in human breast cancer cells is associated with down-regulation and covalent binding at cysteine 303 of β-tubulin. J Biol Chem 289:1852–1865
13. Heyninck K, Lahtela-Kakkonen M, Van der Veken P, Haegeman G, Vanden Berghe W (2014) Withaferin A inhibits NF-κB activation by targeting cysteine 179 in IKKβ. Biochem Pharmacol 91:501–509

Wortmannin

Keywords: [Apoptosis inducer] [PI3 kinase inhibitor]

Structure

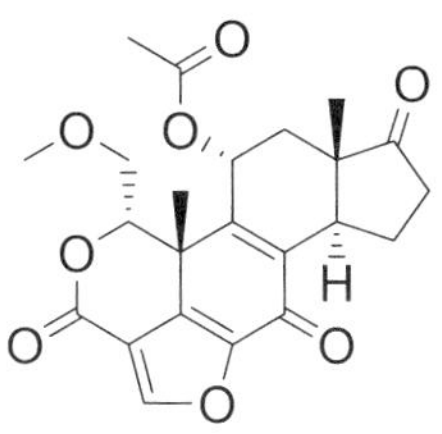

	Wortmannin
Molecular formula	$C_{23}H_{24}O_8$
Molecular weight	428.43
Solubility	DMSO, +++; H_2O, ±; MeOH, ±

Discovery/Isolation

Wortmannin is an antifungal antibiotic isolated from a culture of *Penicillium wortmanni* Klocker [1].

Biological Studies

Wortmannin showed a potent anti-inflammatory activity [2], and inhibited the signal transduction initiated by the bacterial stimulus which leads to the activation of the respiratory burst oxidase [3]. Okada et al. found that wortmannin is a potent and selective inhibitor of phosphatidylinositol 3-kinase (PI3 kinase, PI3K) [4]. Wortmannin inhibits PI3K irreversibly by covalent modification of Lys802, a residue involved in the phosphate transfer reaction [5, 6].

Wortmannin has been used as a bioprobe for investigating the function of PI3P and PI3P signaling pathway including Akt [7, 8] and ATM [9], downstream of PI3K. Wortmannin is also used for cell surviving signals [10–12], the autophagy mechanism [13], and endosome regulation [14].

Biological Activity of Wortmannin

Growth inhibition of *Botrytis allii*, *B. cincerea*, *B. fabae*, *Cladosporium herbarum* and *Rhizopus stolonifer*: 0.4–3.2 μg/ml [1]

IC_{50} for noncompetitive inhibition of all three of the PI3-kinase activities found in the cytosol fraction of guinea pig neutrophils: 5 nM [4]

Inhibition of the phosphorylation of Akt in HEK293 and HeLa cells: 100 nM [12]

IC_{50} for antiproteolytic effect on autophagic sequestration: 30 nM [13]

Inhibition of early endosome fusion in vitro prepared wortmannin cell: 100 nM [14]

References

1. Brian PW, Curtis PJ, Hemming HG, Norris GLF (1957) Wortmannin, an antibiotic produced by *Penicillium wortmanni*. Trans Br Mycol Soc 40:365–368
2. Wiesinger D, Gubler HU, Haefliger W, Hauser D (1974) Antiinflammatory activity of the new mould metabolite 11-desacetoxy-wortmannin and of some of its derivatives. Experientia 30:135–136
3. Baggiolini M, Dewald B, Schnyder J, Ruch W, Cooper PH, Payne TG (1987) Inhibition of the phagocytosis-induced respiratory burst by the fungal metabolite wortmannin and some analogues. Exp Cell Res 169:408–418
4. Okada T, Sakuma L, Fukui Y, Hazeki O, Ui M (1994) Blockage of chemotactic peptide-induced stimulation of neutrophils by wortmannin as a result of selective inhibition of phosphatidylinositol 3-kinase. J Biol Chem 269:3563–3567
5. Norman BH, Shih C, Toth JE, Ray JE, Dodge JA, Johnson DW, Rutherford PG, Schultz RM, Worzalla JF, Vlahos CJ (1996) Studies on the mechanism of phosphatidylinositol 3-kinase inhibition by wortmannin and related analogs. J Med Chem 39:1106–1111
6. Wymann MP, Bulgarelli-Leva G, Zvelebil MJ, Pirola L, Vanhaesebroeck B, Waterfield MD, Panayotou G (1996) Wortmannin inactivates phosphoinositide 3-kinase by covalent modification of Lys-802, a residue involved in the phosphate transfer reaction. Mol Cell Biol 16:1722–1733
7. Burgering BM, Coffer PJ (1995) Protein kinase B (c-Akt) in phosphatidylinositol-3-OH kinase signal transduction. Nature 376:599–602
8. Franke TF, Yang SI, Chan TO, Datta K, Kazlauskas A, Morrison DK, Kaplan DR, Tsichlis PN (1995) The protein kinase encoded by the Akt proto-oncogene is a target of the PDGF-activated phosphatidylinositol 3-kinase. Cell 81:727–736
9. Banin S, Moyal L, Shieh S, Taya Y, Anderson CW, Chessa L, Smorodinsky NI, Prives C, Reiss Y, Shiloh Y, Ziv Y (1998) Enhanced phosphorylation of p53 by ATM in response to DNA damage. Science 281:1674–1677
10. Yao R, Cooper GM (1995) Requirement for phosphatidylinositol-3 kinase in the prevention of apoptosis by nerve growth factor. Science 267:2003–2006
11. Dudek H, Datta SR, Franke TF, Birnbaum MJ, Yao R, Cooper GM, Segal RA, Kaplan DR, Greenberg ME (1997) Regulation of neuronal survival by the serine-threonine protein kinase Akt. Science 275:661–665
12. Ozes ON, Mayo LD, Gustin JA, Pfeffer SR, Pfeffer LM, Donner DB (1999) NF-κB activation by tumour necrosis factor requires the Akt serine-threonine kinase. Nature 401:82–85
13. Blommaart EF, Krause U, Schellens JP, Vreeling-Sindelárová H, Meijer AJ. (1997) The phosphatidylinositol 3-kinase inhibitors wortmannin and LY294002 inhibit autophagy in isolated rat hepatocytes. Eur J Biochem 243:240–246
14. Simonsen A, Lippe R, Christoforidis S, Gaullier JM, Brech A, Callaghan J, Toh BH, Murphy C, Zerial M, Stenmark H (1998) EEA1 links PI(3)K function to Rab5 regulation of endosome fusion. Nature 394:494–498

Xanthohumol

Keywords: [Antiangiogesis] [Antitumor] [Autophagy] [Valosin-containing protein (VCP) inhibitor]

Structure

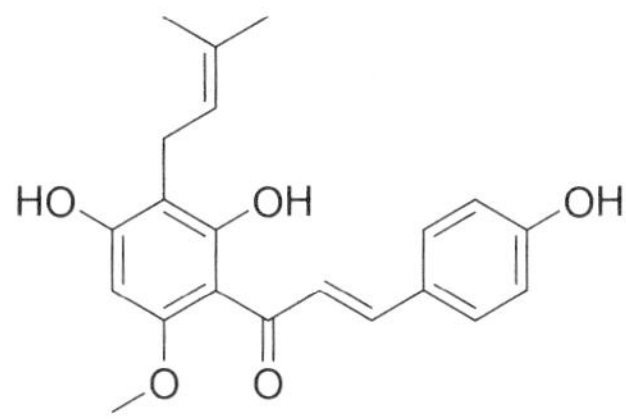

	Xanthohumol
Molecular formula	$C_{21}H_{22}O_5$
Molecular weight	354.40
Solubility	DMSO, ++; H_2O, ±; MeOH, +

Discovery/Isolation/Chemical Study

Xanthohumol is a natural chalcone isolated from hops [1, 2]. The structure was determined by Verzele et al. [3].

Biological Studies

There are several reports about biological activities of xanthohumol as a bone resorption inhibitor [4], diacylglycerol acyltransferase inhibitor [5], farnesoid X receptor agonist [6], angiogenesis inhibitor [7], antitumor compound [8], and so on. It has been reported that xanthohumol represses NF-κB and Akt pathways and induces apoptosis [7, 8].

Sasazawa et al. rediscovered xanthohumol as an autophagy modulator, and identified valosin-containing protein (VCP) as a xanthohumol binding protein by using xanthohumol-immobilized beads [9]. Xanthohumol bound directly to the N domain, which is known to mediate cofactor and substrate binding to VCP. VCP knockdown also induced autophagy, and xanthohumol increased the protein levels of CHOP and GRP78, suggesting that xanthohumol bound to and inactivated VCP. These results may explain how xanthohumol exhibits various biological activities that have been reported previously.

Biological Activity of Xanthohumol

IC_{50} for diacylglycerol acyltransferase: 50.3 μM [5]
Inhibition of endothelial cell chemotaxis and invasion: 5–10 μM [7]
Growth inhibition of prostate cancer cells, LNCaP, C4-2, PC-3 and DU145: 20–40 μM [8]
Autophagy induction in HeLa cells: 10–30 μM [9]

References

1. Power FB, Tutin F, Rogerson H (1913) CXXXV.-The constituents of hops. J Chem Soc Trans 103:1267–1292
2. Stevens JF, Page JE (2004) Xanthohumol and related prenylflavonoids from hops and beer: to your good health! Phytochemistry 65:1317–1330
3. Verzele M, Stockx J, Fontijn F, Anteunis M (1957) Xanthohumol, a new natural chalcone. Bull Soc Chim Belg 66:452–475
4. Tobe H, Muraki Y, Kitamura K, Komiyama O, Sato Y, Sugioka T, Maruyama HB, Matsuda E, Nagai M (1997) Bone resorption inhibitors from hop extract. Biosci Biotechnol Biochem 61:158–159
5. Tabata N, Ito M, Tomoda H, Omura S (1997) Xanthohumols, diacylglycerol acyltransferase inhibitors, from *Humulus lupulus*. Phytochemistry 46:683–687
6. Nozawa H (2005) Xanthohumol, the chalcone from beer hops (*Humulus lupulus* L.), is the ligand for farnesoid X receptor and ameliorates lipid and glucose metabolism in KK-A(y) mice. Biochem Biophys Res Commun 336:754–761
7. Albini A, Dell"Eva R, Vene R, Ferrari N, Buhler DR, Noonan DM, Fassina G (2006) Mechanisms of the antiangiogenic activity by the hop flavonoid xanthohumol: NF-kappaB and Akt as targets. FASEB J 20:527–529
8. Deeb D, Gao X, Jiang H, Arbab AS, Dulchavsky SA, Gautam SC (2010) Growth inhibitory and apoptosis-inducing effects of xanthohumol, a prenylated chalone present in hops, in human prostate cancer cells. Anticancer Res 30:3333–3339
9. Sasazawa Y, Kanagaki S, Tashiro E, Nogawa T, Muroi M, Kondoh Y, Osada H, Imoto M (2012) Xanthohumol impairs autophagosome maturation through direct inhibition of valosin-containing protein. ACS Chem Biol 7:892–900

XAV939*

Keywords: [Antitumor] [PARP/tankyrase inhibitior] [Wnt signaling]

Structure

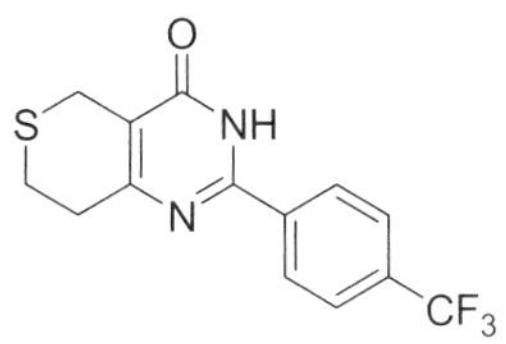

	XAV939
Molecular formula	$C_{14}H_{11}F_3N_2OS$
Molecular weight	312.31
Solubility	DMSO, ++; H_2O, –; MeOH, –

Discovery/Isolation

XAV939 was identifed as a selective inhibitor of β-catenin-mediated transcription by a chemical–genetic screen [1].

Biological Studies

Using a quantitative chemical proteomic approach, it has been revealed that XAV939 stimulated the degradation of β-catenin through the stabilization of axin by inhibiting the poly-ADP-ribosylating enzymes tankyrase 1 and tankyrase 2 [1]. The cocrystal structure of the PARP domain of human tankyrase-2 revealed that XAV939 is anchored in the nicotinamide crevice of the NAD^+ binding site [2].

Because Wnt/β-catenin signaling regulates several biological mechanisms including embryonic development, cell proliferation, cell migration, and so on, several research groups have used XAV939 for investigations. Wang et al. reported that XAV939 robustly induced cardiomyogenesis in mouse ES cells [3], and Fancy et al. found the essential function of Axin2 in remyelination after hypoxic and demyelinating injury using XAV939 [4].

Poly-ADP-ribosylation is involved not only in the Wnt/β-catenin signaling pathway, but also in mitotic spindle formation, sister telomere resolution, and other functions. Li et al. investigated the function of poly-ADP-ribosylation in interphase, and found that poly-ADP-ribosylation is required for efficient replication of herpes simplex virus, and that the tankyrase inhibition by VAX939 decreased viral titers [5]. The other function of tankyrase, the assembly of 26S proteasome through the inhibition of tankyrase-mediated ADP-ribosylation of PI31, a proteaseome regulator, has also been confirmed using XAV939 [6].

Biological Activity of XAV939

IC_{50} values for TNKS1, TNKS2, and PARP1: 11, 4, and 2194 nM, respectively [1]
K_d values for TNKS1, TNKS2, and PARP1: 14 ± 8, 8 ± 3, and 620 ± 130 nM, respectively [2]
IC_{50} for growth of MDA-MB-231 cells in low-serum conditions: 1.5 μM [7]

References

1. Huang SM, Mishina YM, Liu S, Cheung A, Stegmeier F, Michaud GA, Charlat O, Wiellette E, Zhang Y, Wiessner S, Hild M, Shi X, Wilson CJ, Mickanin C, Myer V, Fazal A, Tomlinson R, Serluca F, Shao W, Cheng H, Shultz M, Rau C, Schirle M, Schlegl J, Ghidelli S, Fawell S, Lu C, Curtis D, Kirschner MW, Lengauer C, Finan PM, Tallarico JA, Bouwmeester T, Porter JA, Bauer A, Cong F (2009) Tankyrase inhibition stabilizes axin and antagonizes Wnt signalling. Nature 461:614–620
2. Karlberg T, Markova N, Johansson I, Hammarstrom M, Schutz P, Weigelt J, Schuler H (2010) Structural basis for the interaction between tankyrase-2 and a potent Wnt-signaling inhibitor. J Med Chem 53:5352–5355
3. Wang H, Hao J, Hong CC (2011) Cardiac induction of embryonic stem cells by a small molecule inhibitor of Wnt/β-catenin signaling. ACS Chem Biol 6:192–197
4. Fancy SP, Harrington EP, Yuen TJ, Silbereis JC, Zhao C, Baranzini SE, Bruce CC, Otero JJ, Huang EJ, Nusse R, Franklin RJ, Rowitch DH (2011) Axin2 as regulatory and therapeutic target in newborn brain injury and remyelination. Nat Neurosci 14:1009–1016
5. Li Z, Yamauchi Y, Kamakura M, Murayama T, Goshima F, Kimura H, Nishiyama Y (2012) Herpes simplex virus requires poly(ADP-ribose) polymerase activity for efficient replication and induces extracellular signal-related kinase-dependent phosphorylation and ICP0-dependent nuclear localization of tankyrase 1. J Virol 86:492–503
6. Cho-Park PF, Steller H (2013) Proteasome regulation by ADP-ribosylation. Cell 153:614–627
7. Bao R, Christova T, Song S, Angers S, Yan X, Attisano L (2012) Inhibition of tankyrases induces Axin stabilization and blocks Wnt signalling in breast cancer cells. PLoS One 7:e48670

Xestospongin B and C

Keywords: [Vasodilative activity] [IP3 receptor agonist]

Structure

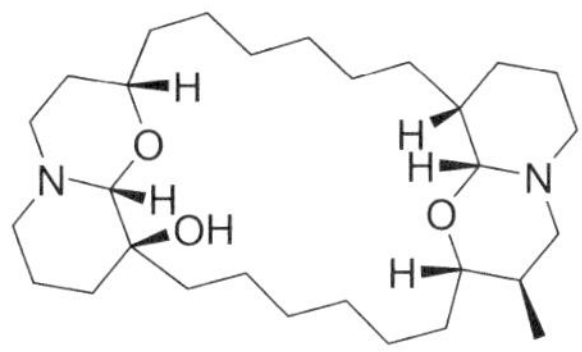

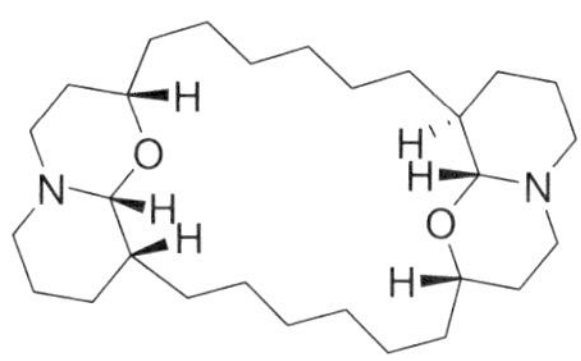

	Xestospongin B	Xestospongin C
Molecular formula	$C_{29}H_{52}N_2O_3$	$C_{28}H_{50}N_2O_2$
Molecular weight	476.75	446.72
Solubility	DMSO, +; H_2O, ±; MeOH, +	DMSO, +; H_2O, ±; MeOH, +

Discovery/Isolation

Xestospongin B was isolated from Australian marine sponge *Xestospongia exugua* as a novel vasodilative compound which induces relaxation of blood vessel in vivo [1].

Biological Studies

By screening of marine sponges for inhibitor of 1,4,5-trisphosphaste receptor (IP_3R), it was found that *Xestospongia* species contained potent blockers of IP_3-mediated Ca^{2+} release including xestospongin A, C, D, araguspongine B, and demethylxestospongin B [2]. Xestospongin C, a most potent IP_3R inhibitor among these five compounds, shows a noncompetitive inhibition mechanism with respect to the IP_3 binding site. Interestingly, xestospongin B is a competitive inhibitor of IP_3-mediated Ca^{2+} release [3]. Both compounds are membrane-permeable blockers of IP_3-mediated Ca^{2+} release in intact cells, and selectively inhibited bradykinin- and carbamylocholine-induced Ca^{2+} efflux from ER.

Xestospongin B induced autophagy in human cells by disrupting a molecular complex of IP_3R and Beclin 1 [4]. ER-targeted Bcl-2 increased the interaction of IP_3R and Beclin 1, resulting in the suppression of xestospongin B-induced autophagy [4, 5]. These results suggest that IP_3R is a new regulator of the Beclin 1 complex involving the autophagy cascade.

Biological Activity of Xestospongin B and C

IC_{50} for IP_3-mediated Ca^{2+} release of xestospongin C in rabbit cerebellar microsomes: 358 nM [2]

K_i for xestospongin B on IP_3 binding for cerevellar membrane: 31 μM [3]

Inhibition of xestospongin B on bradykinin-induced Ca^{2+} signaling in NG108-15 cells: 27 μM [3]

Autophagy induction in HeLa cells: 2 μM [4]

References

1. Nakagawa M, Endo M, Tanaka N, Lee GP (1984) Structures of xestospongin-A, xestospongin-B, xestospongin-C and xestospongin-D, novel vasodilative compounds from marine sponge, *Xestospongia exigua*. Tetrahedron Lett 25:3227–3230
2. Gafni J, Munsch JA, Lam TH, Catlin MC, Costa LG, Molinski TF, Pessah IN (1997) Xestospongins: potent membrane permeable blockers of the inositol 1,4,5-trisphosphate receptor. Neuron 19:723–733
3. Jaimovich E, Mattei C, Liberona JL, Cardenas C, Estrada M, Barbier J, Debitus C, Laurent D, Molgo J (2005) Xestospongin B, a competitive inhibitor of IP3-mediated Ca^{2+} signalling in cultured rat myotubes, isolated myonuclei, and neuroblastoma (NG108-15) cells. FEBS Lett 579:2051–2057
4. Vicencio JM, Ortiz C, Criollo A, Jones AW, Kepp O, Galluzzi L, Joza N, Vitale I, Morselli E, Tailler M, Castedo M, Maiuri MC, Molgo J, Szabadkai G, Lavandero S, Kroemer G (2009) The inositol 1,4,5-trisphosphate receptor regulates autophagy through its interaction with Beclin 1. Cell Death Differ 16:1006–1017
5. Criollo A, Vicencio JM, Tasdemir E, Maiuri MC, Lavandero S, Kroemer G (2007) The inositol trisphosphate receptor in the control of autophagy. Autophagy 3:350–353

ZM447439*

Keywords: [Aurora kinase inhibitor]

Structure

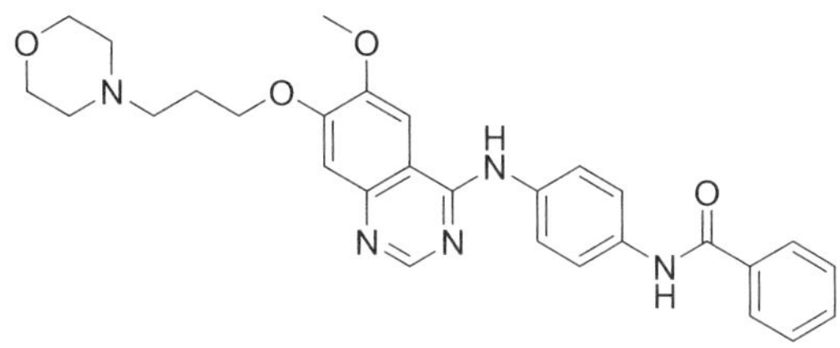

	ZM447439
Molecular formula	$C_{29}H_{31}N_5O_4$
Molecular weight	513.60
Solubility	DMSO, +++; H_2O, +; MeOH, +++

Discovery/Isolation

ZM447439 is one of the first-generation Aurora kinase inhibitors which was developed by AstraZeneca [1]. ZM-447439 inhibits Aurora kinase A (IC_{50} 1000 nM, K_i 100 nM), Aurora kinase B (IC_{50} 50 nM, K_i 0.36 nM) and Aurora kinase C. The compound specifically interacts at the ATP binding site and at the adjacent cleft, found only in Aurora kinases.

Biological Studies

ZM447439 has been shown to induce growth arrest, inhibit histone H3 phosphorylation, and effectively activate apoptosis. In the presence of ZM447439, several properties during mitotic progression such as chromosome alignment, segregation, and cytokinesis during mitosis are perturbed. It is also shown that ZM447439 can override the process of the spindle assembly checkpoint and accelerate meiosis I in mouse oocytes. Based on the use of this inhibitor, Aurora B was found to be important for chromosomal alignment through kinetochore–microtubule attachments in metaphase, activation of the spindle assembly checkpoint, and cytokinesis [2].

Biological Activity of ZM447439

IC_{50} values for Aurora kinase A and Aurora kinase B: 1000 nM and 50 nM, respectively [1]

K_i values for Aurora kinase A and Aurora kinase B: 100 nM and 0.36 nM, respectively [1]

References

1. Ditchfield C, Johnson VL, Tighe A, Ellston R, Haworth C, Johnson T, Mortlock A, Keen N, Taylor SS (2003) Aurora B couples chromosome alignment with anaphase by targeting BubR1, Mad2, and Cenp-E to kinetochores. J Cell Biol 161:267–280
2. Gadea BB, Ruderman JV (2005) Aurora kinase inhibitor ZM447439 blocks chromosome-induced spindle assembly, the completion of chromosome condensation, and the establishment of the spindle integrity checkpoint in *Xenopus* egg extracts. Mol Biol Cell 16:1305–1318

ZSTK474*

Keywords: [Antiangiogeneisis] [Antitumor] [PI3K inhibitior]

Structure

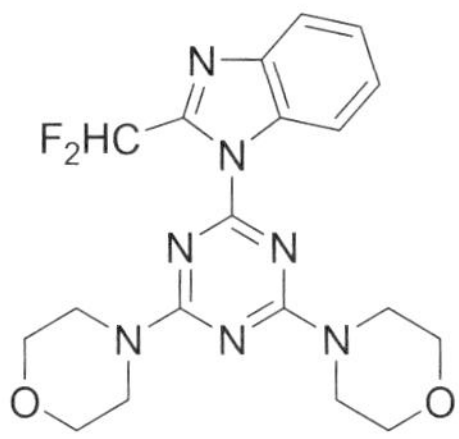

	ZSTK474
Molecular formula	$C_{19}H_{21}F_2N_7O_2$
Molecular weight	417.41
Solubility	DMSO, ++; H_2O, –; MeOH, –

Discovery/Isolation

ZSTK474 was isolated as a strong antiproliferative compound from a chemical library consisting of 1500 compounds of *s*-triazine derivatives [1].

Biological Studies

COMPARE analysis of chemosensitivity measurements from 39 human cancer cell lines suggested phosphatidylinositol 3-kinase (PI3K) as a molecular target for ZSTK474 [1]. An in vitro PI3K kinase assay and molecular modeling revealed that ZSTK474 competitively inhibited class I PI3K kinase by binding to the ATP binding pocket [1, 2]. Oral administration of ZSTK474 resulted in strong antitumor activity against human cancer xenografts without toxic effects in critical organs by cell cycle arrest at G0/G1 phase [1, 3]. ZSTK474 also has potent antiangiogenic activity by inhibition of HIF-1α expression and VEGF secretion from cancer cells [4].

Biological Activity of ZSTK474

Inhibition of cell cycle progression of OVCAR3 in G1 phase: >0.5 μM [1]
K_i for PI3Kδ: 1.8 nM [2]
IC_{50} for inhibiting growth of HUVECs: 146 nM [4]

References

1. Yaguchi S, Fukui Y, Koshimizu I, Yoshimi H, Matsuno T, Gouda H, Hirono S, Yamazaki K, Yamori T (2006) Antitumor activity of ZSTK474, a new phosphatidylinositol 3-kinase inhibitor. J Natl Cancer Inst 98:545–556
2. Kong D, Yamori T (2007) ZSTK474 is an ATP-competitive inhibitor of class I phosphatidylinositol 3 kinase isoforms. Cancer Sci 98:1638–1642
3. Dan S, Yoshimi H, Okamura M, Mukai Y, Yamori T (2009) Inhibition of PI3K by ZSTK474 suppressed tumor growth not via apoptosis but G0/G1 arrest. Biochem Biophys Res Commun 379:104–109
4. Kong D, Okamura M, Yoshimi H, Yamori T (2009) Antiangiogenic effect of ZSTK474, a novel phosphatidylinositol 3-kinase inhibitor. Eur J Cancer 45:857–865

Appendix

ICG-001*

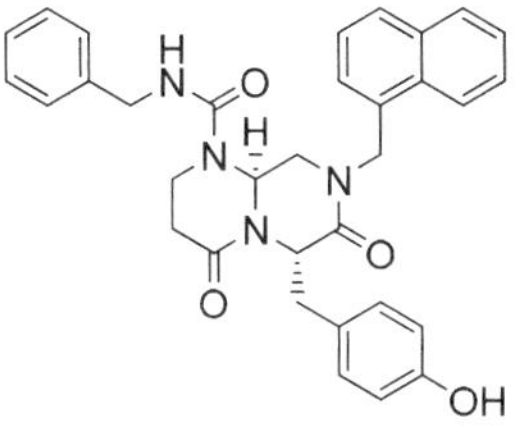

ICG-001 inhibited the β-catenin/TCF-mediated transcription by binding to transcriptional co-activator CBP.

Reference

1. Emami KH, Nguyen C, Ma H, Kim DH, Jeong KW, Eguchi M, Moon RT, Teo JL, Oh SW, Kim HY, Moon SH, Ha JR, Kahn M (2004) A small molecule inhibitor of β-catenin/cyclic AMP response element-binding protein transcription. Proc Natl Acad Sci U S A 101:12682–12687

Origamicin*

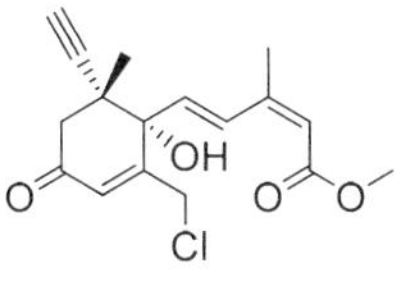

Origamicin decreased HCV replication and nonstructural protein 5B (NS5B) protein contents without cytotoxicity. It is thought that origamicin shows antiviral activity through the inhibition of host proteins involved in protein folding (HSPs and PDI) as an electrophile.

Reference

1. Rakic B, Clarke J, Tremblay TL, Taylor J, Schreiber K, Nelson KM, Abrams SR, Pezacki JP (2006) A small-molecule probe for hepatitis C virus replication that blocks protein folding. Chem Biol 13:1051–1060

Pyrabactin B*, Pyrabactin*

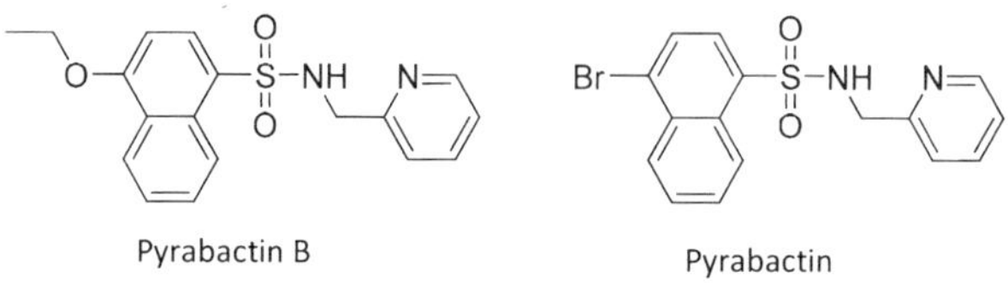

Pyrabactin B was identified as germination inhibitor that acts as agonists of the abscisic acid signal transduction pathway in seeds of *Arabidopsis thaliana*. Park et al. pursed a chemical genetic strategy using pyrabactin, and isolated pyrabactin-resistant mutants named *Pyr1*, which binds and inhibits Mg^{2+}- and Mn^{2+}-dependent serine/threonine phosphatases type 2C (PP2C) in abscisic acid-dependent manner.

References

1. Yang Zhao, Tszfung F Chow, Rachel S Puckrin, Simon E Alfred, Albert K Korir, Cynthia K Larive & Sean R Cutler (2007) Chemical genetic interrogation of natural variation uncovers a molecule that is glycoactivated. Nat Chem Biol 3:716–721
2. Park SY, Fung P, Nishimura N, Jensen DR, Fujii H, Zhao Y, Lumba S, Santiago J, Rodrigues A, Chow TF, Alfred SE, Bonetta D, Finkelstein R, Provart NJ, Desveaux D, Rodriguez PL, McCourt P, Zhu JK, Schroeder JI, Volkman BF, Cutler SR (2009) Abscisic acid inhibits type 2C protein phosphatases via the PYR/PYL family of START proteins. Science 324:1068–1071

Quinostatin*

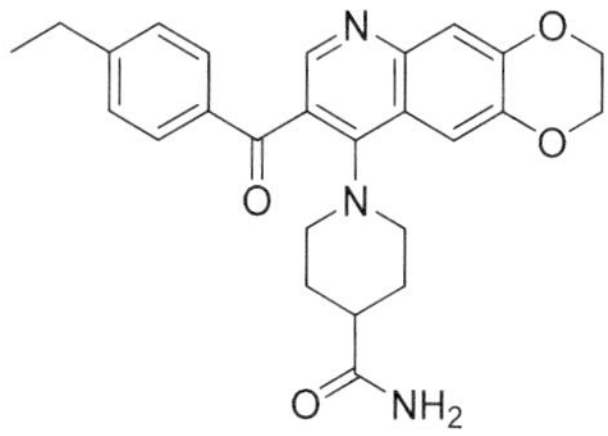

Quinostatin prevented the phosphorylation of S6 in cells through the PI3K/Akt pathway by binding p110 of class Ia PI3K.

Reference

1. Yang J, Shamji A, Matchacheep S, Schreiber SL (2007) Identification of a small-molecule inhibitor of class Ia PI3Ks with cell-based screening. Chem Biol 14:371–377

SMIR4*

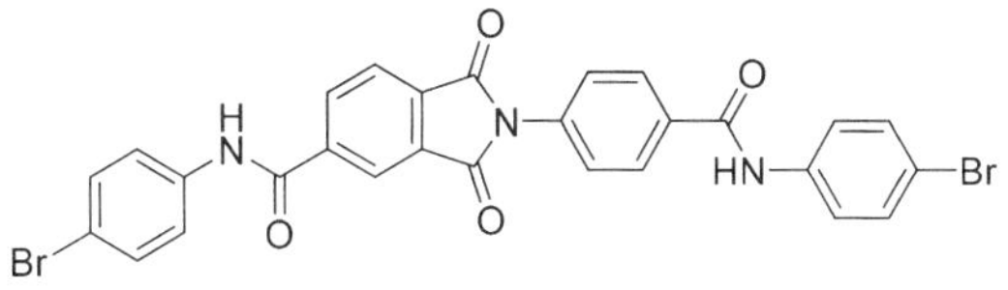

SIMR4 fully and specifically suppressed rapamycin's anti-proliferative effects by binding to NIR1p, a new genetic modifier of mTOR pathway.

Reference

1. Huang J, Zhu H, Haggarty SJ, Spring DR, Hwang H, Jin F, Snyder M, Schreiber SL (2004) Finding new components of the target of rapamycin (TOR) signaling network through chemical genetics and proteome chips. Proc Natl Acad Sci U S A 101:16594–16599

TPh A*

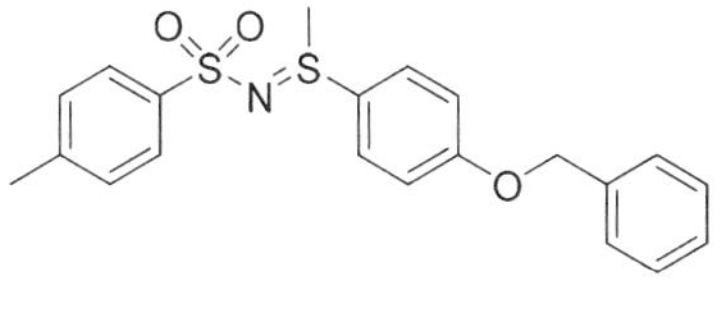

TPh A was identified as a pirin binder, and used for investigating the function of human pirin that is widely expressed in punctate subnuclear structures in human tissues.

Reference

1. Miyazaki I, Simizu S, Okumura H, Takagi S, Osada H (2010) A small-molecule inhibitor shows that pirin regulates migration of melanoma cells. Nat Chem Biol 6:667–673

Authors and Collaborators

Authors

Hiroyuki Osada

Chemical Biology Research Group, RIKEN Center for Sustainable Resource Science,
2-1 Hirosawa, Wako, Saitama 351-0198, Japan
e-mail: hisyo@riken.jp

Nobumoto Watanabe

Bio-Active Compounds Discovery Research Unit, RIKEN Center for Sustainable Resource Science
2-1 Hirosawa, Wako, Saitama 351-0198, Japan
e-mail: nwatanab@riken.jp

Akihiro Ito

[1]Chemical Genetics Laboratory, RIKEN,
[2]Chemical Genomics Research Group, RIKEN Center for Sustainable Resource Science
2-1 Hirosawa, Wako, Saitama 351-0198, Japan
e-mail: akihiro-i@riken.jp

Minoru Yoshida

[1]Chemical Genetics Laboratory, RIKEN,
[2]Chemical Genomics Research Group, RIKEN Center for Sustainable Resource Science,
2-1 Hirosawa, Wako, Saitama 351-0198, Japan
e-mail: yoshidam@riken.jp

© Springer Japan KK 2017

H. Osada (ed.), *Bioprobes*, DOI 10.1007/978-4-431-56529-1

Etsu Tashiro

Department of Biosciences and Informatics, Faculty of Science and Technology, Keio University,
3-14-1 Hiyoshi, Kohoku-ku, Yokohama, Kanagawa 223-8522, Japan

Mitsuhiro Kitagawa

Department of Biosciences and Informatics, Faculty of Science and Technology, Keio University,
3-14-1 Hiyoshi, Kohoku-ku, Yokohama, Kanagawa 223-8522, Japan

Masaya Imoto

Department of Biosciences and Informatics, Faculty of Science and Technology, Keio University,
3-14-1 Hiyoshi, Kohoku-ku, Yokohama, Kanagawa 223-8522, Japan
e-mail: imoto@bio.keio.ac.jp

Takao Kataoka

Department of Applied Biology, Kyoto Institute of Technology,
Matsugasaki, Sakyo-ku, Kyoto 606-8585, Japan
e-mail: takao.kataoka@kit.ac.jp

Takeo Usui

Faculty of Life and Environmental Sciences, University of Tsukuba,
1-1-1 Tennodai, Tsukuba, Ibaraki 305-8572, Japan
e-mail: usui.takeo.kb@u.tsukuba.ac.jp

Hiroyuki Koshino

Molecular Structure Characterization Unit, RIKEN Center for Sustainable Resource Science,
2-1 Hirosawa, Wako, Saitama 351-0198, Japan

Collaborators

Makoto Muroi, RIKEN Center for Sustainable Resource Science
Yasumitsu Kondoh, RIKEN Center for Sustainable Resource Science
Makoto Kawatani, RIKEN Center for Sustainable Resource Science
Yushi Futamura, RIKEN Center for Sustainable Resource Science
Junnosuke Otaka, RIKEN Center for Sustainable Resource Science

Toshihiko Nogawa, RIKEN Center for Sustainable Resource Science
Naoki Kato, RIKEN Center for Sustainable Resource Science
Kazuko Yoshida, RIKEN Center for Sustainable Resource Science
Takeshi Kashiwa, RIKEN Center for Sustainable Resource Science
Lai Ngit Shin, Universiti Sains Malasia
Azhar Rasul, RIKEN Center for Sustainable Resource Science
Amit Subedi, RIKEN Center for Sustainable Resource Science
Kruthi Suvarna, RIKEN Center for Sustainable Resource Science
Tatsuro Kawamura, Max-Planck-Institute of Molecular Physiology
Hiroki Hayase, Japan Science and Technology Agency
Yasushi Takemoto, Kyoto University
Harumi Aono, RIKEN Center for Sustainable Resource Science

H. Osada

N. Watanabe

M. Yoshida

A. Ito

M. Imoto

E. Tashiro

M. Kitagawa

T. Kataoka T. Usui H. Koshino

T. Kashiwa, A. Subedi, Y. Kondoh, T. Nogawa, A. Rasul, J. Otaka, M. Muroi
K. Yoshida, H. Aono, K. Suvarna, M. Kawatani, N. Kato

Y. Futamura N. S. Lai T. Kawamura H. Hayase Y. Takemoto

Index

© Springer Japan KK 2017

H. Osada (ed.), *Bioprobes*, DOI 10.1007/978-4-431-56529-1

MIX
Papier aus verantwortungsvollen Quellen
Paper from responsible sources
FSC® C105338

If you have any concerns about our products,
you can contact us on
ProductSafety@springernature.com

In case Publisher is established outside the EU,
the EU authorized representative is:
Springer Nature Customer Service Center GmbH
Europaplatz 3, 69115 Heidelberg, Germany

Printed by Libri Plureos GmbH
in Hamburg, Germany